EXPERIMENTAL CHEMISTRY

A Laboratory Manual

Second Edition SI units

Peter Rendle, M.A., F.R.I.C.

Michael Vokins, M.A., A.R.I.C.

Peter Davis, M.A.

Edward Arnold

First published 1967
by Edward Arnold (Publishers) Ltd.,
41 Bedford Square
London, WC1B 3DP
Reprinted 1969, 1971
Second edition 1972
Reprinted with corrections 1974
Reprinted 1976, 1978

ISBN: 0 7131 2343 5

Printed in Great Britain by
Fletcher & Son Ltd, Norwich

Preface to the Second Edition

The aim of this book is to cover practical laboratory work for senior forms in schools, for technical colleges and for some university courses. It does not follow any examination syllabus, but covers all that is needed for existing A-level courses. Changes in science teaching since the first edition have laid emphasis on experimental work being conducted in a spirit of enquiry so that students acquire first-hand experience before theoretical interpretations are introduced. This has always been our aim, since in our experience students enjoy problem-solving, and their capacity for understanding is thereby developed.

We have altered units and symbols to agree with the Système Internationale (SI) now being adopted by most examining boards in the United Kingdom. Nomenclature of compounds has proved to be more of a problem, since it does not appear that a uniform system is likely to be adopted in the near future by English-speaking chemists. We have included an appendix explaining our present usage which is inevitably a compromise.

The contents of the book have not been greatly altered, though there has been some influence from the Nuffield Advanced Chemistry course. The opening section, based on considerable practical experience, should enable the inexperienced worker to avoid many pitfalls. The section on inorganic chemistry, based on the periodic table, includes both test tube experiments and preparations on a larger scale. Organic chemistry is no longer divided into 'aliphatic' and 'aromatic' and includes many preparations based on the higher members of homologous series which avoid the problems which beset the beginner trying to deal with very volatile liquids. In the physical and analytical sections we have tried to include experiments giving as wide an experience as possible, though analysis involving the use of expensive spectrometers is regretfully omitted. Although systematic qualitative analysis is waning in popularity, some technical college and A-level courses still require it, and we feel that it will probably also be useful for students in schools working on individual projects.

We would like to thank those who sent in valuable comments on the first edition, and would welcome further suggestions. We hope that the book will continue to be of use as an adjunct to Nuffield and other courses both in the United Kingdom and overseas.

Acknowledgments

We are grateful for permission to use copyright material from Butterworth & Co. (Publishers) Ltd., Cambridge University Press, the McGraw-Hill Book Company, *The Journal of Chemical Education*, *The School Science Review*, and the Oxford and Cambridge Examination Board, and for the assistance of many firms including: British Drug Houses Ltd. (Laboratory Chemicals Division), especially K. K. Hart and Miss B. Savage for permission to use the Emdite analysis scheme developed by them, H. J. Elliott Ltd., Gallenkamp Ltd., Grants Instruments Ltd., Grayshaw Instruments, Griffin & George Ltd., Oertling Ltd., Quickfit & Quartz Ltd.

We also wish to acknowledge permission to use material from courses organized by L. L. Leveson and G. R. Fitch of Bath University of Technology, Dr. J. N. Andrews of the Centre for Nuclear Studies, Bath University of Technology, and L. Davies and A. Dyer of the University of Salford. We should also like to thank warmly Professor H. Irving of the University of Leeds, and E. G. Stroud, Senior Safety Officer of the National Physical Laboratory, for reading portions of the manuscript and making many helpful suggestions.

Finally we would like to thank the members of Group IV at Clifton College who helped in the development of the experiments, Mrs. L. Briggs who typed much of our manuscript, and our Publishers who guided us so ably throughout.

Bristol, 1971 G.P.R., M.D.W.V., P.M.H.D.

Contents

Abbreviations and Symbols

g	grams
mg	milligrams
l	litres (1 litre = 1 dm^3 = $10^{-3}m^3$)
V	volume
cm	centimetres
mm	millimetres
p.p.m.	parts per million
M	concentration (moles per litre)
M_r	relative molecular mass (molecular weight)
A_r	relative atomic mass (atomic weight)
mm Hg	pressure (millimetres of mercury)
P	pressure
T	temperature
K	kelvins
t	time
$t_{\frac{1}{2}}$	half-life (time taken for concentration to be halved)
(s)	solid state
(l)	liquid state
(g)	gaseous state
(aq)	aqueous solution
e^-	electron
F	the Faraday constant (1 mole of electrons)
L	the Avogadro constant ($6{\cdot}023 \times 10^{23}$ mol^{-1})
R	the gas constant ($8{\cdot}314$ J $mol^{-1}K^{-1}$)
ΔH	change in enthalpy
ΔG	change in Gibbs free energy
K_β	equilibrium constant
K_{s_0}	solubility product
e.m.f.	electromotive force
r.p.m.	revolutions per minute
c.p.m.	counts per minute
κ	electrolytic conductivity ($\Omega^{-1}m^{-1}$)
Λ	molar conductivity
Λ^∞	molar conductivity at infinite dilution
[X]	molar concentration of X (mol l^{-1})
M.P.	melting-point
B.P.	boiling-point
pH	$-\log_{10}[H_3O^+]$
A	ampères
V	volts

I Introduction

THIS book is essentially for use in the laboratory. Chemistry is still a practical subject, and the carrying out of experimental work will enable you to gain a greater understanding if you approach it in the right way. Theoretical explanations are omitted deliberately, and you should use textbooks and journals (such as those listed in the Bibliography) in conjunction with this book. If you have learnt nothing new as a result of an hour or two in the laboratory, the time has been largely wasted.

Before starting an experiment, try to find out what its aim is and, as you come to each stage, find out the reason for what you are doing. At the end of the experiment you should have a clear and complete written account of it, including such conclusions as you have been able to draw from it. All numerical readings should be recorded at once, but not on scrap pieces of paper.

The part of the book on laboratory techniques contains a great deal of practical advice which will help you in all types of work in the laboratory. Before carrying out an experiment which involves a technique with which you are unfamiliar, read through the appropriate section. You should in any case read through the sections on safety, weighing, and setting up apparatus before doing any of the experiments. Also in Chapter II you will find a set of diagrams of assemblies of apparatus which are used in many different preparations throughout the book. These diagrams (Figs. A–N) are printed on coloured paper to enable you to refer to them easily.

The arrangement of the experiments in the book is according to subjects. It is not intended as a systematic course to be followed from start to finish. You are expected to be selective and to carry out some of the experiments in each group.

In the analytical section a new method of cation analysis, which avoids the use of hydrogen sulphide, is included.

Names of chemical compounds

The system of naming used is that recommended by the Chemical Society, based on the internationally agreed I.U.P.A.C. rules. You may find some of the names unfamiliar, but they are all logical and easy to understand. Inorganic compounds are usually named on the Stock notation: the oxidation number (valency) is put in Roman numerals after the name of the element. For example, ferrous sulphate becomes iron (II) sulphate, ferric chloride becomes iron (III) chloride and potassium ferrocyanide becomes potassium hexacyanoferrate (II). If the element has only one common oxidation number, it is omitted from the name. This notation is rarely used for compounds between non-metals.

Remember that, in aqueous solution, the hydrogen ion is always hydrated (i.e. H_3O^+) even when, for convenience, it is written H^+.

Quantities

Quantities are usually given in terms of moles, and concentrations in terms of moles per litre (molarity, M). Equivalents and normalities are never mentioned; this avoids ambiguity when the same compound can react in different mole ratios in different types of reaction.

A mole is the amount of a substance which contains the same number of molecules (or atoms, or ions, or electrons as the case may be) as there are carbon atoms in exactly 0·012 kilograms of carbon-12.

Reference: Guggenheim, E. A., *J. Chem. Educ.*, 1961, **38**, 86.

Examples

Iodine, I_2: one mole weighs 253·8 g

Copper (II) sulphate pentahydrate, $CuSO_4.5H_2O$: one mole weighs 249·69 g

A molar solution of copper (II) sulphate would therefore contain 249·69 g per litre of the hydrated crystals.

The convention shown in Table 1 has been adopted in the book to express the accuracy with which quantities should be measured.

It is very convenient to know the volume capacity of various test tubes in common use, and it is also helpful to mark all watch-glasses with their weights (use a diamond writer).

Table 1

Quantity	*Method of measurement*
10·0 cm^3 aliquot	Use a pipette
Exactly 10 cm^3	Use a pipette or a burette
10 cm^3	Use a measuring cylinder
About 10 cm^3	Use a test tube (e.g. half full)
Weigh accurately a sample of roughly 10 g	Use a rough balance to 10 g ± 0·1 g, then reweigh on an analytical balance to ±0·001 g
10 g	Use a rough balance to ±0·1 g
Roughly 10 g	Use a rough balance to ±0·2 g

Journals

Educ. Chem.	Education in Chemistry
J.A.C.S.	Journal of the American Chemical Society
J. Chem. Educ.	Journal of Chemical Education
J. Chem. Soc.	Journal of the Chemical Society
J. Roy. Inst. Chem.	Journal of the Royal Institute of Chemistry
Proc. Chem. Soc.	Proceedings of the Chemical Society
S.S.R.	School Science Review

References are given in the following style: Name of author(s) and initials, *abbreviated name of journal.* year of publication, **volume number**, page number.

BIBLIOGRAPHY

1. Books

General

BJERRUM, J., SCHWARZENBACH, G. and SILLEN, L. G., eds. *Stability Constants of Metal-ion Complexes*. Chemical Society, London, 1964.

CAHN, R. S. and CROSS, L. C., eds. *Handbook for Chemical Society Authors*. Chemical Society, London, 1960.

HODGMAN, C. D. et.al., eds. *Handbook of Chemistry and Physics*. Chemical Rubber Publishing Co./ Blackwell Scientific, Oxford. (Annual publication.)

KIEFFER, W. F. *The Mole Concept in Chemistry*. Reinhold, New York; Chapman & Hall, London, 1962.

KNEEN, W. R., ROGERS, M. J. W., and SIMPSON, P., *Chemistry: Facts, Patterns and Principles*. Addison-Wesley, London, 1972.

NUFFIELD ADVANCED SCIENCE *Book of Data*. Penguin, London, 1972.

Inorganic

ADAMS, D. M. and RAYNOR, J. B. *Advanced Practical Inorganic Chemistry*. Wiley, New York and London, 1965.

COTTON, F. A. and WILKINSON, G. *Advanced Inorganic Chemistry*, 3rd edn. Interscience, New York and London, 1972.

HESLOP, R. B. and ROBINSON, P. L. *Inorganic Chemistry*, 2nd edn. Elsevier, Amsterdam, 1962.

Inorganic Syntheses McGraw-Hill, New York and Maidenhead, 1939–1963. BOOTH, ed., vol. 1, 1939; FERNELIUS, ed., vol. 2, 1946; AUDRIETH, ed., vol. 3, 1950; BAILAR, ed., vol. 4, 1953; MOELLER, ed., vol. 5, 1957; ROCHOW, ed., vol. 6, 1960; KLEINBERG, ed., vol. 7, 1963. Now annual.

MOODY, B. J. *Comparative Inorganic Chemistry*, 2nd edn. Edward Arnold, London, 1969.

PALMER, W. G. *Experimental Inorganic Chemistry*. University Press, Cambridge, 1954.

REMY, H. *Treatise on Inorganic Chemistry*, 2 vols. Elsevier, Amsterdam, 1956.

Organic

CLARK, N. G. *Modern Organic Chemistry*. Oxford University Press, London, 1964.

FINAR, I. L. *Organic Chemistry*, vol. 1; vol. 2. Longmans, London, 1973; 1968.

Organic Syntheses, collective vols I–IV. Wiley, New York; Chapman & Hall, London, 1932–1955.

VOGEL, A. I. *Elementary Practical Organic Chemistry*, 3 vols. Longmans, London (1957–1958).

Physical

ADAM, N. K. *Physical Chemistry*. Oxford University Press, London, 1956.

CHASE, G. D. et al. *Experiments in Nuclear Science*. Burgess, Minneapolis, 1964.

CHOPPIN, G. R. *Nuclei and Radioactivity*. W. A. Benjamin, New York, 1964.

LADD, M. F. C. and LEE, W. H. *Practical Radiochemistry*. Cleaver-Hume: Macmillan, London, 1964.

LATHAM, J. LIONEL *Elementary Reaction Kinetics*. Butterworth, London, 1962

MACINNES, D. A. *The Principles of Electrochemistry*. Dover, New York; Constable, London, 1961.

SALMON, J. E. and HALE, D. K. *Ion exchange: A Laboratory Manual*. Butterworth, London, 1959.

STOCK, R. and RICE, C. B. F. *Chromatographic Methods*. Chapman & Hall, London, 1963.

TRUTER, E. V. *Thin Layer Chromatography*. Cleaver-Hume, London, 1964.

Analytical

BELCHER, R. and WILSON, C. L. *New Methods in Analytical Chemistry*, 1st and 2nd edns. Chapman & Hall, London, 1955 and 1964.

KOLTHOFF, I. M. and SANDELL, E. B. *Textbook of Quantitative Inorganic Analysis*, 3rd edn. Collier–Macmillan, New York and London, 1952.

LEVESON, L. L. *Introduction to Electro-analysis*. Butterworth, London, 1964.

MOELLER, T. *Qualitative Analysis*. McGraw-Hill, New York and Maidenhead, 1958.

OPENSHAW, H. T. *A Laboratory Manual of Qualitative Organic Analysis*, 3rd edn. University Press, Cambridge, 1955.

SCHWARZENBACH, G. *Complexometric Titrations*. Methuen, London, 1957.

SHRINER, R. L. and FUSON, R. L. *The Systematic Identification of Organic Compounds*, 5th edn. Wiley, New York and London, 1964.

VOGEL, A. I. *A Textbook of Macro and Semimicro Qualitative Inorganic Analysis*, 4th edn. Longmans, London, 1954.

VOGEL, A. I. *A Textbook of Quantitative Inorganic Analysis*, including elementary instrumental analysis, 3rd edn. Longmans, London, 1961.

2. Journals

Education in Chemistry (bimonthly)
Royal Institute of Chemistry, 30 Russell Square, London, W.C.1.

Journal of Chemical Education (monthly)
20th and Northampton Streets, Easton, Pa., U.S.A.

School Science Review (quarterly)
John Murray Ltd., 50 Albemarle Street, London, W.1.

Satis (bimonthly)
Project Technology, College of Education, Loughborough.

II Laboratory Technique

1. Safety

Too often safety in laboratories consists of precautions for a particular experiment, a fire extinguisher fixed to the wall and the occasional adhesive plaster used for minor cuts.

Nevertheless, major accidents do occur even in well run laboratories, and it is essential that preparation should have been made to cope with any emergency. In this section only major points, which should be soundly learnt by everyone, are discussed, but we hope the topic will be enlarged by special lectures on safety and reading from the following books:

National Chemical Laboratory, *Safety Measures in Chemical Laboratories,* H.M.S.O., London, 1964.
Gray, C. H., ed., *Laboratory Handbook of Toxic Agents,* Royal Institute of Chemistry, London, 1965.
Guy, K., *Laboratory First Aid,* Macmillan, London, 1965.

In addition, there are two wall charts available from British Drug Houses Ltd.: *Spillages of Hazardous Chemicals* and *Laboratory First Aid.*

Precautions to take

The fundamental precaution is to be safety-conscious; it is too late to think about safety when something starts to go wrong. ***You must know IN ADVANCE what action you are going to take if an accident occurs;*** make sure that you are aware of the hazards associated with each phase of the experiment, and take all the proper precautions. Is it safe to sit down near an ether distillation? If a flask falls, do you attempt to catch it or step well away?

In chemical laboratories the main dangers are the combustion of organic solvents, cuts from broken glass and poisoning by hazardous chemicals.

Fire precautions consist of anticipating the possibility of and consequences of fire.

Splints, not waste paper, should be used for lighting Bunsen burners, and the splint must be properly extinguished before being discarded.

Extra care is necessary when using low-boiling solvents (ether, propanone, ethanol, petroleum spirit). To distil or evaporate these solvents a preheated water-bath is used, and no burners should be alight nearby.

If distilling large volumes of inflammable solvents, stand the apparatus in a tray capable of containing the entire volume of liquid in case a flask breaks.

For occasions when fires do occur, you should know where the extinguishers are kept and how to operate them. A small fire in a beaker can often be extinguished by covering it with a watch-glass.

Broken glass is perhaps the major cause of laboratory accidents.

Learning the correct techniques is the way to avoid breaking apparatus, but always inspect apparatus for cracks or deep scratches before use.

Protect your hand with a cloth when fitting glass tubing and thermometers to corks; the use of standard ground-glass joint apparatus and screw-cap adaptors almost eliminates the need for cork boring.

The consequences of explosions can be minimized by the use of safety screens. Safety screens are placed in front of vacuum desiccators (which should also be covered with a cloth) and vacuum distillations; they are also needed in all experiments where faulty technique could lead to an explosion.

If chemicals are known to be dangerous, avoid them whenever possible; oxygen can be prepared from hydrogen peroxide, dinitrogen monoxide from the oxidation of hydroxyammonium chloride in solution.

It cannot be emphasized too strongly that you should in no circumstances carry out experiments with highly oxidizing or explosive-forming materials unless under the strictest supervision.

Hazardous chemicals are more numerous than is often appreciated because in many cases serious danger only arises with repeated exposure. Table 2 lists the commonest chemicals known to be hazardous.

Table 2

Hazardous chemicals	*Comments*
Salts of Ag, As, Ba, Be, Cu, Hg, Ni, Pb, Sb, Tl, V, Cr; $C_2O_4^{2-}$, F^-, MnO_4^-, CN^-	Most of these are very dangerous, but only if swallowed; however, As, Be and Tl can be absorbed through the skin; $AgNO_3$ causes caustic burns; Hg vapour is significant even at room temperature
Radioactive chemicals	See section on Radioactivity, p. 140
H_2S	Almost as poisonous as HCN, exposure dulls the sense of smell
SO_2, NO_2, Cl_2, Br_2, I_2, HNO_3, H_2SO_4, HF	All are dangerous as well as unpleasant. When concentrated, all cause rapid destruction of the skin; HF is especially dangerous
Na, K and oxides	Handle with care
P, oxides and chlorides	Handle with care
$HClO_3$, $HClO_4$ and their salts	Highly oxidizing
Chlorinated alkanes	Most of these are narcotic, causing mental confusion
Aniline, and aromatic amines	Toxic by vapour and skin absorption, may be carcinogenic
Benzene	Toxic vapour causing dizziness; if you can detect the odour the concentration is above the safe limit
Benzoyl chloride	Very irritant
Dimethyl sulphate	Very irritant and corrosive
Ether (Ethoxyethane)	Very readily inflammable (e.g. on hot metal)
Ethane diamine	Irritant and harmful by skin absorption
Hydrazine	Corrosive
Nitrobenzene	Toxic by vapour and skin absorption
Phenol and cresols	Burn the skin

To avoid mishaps with chemicals it is essential to check bottle labels before using a chemical, measure samples with a safety pipette if solutions are concentrated or volatile, and never eat in a laboratory or taste chemicals (even sucrose, for it could be contaminated).

Fume cupboards must always be used if experiments involve hazardous volatile chemicals; the primary use of fume cupboards is to remove gases which poison, not gases which smell.

Unexpected accidents have occurred when opening containers of volatile chemicals; before attempting to open bottles of '0·880' ammonia or anhydrous aluminium chloride, a cloth should be wrapped round the neck to absorb fumes. Such bottles should be kept cool and never stood near radiators or in sunlight.

If a bottle lacks means of identification of the contents, then the compound must be disposed of by a responsible person.

First aid emergency treatment

There is no substitute for professional medical examination at the earliest possible moment, and the emergency procedure should be clearly stated on a prominent notice outside the laboratory and also alongside the nearby telephones.

In the event of an accident, however minor, a crowd of onlookers is most undesirable.

Burns caused by heat or scalding should be treated by immersion of the affected part in cold water and then applying a cold wet dressing.

If clothing catches fire, the victim should lie on the floor, and the fire should be smothered with a blanket.

The treatment of chemical burns varies with the chemical responsible, but washing with plenty of water is the first remedy. Special washes, for use according to Table 3, should be available in the First Aid cupboard.

Table 3

Chemical	*Neutralizing wash*
Acids	2M ammonium carbonate (leaves no residue on clothes)
Alkalis	M acetic acid (ethanoic acid)
Bromine	2M ammonia
Hydrofluoric acid	As for acids, then hospital treatment
Phenol	Ethanol, then hospital treatment
Phosphorus	0·1M copper sulphate
Sodium	Ethanol on a cotton-wool pad

If ***poison is swallowed,*** give plenty of water at once if the patient is conscious. For corrosive poisons give calcium hydroxide solution (lime water) as soon as possible and ***do not give an emetic.*** An emetic, or the specific antidote, is only given in the case of non-corrosive poisons.

In case of gassing a knowledge of artificial respiration methods may be required, and the correct person(s) to summon should be listed in the emergency procedures.

Cuts. The patient should be seated for treatment as some people faint even with a minor cut. The wound is washed well with cold water, inspected and any foreign bodies removed. Then an antiseptic cream of Cetrimide is applied, and the affected part protected with a suitable dressing.

If bleeding is severe apply firm pressure round the wound, and cover with a pad and firm bandaging; in such cases a *loose* tourniquet is put in place as a precaution.

Electrical mishaps are usually avoidable by the correct installation of equipment. If electrical equipment is involved in a fire, first switch off the current and then use a carbon dioxide extinguisher on the flames. A person rendered unconscious by shock must be drawn away from apparatus, using an insulating material (dry wool) over the hands. Artificial respiration should be given.

If this discussion makes safety seem rather grim, remember that true caution only comes through full knowledge; it is foolhardy to rely on luck when simple precautions can be learnt.

2. Balances and weighing

A great many experiments in chemistry involve weighing at some stage. Much time can be wasted over weighing procedures, and one of the biggest time wasters is the habit of weighing to a degree of accuracy in excess of the requirements of the experiment.

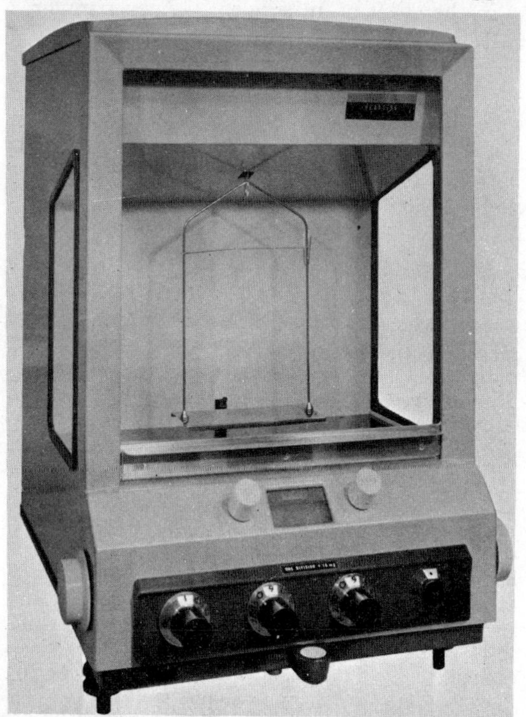

Fig. 1. The Oertling R.10 analytical balance.

For qualitative experiments, like preparing inorganic salts or organic substances, weighing to 0·1 g is quite sufficient.

For making up small quantities, of the order of $200\ cm^3$ of solutions for titrations, it is necessary to weigh to 0·01 g.

For gravimetric work, an accuracy of 0·001 g is usually adequate.

Even if weighing is only carried out to the required degree of accuracy, time can be wasted in the actual process, and unless some method is used whereby weighing is carried out rapidly, many experiments cannot be done in the time normally available.

If possible an automatic balance (see Fig. 1) should be used for accurate work, i.e. weighing to 0·01 g or 0·001 g, and a rough laboratory balance (see Fig. 2) for weighing to 0·1 g.

Small, plastic-topped specimen tubes (see Fig. 3) are available which are cheap and make excellent weighing bottles. These tubes have the advantage that, being mass produced, they are remarkably similar in weight. Ideally each student should have a matched pair of these. Any two tubes can be matched merely by adding grains of sand to the lighter one until they are of equal weight. The

Fig. 2. A convenient rough balance.

tube used for weighing the substance should have its weight scratched on by a diamond pencil.

The matched tubes should never be used for any other purpose than weighing.

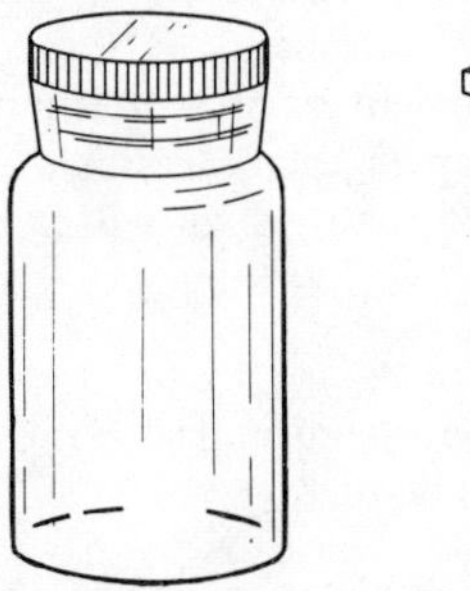
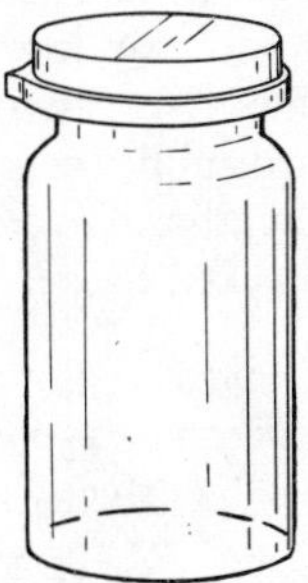

Fig. 3. Weighing bottles.

Suggested rough weighing technique

The pans of the rough balance can be conveniently protected by enclosing them in polythene bags.

To weigh a rough amount of solid (i.e. to the nearest 0·1 g) place equal-sized pieces of scrap paper on each pan and adjust the riders to about 10% less than the final weight required. Using a spatula, add solid to the left-hand pan until it falls; now adjust the riders to their final position and continue to add solid, but more cautiously, until an equipoise is obtained.

Suggested accurate weighing technique

The actual manipulation of accurate balances is not described as each type requires its own special instructions, usually provided by the manufacturer. The drill to be observed will depend to some extent on facilities; each laboratory should have its own 'house rules'.

To weigh an accurate amount of solid (i.e. to the nearest 0·01 g or better) place a weighing bottle on one pan of the rough balance and its tare weight on the other pan, and weigh out a rough amount of solid as close as possible to the accurate weight required. If the solid contains large crystals or lumps it should be lightly ground in a mortar before weighing.

The weighing bottle with contents is now capped, wiped clean and weighed using the correct procedure on an accurate balance, the weight being recorded immediately in a notebook.

Next the solid is tipped out into a beaker or whatever is suitable, no attempt being made to remove the traces of solid which will cling to the weighing bottle. Reweigh the nearly empty bottle accurately. The loss in weight is the accurate weight of solid taken. This avoids the rather awkward process of washing out all the solid from the bottle and is quicker and more accurate.

The method is often used, as it is rarely necessary to weigh out an exact amount. It is bad practice to weigh out, for example, 1·26 g of a solid to make an exact 0·10M solution. It is better to use the above method, finish up with a weight of 1·32 g and express the solution as:

$$(1{\cdot}32/12{\cdot}6)\text{M} = 0{\cdot}105\text{M}$$

This avoids the very messy practice of adding and removing odd crystals to try to get a weight exact.

By using the above method it is never necessary to have any loose chemicals near an accurate balance. The preliminary transfer is done at a rough balance, and only a closed bottle is used on the main balance.

3. Setting up apparatus

The number of assemblies required for the preparations in this book has been made as small as possible, so that most of them can be built up from a set of apparatus with interchangeable ground-glass joints. The 27 BU set manufactured by Quickfit and Quartz Ltd. is recommended, but any similar set can be used. In addition to the set, the following extra pieces of apparatus are needed for some of the experiments: a 250 cm^3 flask, a vented receiver adaptor, a multiple adaptor, a fractionating column, an air condenser, a drying tube and, particularly useful, a screw-cap adaptor.

If apparatus with ground-glass joints is not available, the experiments can nearly all be carried out satisfactorily using corks or rubber stoppers; the only exceptions are those which involve corrosive vapours, such as nitric acid or sulphur trioxide.

When ground-glass joints are used, it is not necessary to lubricate them except when high temperatures are involved. If a joint becomes seized, try the following methods of loosening it: (*a*) rock the cone in the socket, (*b*) tap the joint gently with a block of wood, (*c*) warm the joint in a small flame, then tap gently, (*d*) soak the joint in penetrating oil, then try tapping.

A common cause of seizure is a caustic alkali. Try to keep alkalis off the ground-glass, and if they do get on it, wash thoroughly as soon as possible. *Seizures can usually be avoided by dismantling the apparatus immediately after use.*

Care should always be taken, when glass apparatus is set up, to avoid strain. It is best to start with one piece, and build up from there. To take the apparatus for distillation as an example:

(*a*) lightly clamp the flask at a height convenient for heating,
(*b*) attach the still-head, screw-cap adaptor and thermometer (no more clamps are needed for these),
(*c*) attach the rubber tubing to the condenser, then position a clamp and stand so that the condenser will rest on the lower, fixed, side of the clamp. Attach the condenser to the still-head, and clamp lightly,
(*d*) attach and support the receiver adaptor and the receiver.

A similar procedure should be followed for the other assemblies.

Notes on individual assemblies (Figs. A-N)

Reflux. Clamp the flask and the condenser. If an air condenser is used, clamp it at the top.

Distillation. Use a vented receiver adaptor in the following circumstances:

(*a*) if a noxious gas or vapour is given off, and must be led off by rubber tubing to an absorption apparatus or a sink,
(*b*) if an inflammable vapour is given off (for example in ether distillation), and must be led off by rubber tubing to below bench level.

Where an air condenser is specified, it is frequently adequate to attach the receiver adaptor directly to the still-head.

Fractional distillation. Clamp the fractionating column only at the top. If a column is not available, a vertical air condenser or an ordinary condenser with an empty jacket can be used instead, though it will be less efficient.

Gas evolution. It is usually necessary to use apparatus larger than that in the set. A 250 cm^3 flask, with a B24 joint, and a 100 cm^3 dropping funnel are satisfactory for most purposes. If these are not available, it is convenient to prepare a number of standard rubber stoppers, each carrying a dropping funnel and a delivery tube, which will fit 250 cm^3 wide-necked flasks.

Gas drying. If ground-glass jointed apparatus is not available, a 250 cm^3 conical flask with a rubber stopper is perfectly adequate.

Gas absorption. If it is necessary to dissolve a gas in a liquid, the best method is to use a Buchner flask fitted with a wide glass tube in a rubber stopper. This overcomes the 'suck-back' problem by equalizing the internal pressure with that of the atmosphere.

Heating a solid in a gas. The use of apparatus with ground-glass joints is not recommended, since the combustion tube is often irreparably damaged by the heat of the reaction. If the solid to be heated is in wire form, it can be put in the tube as it is, but if it is finely divided, it is best put into a porcelain boat first.

Soxhlet extraction. Details will be found on p. 12.

Use of corks

Even when apparatus with ground-glass joints is normally used, there are still occasions when corks are required. For efficiency corks must be rolled before use, and bored with care.

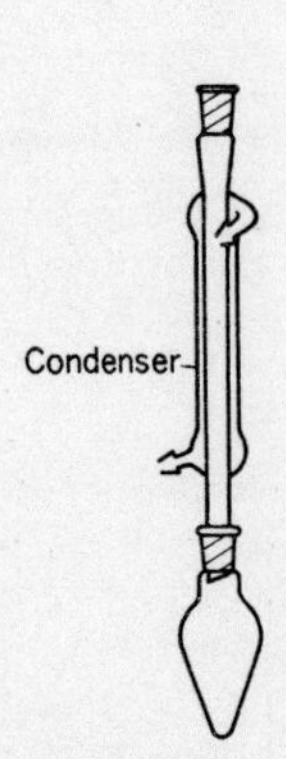

Fig. A. Heating under reflux.

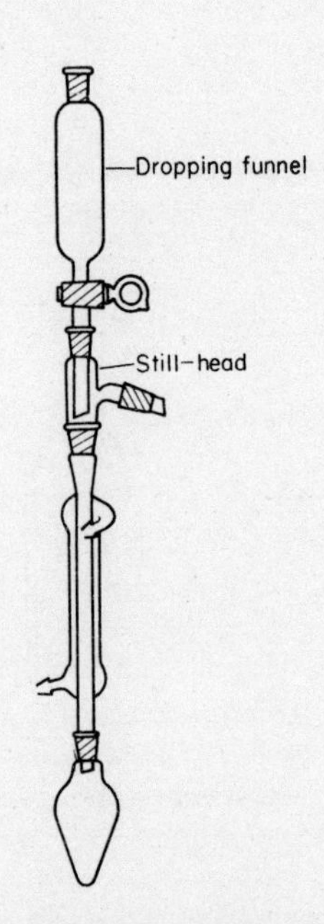

Fig. B. Heating under reflux with addition.

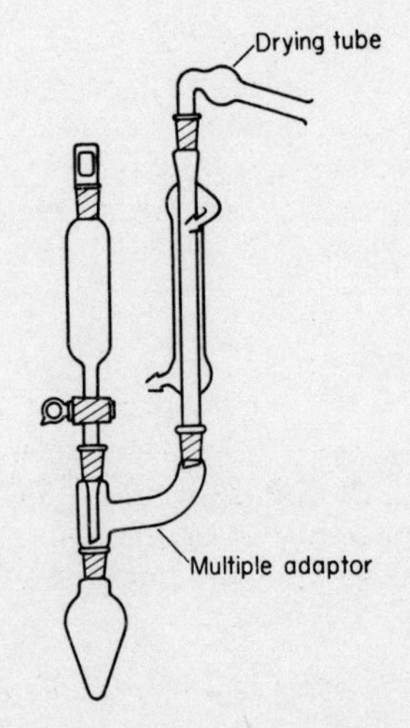

Fig. C. Heating under reflux with addition and exclusion of water.

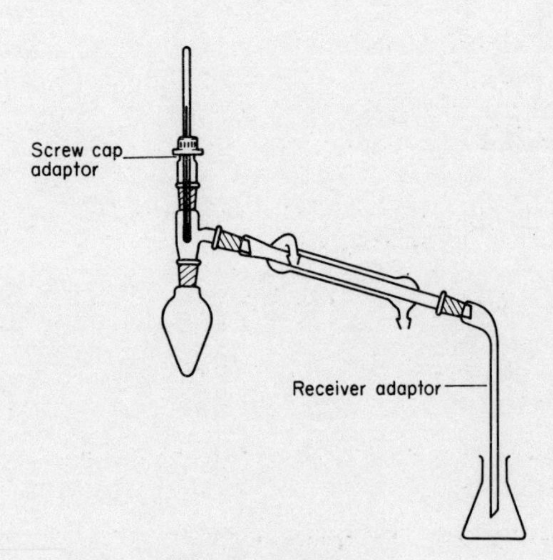

Fig. D. Distillation.

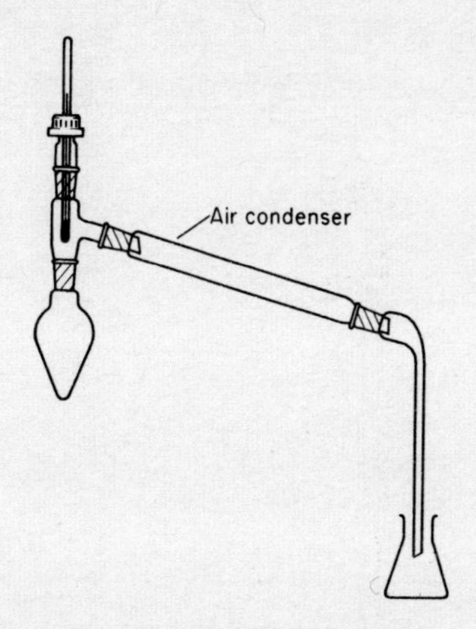

Fig. E(i). Distillation with air condenser.

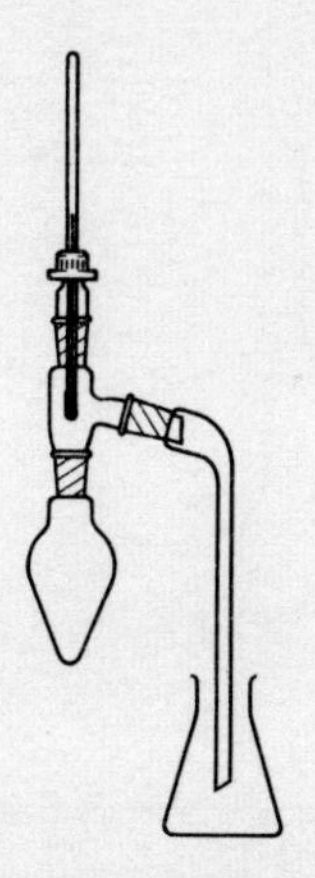

Fig. E(ii). Distillation with receiver adaptor as air condenser.

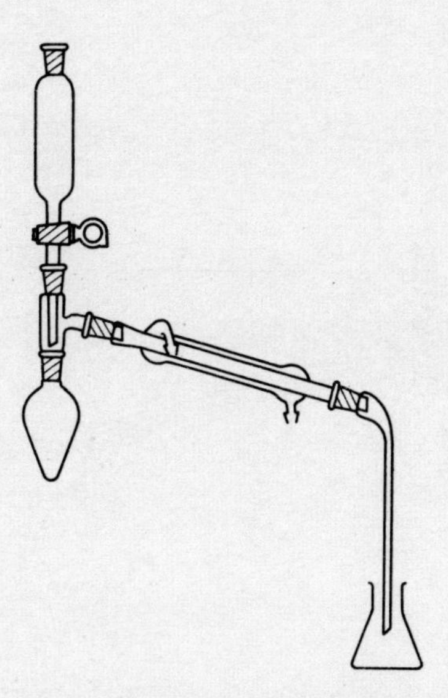

Fig. F. Distillation with addition.

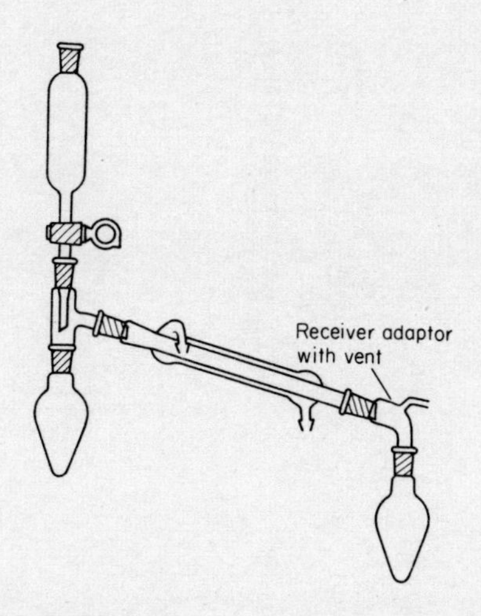

Fig. G. Distillation with vent, with addition.

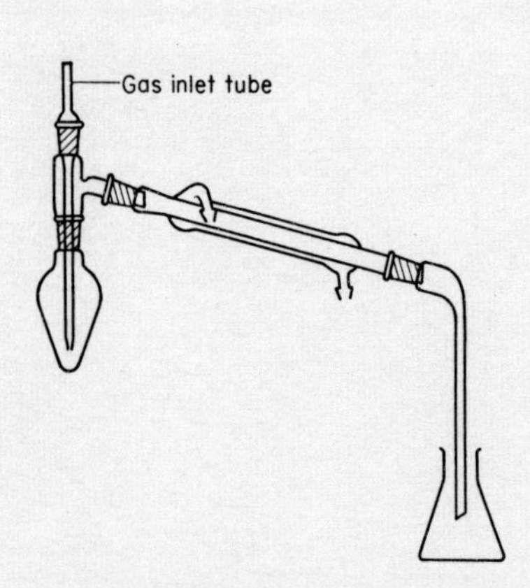

Fig. H. Distillation with gas inlet, or steam-distillation.

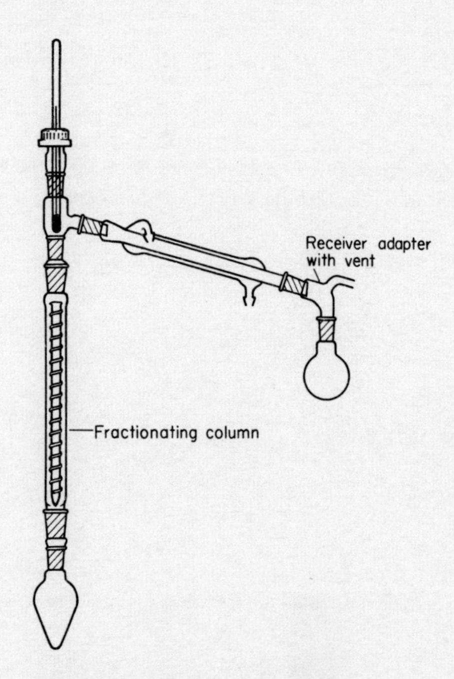

Fig. I. Fractional distillation.

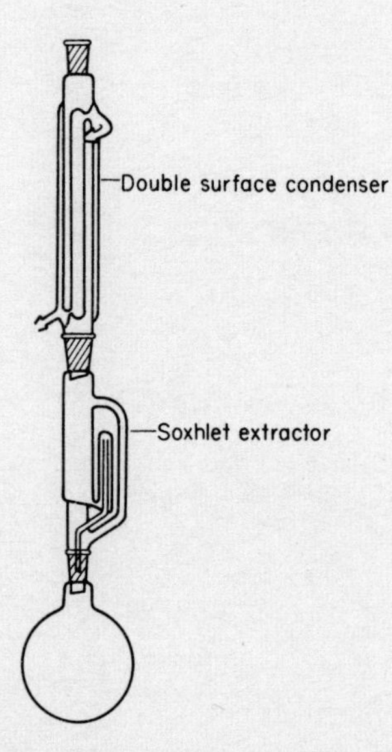

Fig. J. Extraction using Soxhlet apparatus.

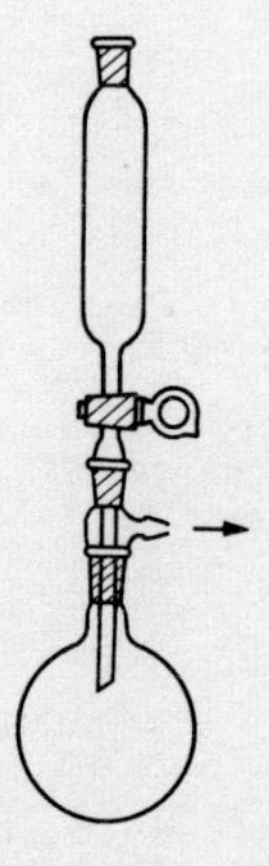

Fig. K(i). Gas evolution (ground-glass joints).

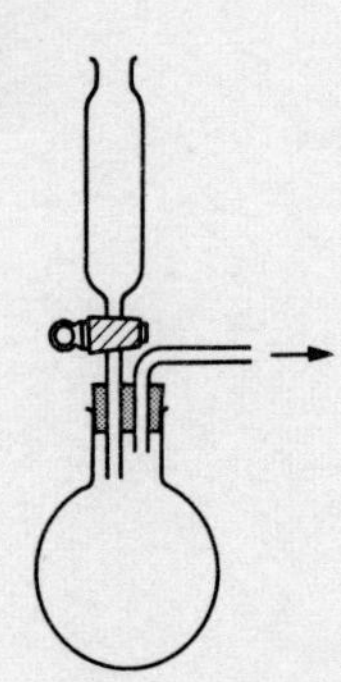

Fig. K(ii). Gas evolution.

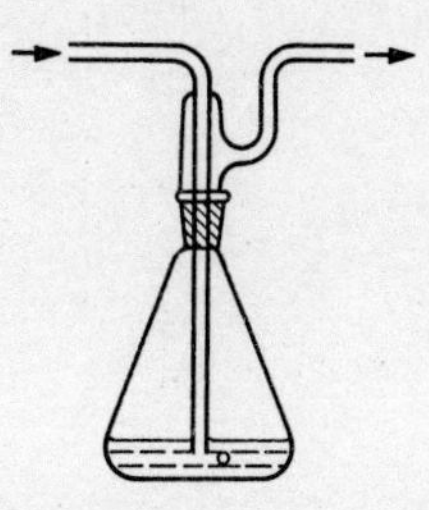

Fig. L(i). Gas drying (ground-glass joints).

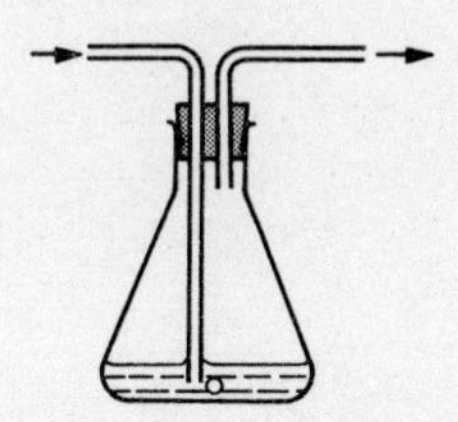

Fig. L(ii). Gas drying.

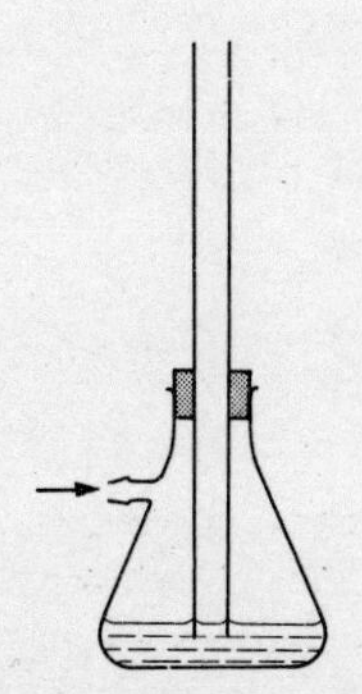

Fig. M. absorption for dissolving a very soluble gas, or as a trap for a poisonous gas such as chlorine.

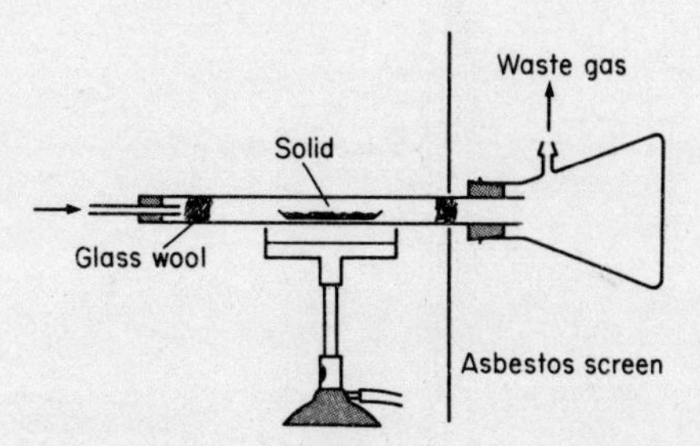

Fig. N. Heating a solid in a stream of gas and collecting a sublimate.

A cork of the correct size should only just go into the neck of the flask. Soften it by rolling between the fingers, or between sheets of paper on the bench. Never try to roll a cork which already has a hole in it; it will almost certainly split.

To bore a cork, choose a sharp borer slightly smaller than the tube or thermometer which is to go into the hole. Hold the cork in the hand, and push and rotate the borer until the hole is approximately half way through it. Now reverse the cork, and continue boring from the other end until the holes meet in the middle. Now use a rat-tailed file to increase the size of the hole until the tube or thermometer fits it with gentle pushing, but with no strain. Place the cork on the file, and rotate it with the hand or on the bench; do not use a sawing action as this will cause an eccentric hole which is likely to leak.

When inserting tubes or thermometers into holes in corks, it is an advantage to moisten them with a little ethanol as a temporary lubricant. If a cork becomes stuck to a tube or a thermometer during use, it is best to cut it off, rather than risk breakage. The majority of cuts which occur in the laboratory happen when pushing tubes through, or removing them from, corks.

4. Methods of heating

A variety of heating techniques may be used, depending on the apparatus and period of heating.

A direct burner should only be used if held in the hand for a simple distillation, when the operator has to observe the temperature of the vapour, and needs more delicate control of the source of heat.

Wire gauze and tripod are commonly used for most experiments and all other distillations. The flame should not be too high, certainly not roaring.

It is very convenient if, in distillations, the tripod and gauze are arranged so that the flask is a few mm above the gauze. This means that the tripod and gauze can be removed without disturbing the distilling flask or having to dismantle the apparatus unnecessarily. The method is especially useful in reflux work.

For accurate work with crucibles a clean silica triangle is used in place of the gauze.

A sand-bath is a shallow tray of metal containing a thin layer of sand. This is a useful alternative to the gauze when rather higher temperatures are employed. The heating is more uniform, and fracture of the glassware is far less likely. It is essential to avoid too thick a layer of sand. Being a bad conductor of heat, a thick layer of sand will result in a lot of time wasted while it warms up.

A water-bath is essentially a saucepan of water, usually fitted with a series of rings which cut down evaporation. Beakers are stood on top of the rings, but flasks are best inserted in the water up to about half way, and the temperature of the water is controlled with a thermometer directly in the water. This procedure is used for low-boiling liquids.

If, like ether, the solvents are also dangerously inflammable, no flame is used at all, the water being heated somewhere else. For small-scale work an ordinary beaker of appropriate size can be used, and the solvent allowed to evaporate freely *in a fume cupboard.*

Thermostatically controlled water-baths are valuable in many physical chemistry experiments.

An oil-bath is used in much the same way as the water-bath, but for reaction mixtures which have to be held for a time at temperatures rather in excess of 100°C. Oil-baths are used in the range 100°C to about 180°C. They become inconvenient at higher temperatures.

Mineral oil of some type is usually employed in large oil-baths, but for small-scale work a beaker of liquid paraffin or one of the phthalic esters may be used.

Metal-baths are rather expensive, but consist of a low-melting alloy and are only used for temperatures in excess of those reached comfortably with an oil-bath.

An air-bath is essentially a small oven and is usually improvised by adapting a suitable tin. It is useful in destructive distillation where direct heating of solid matter leads to extensive charring. Depolymerization of Perspex is a good example of an experiment in which the air-bath is used (see p. 84).

Electrical methods, e.g. the heating mantle and hotplate, are of value in the handling of inflammable liquids of low boiling-point, but usually a water-bath can be used for the same purpose. These methods are also useful for long periods of heating.

5. Reflux and distillation

Unlike ionic reactions, which are frequently extremely rapid, reactions between covalent substances tend to be slow. Particularly in organic chemistry it may be necessary to keep a reaction mixture hot for a matter of hours. This, coupled with the fact that volatile and inflammable solvents have to be employed, makes it necessary for special equipment to be used.

Reflux

The use of a reflux condenser is often necessary. It is used whenever a reaction mixture has to be kept boiling for an appreciable time and the solvent is volatile. A water condenser may be used for solvents boiling up to about 130°C, and for higher boiling-point solvents an air condenser is adequate. The flask must never be filled more than half way, the size of flask is chosen by consideration of the total volume of the reaction mixture. A piece of boiling stone, porous pot or similar substance to promote even boiling is necessary for all reflux procedures. The object of the apparatus is to keep the solution hot while the reaction is proceeding, without loss of solvent. It is pointless to boil violently, and the heating should be controlled so that the solution is merely simmering. The flask may be heated by any of the usual methods except by that of a direct burner.

Distillation

The purpose of distillation is to purify a liquid, or to remove a solvent from a solution. The flask must never be more than half full, anti-bumping granules should always be used, and the choice of condenser is the same as for reflux work. The heating of the flask may be by any of the usual means, and a direct burner can be used, provided it is held in the hand and the flame moved around. As a guide, the length of the flame should be about the same as the diameter of the flask. A larger flame may cause excessive heating of the upper part of the flask and lead to fracture. Distillation of a liquid to purify it should be at such a rate that no more than 2 drops per second of distillate are obtained.

Removing a large quantity of solvent may be done much more rapidly.

Fractional distillation

The purpose of fractional distillation is to separate two liquids of different boiling-point. As with other forms of distillation, the flask must never be more than half full, and anti-bumping granules should always be used. To get a good separation of the liquids, *it is essential that the distillation be carried out very slowly*. The slower the distillation the better the separation. A rate of 1 drop per second of distillate should be the aim.

Since the efficiency of the process depends on the fractionating column reaching thermal equilibrium (that is, there should be a gradual increase in temperature from the top to the bottom of the column), best results are obtained if draughts are excluded. The source of heat should be steady, and not intermittent, therefore the use of a burner held in the hand is not recommended. A water-bath or a wire gauze is suitable.

Steam distillation

Some high-boiling liquids, insoluble in water, may be distilled by this method. The distillate consists of two layers; one is water and the other is the required liquid which, except for being wet, is usually pure.

There are two procedures which may be used in steam distillation. Whichever method is chosen, the flask must not be more than half full; anti-bumping granules should be used in the second, but are not necessary in the first.

Method 1

A separate flask or can is used, containing water which is boiled to provide the steam. This is led into the distilling flask, and when the mixture reaches boiling-point, a mixture of water and the required substance distils over. Due to local cooling the distilling flask gradually fills. If the flask becomes more than half full, impurities are likely to be splashed over with the distillate. To avoid this the distilling flask should be heated directly on occasions to reduce the volume. This should be done without stopping the supply of steam. A water condenser is used for all steam distillation work, and the rate of distillation is controlled by the efficiency of this condenser. It is possible to steam distil some low-melting solids such as *o*-nitrophenol. To avoid the condenser becoming clogged in these cases, the flow of water is restricted in the condenser so that the solid remains molten.

It is very important at the end of the distillation, which can be observed as that time when only water is arriving in the receiver, to disconnect the source of steam before removing the burner from the steam generator. If this is not done, there is a tendency for the residues to be forced back into the steam generator.

Method 2

No steam generator is used in this method, but water is added to the distilling flask, making certain that the total volume is less than half that of the flask. Anti-bumping granules are used, and the mixture distilled in the ordinary way. It is convenient to attach a dropping funnel to the distilling flask so that more water can be added if it becomes necessary. This method is quicker and more convenient if only small quantities of material are being handled.

6. Use of the separating funnel

The separating funnel is used for several important processes. Unless care is taken, its use can be one of the major causes of mechanical loss. The choice of size is particularly important and, as with flasks in distillation, the smallest which will do the job is best.

Separating two immiscible liquids

The liquid mixture is poured into the funnel, and the funnel is gently agitated to assist in the separation into layers. The funnel should always be stoppered but if a particularly volatile substance, such as ether, is present, the stopper should be removed occasionally to avoid the possible build-up of pressure.

When separation into layers has occurred, the stopper is removed and the lower layer run off into a small flask. Swirling the funnel and allowing separation to occur again frequently provides a further small sample of the lower layer.

The top layer is poured from the top of the funnel into a second flask. It is a wise precaution always to keep both liquids, even if one of them is to be discarded. It is surprising how often the wrong layer is thrown away.

Washing a crude liquid

One of the commonest procedures consists in shaking up a crude liquid product with an aqueous solution to remove some of the impurities present. The reagents should always be used in small quantities, and the process repeated if necessary. Mechanical loss is always greater when large volumes of washing solutions are used.

Gases are often formed in considerable quantities during the cleaning process, and it is essential to release the pressure frequently. This is best done by inverting the well-stoppered funnel and opening the tap.

If the required substance is the top layer, then running off the bottom layer is quite simple. The whole of the bottom layer of waste should not be run off each time. It is better to leave a little of the aqueous solution, and add further fresh reagent. The careful separation is only done when running off the last of the various washing solutions. This avoids the risk of inadvertently letting out a few drops of the product being treated.

When the required substance happens to be the bottom layer, avoiding mechanical loss becomes more difficult. If the product is run off between each wash and then returned to the funnel for the next, the loss becomes very great. The best compromise is obtained by using rather larger volumes of washing solutions, and decanting the spent solution from the top of the funnel. In this way the product never leaves the funnel until the final wash is over. It is then run out into its receiver, leaving the final washing solution in the funnel.

Liquid extractions

The separating funnel is often used to extract a solute from one solvent by means of a second solvent immiscible with the first. The removal of a solute from water by means of ether is one of the commonest applications.

The size of the funnel is chosen to accommodate the whole of the aqueous solution if possible. This saves a lot of time which would be spent in repetition. A series of extractions with a small quantity of ether is more effective than one with a large amount of ether. In practice the volume used is that which gives the smallest manageable top layer, bearing in mind that the ether solution must be decanted from the top of the funnel. If the layer is too small, decantation becomes difficult. The solution is usually extracted about three times with fresh quantities of ether, and all the ether extracts are decanted into one flask. After the final extraction the aqueous layer is run off, and the last ether layer decanted completely into the flask. The ether solution is then dried, and the ether removed by distillation to obtain the solute.

7. Use of the Soxhlet extractor

The extractor which is most generally useful is the one designed by Soxhlet to extract, by means of a volatile solvent, soluble substances from a solid (see Fig. J). Other extractors are available for the extraction of one immiscible solvent by another, but they are not included here.

Setting up the extractor

The flask must have at least twice the capacity of the extractor since, just before siphoning, the extractor will be full.

An efficient condenser, preferably double-surfaced, must be used, and porous pot or similar device must be put in the flask to ensure smooth boiling.

Paper thimbles are available to fit the extractor, but their use is not essential unless the solid being extracted is rather fine. It is usually more convenient to use a piece of metal gauze, shaped to fit the bottom of the extractor, to keep the siphon tube clear. The gauze should be covered with a thin layer of cotton-wool.

The solid to be extracted should come not higher than about 1 cm from the top of the by-pass tube, to avoid any interference with the free flow of vapour. If the solid is likely to float, it should be covered with a layer of cotton-wool.

The weight of the apparatus is quite considerable, and three separate clamps should be used for the flask, the extractor and the condenser. The whole apparatus should be supported on a gauze, water-bath, sand-bath or heating mantle. If a hot-plate is used, a small square of asbestos should be placed between the flask and the hot-plate.

If the apparatus is to be left unattended, electrical heating should be used.

These precautions are important because a large volume of solvent, often inflammable, is being used, and risk of cracking glassware must be kept to a minimum. Standing the whole apparatus in a large metal dish helps to avoid the spread of any spilt solvent.

The working of the extractor

When under reflux, the solvent in the flask boils and the vapour goes through the by-pass tube into the condenser. Here it condenses and drops on to the solid in the extractor. When the solvent level in the extractor reaches the level of the top of the siphon tube, the whole volume, containing dissolved substances, siphons into the flask. When functioning correctly, the column of solution in the siphon tube breaks when the extractor is nearly empty; the siphoning does not start again until the extractor has refilled.

The process continues automatically, and the extraction is continued until the soluble substances have been dissolved completely. Since these are often coloured, the pale colour of the siphoning liquid will give an indication of when to stop.

When it is necessary to recharge the extractor with a fresh batch of solid during an extraction, the source of heat is taken away just as the siphoning starts. This empties the extractor, and also cools the solvent in the flask enough to stop it boiling. When the extractor is refilled, it is useful to add a little fresh porous pot to the solvent at the same time.

If the solution is going to be evaporated, it is an advantage to stop the heating when the extractor is nearly full of solvent. In this manner half the solvent is already removed.

If it is found during the extraction that the siphon is working continuously, the solid is probably packed badly. Either it is too tightly compressed, or a little has got beyond the gauze and is blocking the siphon tube. The only solution is to stop the extraction and repack the solid.

Cleaning the extractor

Sometimes, especially when using plant material, the extractor becomes very dirty, particularly near the bottom of the siphon tube. If it is not possible to clean it by rinsing, then it can be cleaned thoroughly and without danger of breaking the delicate siphon tube as follows:

Fit a 250 cm^3 flask to the extractor, and attach a condenser to the top. Place 30 cm^3 of concentrated nitric acid and a little porous pot in the flask, and heat gently to boiling. Continue to boil the acid gently until about half of it is in the extractor. By this time the stains should be dissolved away or oxidized and, when the apparatus is cooled and dismantled, they can be washed away with water. The method is only suitable if apparatus with ground-glass joints is used, as the nitric acid will attack corks or rubber stoppers.

8. Filtration methods

There are a variety of techniques used for the separation of a liquid or solution from a solid.

Simple filtration

The use of a filter funnel and a piece of filter paper folded into four is usually reserved for inorganic substances. Precipitates obtained in qualitative analysis and inorganic problem work are often rather fine, and cannot be efficiently filtered at the pump. But organic solids are usually to be separated from a volatile solvent, and the comparative slowness of simple filtration brings in complications caused by evaporation.

It is essential in simple filtration to ensure that the paper is really carefully folded. The paper must be fitted carefully into the funnel and wetted thoroughly with water, or the appropriate solvent, before filtration is started.

The contents of the filter paper should not reach within half an inch of the top of the paper. These simple precautions can make all the difference to the time taken for a filtration to reach completion, and should never be neglected.

Filtering of organic liquids

This is usually done to remove solid impurities which are not in a very fine state of subdivision. A normally folded filter paper will do for this, but the 'fluted' filter paper gives a faster rate of filtration. Basically a fluted filter paper is one that is folded to give a corrugated effect which allows

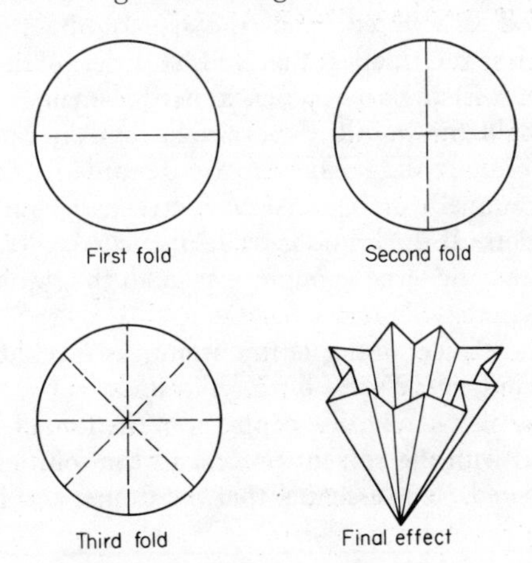

Fig. 4. Folding a fluted filter paper.

the whole of the paper to be active rather than half as is the case with simple filtration.

There are a variety of ways of folding such a paper; one of the easiest is as follows:

The paper is carefully folded in half, opened

out, and then folded in the same direction at right angles to the original fold. The paper is then folded twice more, the folds being all in the same direction and mutually at 45°. Each section is now individually folded in the opposite direction. The result is a fluted paper with sixteen faces (see Fig. 4). This is placed in a suitable sized funnel, and pushed down so that the ridges all touch the side of the funnel. Since all the paper is being used, only one layer thick, filtration is appreciably faster.

For filtering a small amount of liquid to free it from a drying agent it is better to use a very small piece of cotton-wool, pushed lightly into the top of the funnel stem. The mechanical loss entailed by absorption on a filter paper is thus obviated, and much higher yields of product obtained.

The Buchner funnel and filter pump

This system of filtration is the most widely used when dealing with recrystallized substances. The Buchner funnel may be attached to the flask by means of a cork, but a much more useful device consists of a flat piece of rubber, cut from a car tyre, with a hole in the centre capable of receiving the funnel stem and making a good seal.

The disc of rubber allows, within reason, any size funnel to be fitted to any size flask. If this method is adopted, then the size of the funnel chosen is the smallest that will hold the solid, and the flask is similarly chosen to be the smallest that will hold the liquid, if both solid and liquid are required. If the solid is to be discarded, then a large funnel can be used to increase the rate of filtration. If the liquid is to be discarded, then the flask may be large enough to take all the liquid as well as the washings.

This choice of size is important, as mechanical loss can be very great during filtration.

The filter-paper disc is placed in the funnel, and wetted with the solvent present in the solution to be filtered. It is essential that the funnel and flask be perfectly dry. If the solvent concerned is ethanol, then the paper may be wetted with water. The pump is turned on and the paper pressed into place. During filtration the pump must never be turned off, as this may cause water from the pump to be drawn back into the filtrate. When all the material has been filtered, the pump is disconnected from the flask while the pump is still running. If some of the solid has not been transferred to the funnel, a portion of filtrate is retrieved and used for swilling the residue into the funnel. The solid is washed free of filtrate by pouring *a small portion* of chilled fresh solvent into the funnel while the pump is disconnected. Finally, the solid is drained as dry as possible by suction from the pump and pressure from a clean glass stopper.

Gravimetric filtration

In quantitative work it is essential that all the solid be transferred, and retained in the filter funnel. A filtering crucible with a porous sintered-glass bottom is the most convenient apparatus to use. Porosities from 0 (coarse) to 5 (very fine) are available, although for most purposes porosity 3 is best; a few fine precipitates will require porosity 4.

The sintered-glass crucible is dried in an oven and accurately weighed before use.

To collect the solid the pre-weighed crucible is set in the mouth of the Buchner flask by means of a firm rubber cone. The pump is turned on, and as much supernatant liquid as possible is decanted off through the crucible. The liquid should be directed into the crucible down a glass rod.

The solid is then transferred, using a gentle jet of water to swill out all particles. If solid clings to the apparatus, it can be rubbed off using a glass rod protected with a rubber 'policeman'. The pump suction at this stage should be as gentle as possible, otherwise the porous glass may clog. Finally, the solid and crucible are washed repeatedly to remove all soluble materials, and dried to constant weight.

9. Drying methods

The drying of liquids

In the majority of cases with organic liquids extreme drying is not usually necessary, and drying agents like *anhydrous calcium chloride* or *anhydrous sodium sulphate* are adequate. Of the two, calcium chloride is the more efficient, but also the more messy.

As calcium chloride will remove water and

ethanol, it is employed when both need removing but, if the drying is only to remove water, anhydrous sodium sulphate is generally employed. Sodium sulphate will only work at temperatures below 30°C and should be used at room temperature. It is capable of removing its own weight of water, and the use of too much drying agent should be avoided at all times. The drying agent will be 'wetted' with the required product, and a large mechanical loss will be entailed.

In order to dry an organic liquid, whether a starting material or a product, it is placed in a suitable sized conical flask, fitted with a good stopper or cork, and the drying agent added. The corked flask is shaken at intervals, and left for at least five minutes, preferably longer.

If sufficient drying agent has been used some should remain unchanged in appearance: an opaque powder of sodium sulphate or firm granules of calcium chloride.

The drying of solids

There are various methods of drying solid materials. When deciding which method to use it is important to know something of the physical properties of the material. If dehydration of a hydrate or melting of an organic solid occurs, recrystallization has to be repeated with further loss of time and material.

For *steam drying* the substance to be dried is placed on a watch-glass, and this is supported on top of a water-bath by means of an appropriately sized ring. The water is kept boiling gently, and the steam escapes round the side of the watch-glass. It is inadvisable to attempt to dry any substance of melting-point lower than about 110°C in this way. Care must be taken to avoid getting water from the bottom of the watch-glass on the substance in the subsequent handling.

A *thermostatic oven*, kept at a few degrees above 100°C, is one of the most convenient methods of drying the higher melting solids. Since such an oven will be a communal one it is essential that the specimen to be dried is clearly labelled.

Although the method of *air drying* takes longer than the others, it is one of the safest for non-deliquescent solids. The damp solid, drained as dry as possible on the filter, is transferred to a watch-glass and spread out evenly. This can be left to dry overnight in some dust-free place. If possible a second, larger watch-glass should be arranged over the product as a precaution against dust, but this cover should allow free evaporation.

Though the *desiccator* is ideal for many solids, care must be taken when drying hydrates. It is quite possible to lose some water of crystallization if the dehydrating agent is very effective.

Samples to be dried should be spread out on a watch-glass and labelled with their name and date.

The desiccator must be regularly recharged with fresh desiccant, and the ground-glass seal kept greased with the minimum of Vaseline, so it appears transparent. A few desiccants are listed in Table 4 with comments on their relative usefulness.

Table 4

Desiccant	*Remarks*
Phosphorus (V) oxide	Expensive, fast and efficient
Concentrated sulphuric acid	Cheap, hazardous, fast and efficient. If $BaSO_4$ is dissolved in the acid, it precipitates when the drying capacity is exhausted
Calcium chloride	Cheap, moderate effectiveness. Use if ethanol was the solvent
Soda-lime	Use if acidic vapours need to be absorbed
Silica gel	Readily regenerated, limited effectiveness. Changes colour when exhausted if stained with $CoCl_2$

It is important to remember that after opening a desiccator takes at least two hours to re-establish a dry atmosphere.

A *vacuum desiccator* is used to speed the drying of a sample. The sample must be covered with a second watch-glass and the desiccator de-evacuated slowly to avoid blowing the sample about.

A vacuum desiccator must be covered with a cloth when being evacuated and de-evacuated to guard against an implosion.

10. Recrystallization and purification of solids

Organic solids, when first prepared, are rarely pure. The original solid must be recrystallized from an appropriate solvent. If the solvent is an inflammable liquid, as it often is, it is better to carry out a recrystallization under reflux, until experience has been gained. With ethanol, a very common solvent, it is quicker and neater to use a conical flask, but this does entail a risk of fire.

Reflux method

The solid is placed in a suitable sized flask, preferably a conical flask as it can be easily put aside to cool, and a condenser attached. A little solvent is poured down the condenser and the mixture raised to boiling point. If all the solid has not dissolved, a little more solvent is added after removing the Bunsen burner, until the solid just dissolves at the boiling-point. If there are no insoluble solid impurities, the solution will be clear. The flame is extinguished, the flask disconnected and the solution cooled by swirling the flask under a gentle stream of water from a tap. The solid usually crystallizes at once, but, if slow to start, scratching the inside of the flask with a glass rod frequently helps crystals to form. The flask should be cooled at least to room temperature, or preferably rather lower by placing the flask in iced water or in a refrigerator.

The pure product is filtered off at the pump. It is essential for the filter flask and funnel to be clean and dry, except for the solvent concerned. The mixture to be filtered is poured on to the filter paper and the solid remaining in the flask is washed out with the filtrate. This is important. The filtrate is, of course, a saturated solution of the required solid, and so the filtrate cannot reduce the yield by dissolving some of the crystals.

The filtrate is used for washing out the flask several times until all the solid has been transferred to the filter. On no account should fresh solvent be used for transferring the solid to the filter.

To wash the solid on the filter to remove traces of the solution, the pump should be disconnected, enough fresh solvent just to cover the solid poured on to the solid in the funnel, and then the pump reconnected. In this way the mechanical loss is reduced.

The recrystallized solid is then dried in a suitable manner, bottled and labelled.

Open flask method

This is essentially the same as the previous method, but is carried out over a very small flame with an open conical flask. The solvent is only just allowed to come to the boil and then the flask is removed from the flame. It is possible to see the vapour condensing inside the flask, and there should not be a risk of fire if care is taken. The obvious advantage of this method is its speed.

The method is not suitable for low-boiling solvents such as ether.

Recrystallization requiring hot filtration

If, during a recrystallization, there is an insoluble solid impurity, it becomes necessary to filter the hot solution. Care must be taken that no crystallization occurs during the process as this would block up the filter funnel and cause great difficulty. To avoid this, the following procedure is used:

The crude solid is dissolved in the solvent in the normal way, and when all the solute has just dissolved at the boiling-point, a further small quantity of solvent is added. This ensures that the solution is not saturated. This solution is kept hot while a separate sample of solvent is heated to boiling and then poured through the prepared Buchner funnel. This procedure heats up the funnel and flask. The filter paper, which must be in position, is held in place by a glass rod. The funnel chosen should be reasonably large. Not only does this retain the heat better, but filtration will be faster.

The hot solution is now filtered rapidly with the pump full on. As soon as all the solution is through the funnel, the pump is disconnected and the funnel removed.

At this stage the solute will almost always have begun to crystallize out in the Buchner flask. To save mechanical loss, the solution should be kept in the Buchner flask and cooled under the tap in the normal way. The final filtration to collect the crystals therefore requires another Buchner flask.

The use of activated charcoal

Sometimes there are coloured impurities present in the crude material to be recrystallized. These are removed from the solution while hot by adsorption on activated charcoal.

The recrystallization is carried out normally until the crude material is dissolved. A little extra solvent is added, and the mixture cooled slightly. A small amount of activated charcoal is added to the cooled material. *It is important to cool the solution before adding the charcoal*, as this material tends to promote boiling. Often the whole solution will boil over violently on the addition of charcoal if insufficiently cooled.

The mixture with the charcoal is allowed to boil gently for a few minutes, and is then filtered hot, using the method described above. It is important to ensure that the paper is well fitted or charcoal may get round the edges and spoil the product. As in hot filtration, the funnel should be large so that the filtration is as rapid as possible. The flask should be of a suitable size for the volume of purified solution obtained.

The use of a chromatographic column for purification

Some impure organic solids, especially those appreciably soluble in ether in the cold, can be purified very effectively by the use of a column. Suitable examples of this type of substance are benzene-azo-2-naphthol and 2-methoxynaphthalene. These substances are difficult to purify satisfactorily by other means.

An alumina column is prepared using ether, the column of alumina being 5 to 10 cm long for a few grammes of impure material. All the impure solid is placed on top of the alumina, and ether is run through the column. A pure dilute solution of the required product is obtained from the column, and dark bands of impurity are formed on the column.

If ether is being used as the solvent, the end of the column should be *inside* the collecting vessel. Otherwise, the evaporation of the ether at the end of the column is often accompanied by oxidation of the purified solute. This is especially true of benzene-azo-2-naphthol.

For further details see p. 115.

11. Determination of melting-points

The determination of melting-points is used extensively and is of greater importance than boiling-points as a criterion of purity. Melting-points of solids are very sensitive to traces of impurity, being lowered by many degrees. Also, solids are easily recrystallized and can be obtained in a state of purity not normally obtained with liquids.

Because of this fact, many liquids are characterized by preparing from them solid derivatives and determining their melting-points.

The apparatus

There is a great variety of pieces of apparatus designed for the determination of melting-points. Apart from electrical methods, they all consist of some type of liquid-bath, a thermometer and a melting-point tube.

The liquid-bath can be a simple beaker, or better a 100 cm³ Kjeldahl flask fitted with a small side arm (see Fig. 5). If possible the beaker or flask should be permanently mounted with a gauze on a stand.

The choice of liquid is not easy. The liquid employed should be clear, capable of being heated without boiling or catching fire up to at least 250°C, and safe to use. Such a liquid is difficult to find.

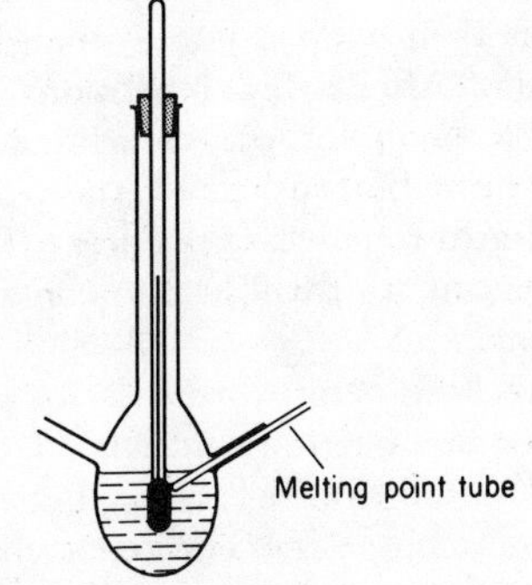

Fig. 5. Apparatus for the determination of melting-point.

Concentrated sulphuric acid is probably the best on the first two points for, if a crystal of potassium nitrate is occasionally added, it remains quite clear.

It suffers the obvious disadvantage of danger, and also is hygroscopic.

Liquid paraffin is safe, but darkens easily and gives off irritating vapours at the higher temperatures. It may also catch fire.

A variety of *high-boiling esters* are available and these seem to offer the best compromise. Dibutyl phthalate is suitable.

The thermometer should have an internal scale, as the graduations of an external scale tend to lose their paint and become difficult to read. A small rubber band round the thermometer may be used to keep the melting-point tube in place when the beaker is being employed, but usually the tube will remain in place by capillarity. A circular glass stirrer is required for the beaker.

If the Kjeldahl flask is being used, no stirrer is required, and the melting-point tube is inserted through the side arm so that the filled end rests against the bulb of the thermometer.

The melting-point tube consists of a thin glass tube, about 0·1 cm internal diameter and about 7 cm long. It is sealed at one end. Occasionally these tubes are supplied open at both ends. If one end is held just inside a small flame it seals up very easily.

The method

An unglazed porous plate can be used to dry a freshly filtered solid in order to determine its melting-point. A sample of the damp solid is crushed on to a small area of the plate with a metal spatula. It is then scraped into a small heap and crushed again. After a few repetitions the solid will be dry enough for the determination. All excess solid must be brushed off the plate and a pencil line drawn round the used area. This must not be used again, as it will lead to contamination of future samples.

To fill a melting-point tube, it is necessary for the solid to be in a fine state of division, and large crystals should be crushed with the end of a spatula on a watch-glass. The open end of the tube is pushed through the pile of solid against the spatula until about a mm length of solid has entered the tube. The tube is now tapped vertically against a hard surface, such as an inverted watch-glass, and the solid will fall to the bottom. The filling and tapping process is repeated until a total length of from 0·5 to 1 cm of solid is compacted into the bottom of the tube.

At least three tubes should be filled in this way for a determination.

To determine a melting-point takes some time until experience is gained. Time can be saved if, at the beginning, a small sample of the solid is placed on the outside of a test tube containing boiling water. If the substance remains unchanged it melts above 100°C. If it does melt, the rate of melting gives an idea of the approximate melting-point; e.g. if it melts slowly, the melting-point is probably above 70°C.

A tube is now placed in the apparatus and adjusted until the crystals of solid can be easily seen. Heating is carried out with a micro-burner, with a flame not more than 3 cm high. The burner should be left under the gauze without being disturbed. The temperature rise per second then becomes fairly uniform. As soon as signs of melting are seen (these are usually a contraction of the solid followed by a damp appearance), the burner is removed and the temperature recorded. This figure is a rough one, but gives a guide for the second determination.

The old tube is removed, and the temperature allowed to drop at least 15°C before the second tube is positioned. Now the burner is applied with care so that the rise in temperature can be controlled and, if necessary, kept at a given point. The formation of a visible meniscus in the melting-point tube is taken as a sign of melting. The temperature is recorded. The tube is removed, the temperature allowed to drop at least 10°C and the third tube inserted. A further accurate determination is now made with this third tube.

Pure solids should melt over a range of only 0·5–1°C.

12. Determination of boiling-points

A pure liquid will distil at a constant temperature, at a given pressure. This fact is used as the basis for boiling-point determination for checking the purity of liquid products.

Whenever a liquid substance has been prepared its boiling-point should always be determined.

A small distilling flask is used (if possible 10 cm³) and about 4 cm³ of test liquid should be used. Less liquid than this tends to result in overheating and a variable boiling-point even if the liquid is pure.

The sample is distilled, holding the Bunsen burner in the hand. A small flame must be used, and only that part of the flask containing the liquid should be heated. Porous pot is essential.

Distillation should be slow, not more than 2 drops per second, and must be continued once started. If the liquid is allowed to stop boiling, subsequent distillation will be less smooth. The temperature is observed carefully throughout, and the first fraction is discarded. When the temperature remains steady, the distillate is collected and the temperature recorded. The last 0·5 cm³ of liquid is allowed to remain in the flask undistilled. There is a danger of overheating if distillation of this last fraction is attempted.

Using a clean set of apparatus, the middle fraction collected in the first stage is now redistilled, the first fraction being discarded until the temperature remains constant. The liquid now obtained as the distillate is quite pure, and the boiling-point is recorded.

Unless the liquid is required exceptionally pure, the second distillation may be omitted and the boiling-point recorded as the temperature at which the middle fraction distilled.

13. Yields and their calculation

However carefully a reaction may be carried out, yields in organic chemistry rarely approach 100%. This is due to side reactions, and is unavoidable.

A further big source of loss in all preparations is mechanical loss incurred by evaporation, and by the wetting of the walls of the containing vessels. It is important at all times to think of evaporation, and the loss entailed by the wetting of containers, when organic liquids are being handled. A volatile liquid should never be left in an unstoppered flask.

During all manipulation from the start of the experiment until the final product is safely bottled and labelled, the smallest possible containers should be used. Whether flask, filter funnel or dropping funnel, the article should be chosen as the smallest that will contain the amount of liquid being used. Filter papers are by far the worst offenders; the amount of liquid absorbed by filter paper is large, and it is possible to lose all of a product by using too large a piece of paper.

The final quantity of product obtained is expressed as a percentage of the theoretical yield possible, as calculated from the equation for the reaction. It is not often necessary to use a completely balanced equation, but merely the relative weights of the original material and the product.

Some reagents are used in slight excess in nearly all preparations, and their weights need not be considered in the calculation.

Example 1: The preparation of 1-iodobutane

butan-1-ol (C_4H_9OH, M_r 74) ⟶ 1-iodobutane (C_4H_9I, M_r 184)

Hence 74 g of butan-1-ol should give 184 g of 1-iodobutane.

Suppose the preparation was carried out with 14·8 g of butan-1-ol, and the final yield was 31·2 g of pure 1-iodobutane,

then the theoretical yield was: $\frac{14{\cdot}8}{74{\cdot}0} \times 184 = 36{\cdot}8$ g

so the percentage yield was: $\frac{31{\cdot}2}{36{\cdot}8} \times 100 = 85\%$

The result is quoted to the nearest whole number.

Example 2: The preparation of ethyl acetate

In this case two organic substances are used, ethanol and acetic acid. Which should be used for the calculation of the yield? As an isolated case, it is difficult to see which is the more important, but if the preparation were one of a series of investigations, the important one would be obvious.

Suppose a series of experiments were carried out to find the amount of ester obtained when aliphatic acids were treated with ethanol and concentrated sulphuric acid. Clearly in this case the acid would be the important item, and the calculation of yield would be based on it.

It is sometimes of interest in a case like this to work out the yield on both the substances. When calculating the yield based on the ethanol, that used in the preliminary stage with the sulphuric acid must also be included. (See preparation of ethyl acetate on p. 60.)

Suppose 12 g ethanol and 8 g acetic acid were used, and the yield was 10 g of ethyl acetate.

$$\underset{M_r\ 46}{C_2H_5OH} + \underset{M_r\ 60}{CH_3CO_2H} \longrightarrow \underset{M_r\ 88}{CH_3CO_2C_2H_5} + H_2O$$

Hence 46 g ethanol with 60 g acetic acid should give 88 g of ethyl acetate.
Based on the acid the theoretical yield was:

$$\frac{8\cdot0}{60\cdot0} \times 88 = 11\cdot7 \text{ g}$$

so the percentage yield was:

$$\frac{10\cdot0}{11\cdot7} \times 100 = 85\%$$

Based on the alcohol the theoretical yield was:

$$\frac{12\cdot0}{46\cdot0} \times 88 = 23\cdot0 \text{ g}$$

so the percentage yield was:

$$\frac{10\cdot0}{23\cdot0} \times 100 = 43\%$$

In this case the much lower yield is due to two factors. Firstly, the total weight of alcohol used was in fact greater than the theoretical quantity needed for 8 g of acetic acid, and secondly, the alcohol is extensively decomposed by the sulphuric acid. However, if the alcohol was the important factor in a series of reactions it would be necessary to count all the 12 g used, because the addition of the alcohol to the sulphuric acid is an essential part of the procedure.

The result of these calculations shows that the method is good for forming ethyl esters of aliphatic acids, but gives poor yields as a method for making acetates of simple alcohols.

In research work in organic chemistry, if a yield is very low (e.g. 10%) the important original material is recovered, and the percentage recovery recorded.

The calculation of percentage yield is of extreme importance when a series of reactions is being carried out, each starting material being the product of the previous stage. If yields are low, the material will effectively disappear before the series is complete.

Suppose a five-stage preparation is carried out and the yield in each stage is 80%. This seems good, but the overall yield of final product is given by the equation, for n stages:

$$(\text{Fractional yield})^n \times 100 = \text{percentage yield overall;}$$

e.g. $$(0\cdot80)^5 \times 100 = 33\%$$

14. General cleaning

One of the most important, and also one of the most neglected, aspects of practical work is the cleaning of apparatus. Once allowed to get really dirty, glassware becomes much more difficult to clean, and the cleaning can entail the use of expensive chemicals, whereas, if used at the right time, water would have been sufficient.

The right time to clean apparatus is directly after use.

At the end of an experiment, and to a certain extent during an experiment, all used glassware should be rinsed out with tap water. This procedure will remove a surprisingly large number of substances if done as soon as possible. However, in organic work, flasks are frequently made dirty by substances insoluble in water, but easily soluble in organic solvents. For each worker to wash out glassware individually with solvent becomes extremely expensive, so a communal washing place should be available.

The communal washing-up place is provided with an assortment of beakers, a few filter funnels, a bottle of suitable solvent and a residue bottle into which all washings are poured. For most purposes, ethanol and propanone are among the best solvents. Only one solvent should be provided, thus avoiding the possibility of mixing the residues.

It is possible to recover the washings by distillation or column chromatography and use them over again. The actual loss of solvent is surprisingly small.

Particular attention must be paid to Buchner funnels. Being opaque, these must be washed out thoroughly with solvent to remove substances which become lodged in the holes.

Inevitably some apparatus will not become clean by treatment with solvent alone. Hot water and detergent, or even hot liquid detergent alone may have to be used.

For flasks which become stained with tarry deposits and other difficult materials, concentrated nitric acid is useful. A special supply is kept for the purpose. The flask to be treated is set up under a reflux condenser, porous pot and about 15 cm^3 of concentrated nitric acid added. This is slowly heated to boiling. *A fume cupboard (hood) should be used* for this apparatus, since nitrogen dioxide is evolved during the oxidation of the impurities. The same method can be used for cleaning condensers, especially double-surface condensers, fractionating columns, etc. The dirty piece of apparatus is attached to the flask, the acid boiled carefully and the process stopped when the condensing acid just reaches the top. In all these cleaning processes with nitric acid the assembly is allowed to cool down to near room temperature, dismantled and the acid returned to the cleaning stock. This method can only be used, of course, with glassware fitted with interchangeable ground-glass joints.

For small pieces of glassware which are really difficult to clean, an acid-bath can be used. The most convenient mixture consists of concentrated sulphuric acid with a little nitric acid added. More nitric acid is added at intervals as the previous amount gets used up in oxidizing the impurities. Since the mixture is rather dangerous it should not be kept in the laboratory, but in some suitable place in a dispensary or store. Ideally the acid-bath should consist of a large desiccator. So long as the top is well greased it will keep the acid from absorbing water and can easily be removed. Small pieces of apparatus can be put into, and removed from, this acid-bath using an old pair of crucible tongs. When removed, the articles are rinsed under the tap and either set aside to drain dry or placed in an oven.

If an acid-bath is being used regularly, all Buchner funnels which will fit in should be cleaned this way fairly frequently.

One of the most useful pieces of apparatus is the metal spatula of the Chattaway pattern. It is essential for handling solids during filtration processes. A spatula, which is usually made of nickel or stainless steel, should never be cleaned with acids or abrasive material. Even fine steel wool causes scratches which hold impurities. The safest material for cleaning a spatula is the wadding used for cleaning chromium plate. A spatula needs frequent cleaning.

15. Recovery processes

Silver

A number of materials used in the laboratory can quite easily be recovered, with a consequent saving of money so long as the process does not involve expensive materials or take too long.

Silver nitrate is the most expensive of the materials used frequently in the laboratory. A certain saving is made by using 0·05M solutions for most purposes, but when a large amount of solution is being used, as in titrations, it is still worth while recovering it.

The used solution should be collected in a communal container of several litres capacity. Add dilute sodium chloride solution to the container before the collection and, when the residues are all collected, add a slight excess of ammonia solution and of granulated zinc.

The silver is gradually displaced by the zinc, and the complete reaction takes several days, with occasional stirring.

Decant off the supernatant solution, wash the metallic precipitate with water by decantation, and then add an excess of 2M hydrochloric acid to dissolve the zinc.

Decant off the acid, wash with water again, and then filter off the crude silver at the pump. Wash thoroughly with water and then dry in an oven, at over 100°C.

Even after this treatment there is still zinc present, but the silver is pure enough to be made into silver nitrate solution for use as an ordinary laboratory reagent for testing for halide ions.

To purify it further, the residual zinc can be burnt off. Place the silver in a suitable ceramic container, and heat it strongly in an oxy-gas flame. The silver will melt and the zinc will burn. The residual block is nearly pure silver. To make up a solution from it, add insufficient concentrated nitric acid to dissolve the weighed block, and, when the reaction is over, remove and weigh the residual silver. The weight of silver nitrate in the solution can be calculated, and the solution can be diluted to the concentration required.

Iodine

Although iodine is not as expensive as silver, a great deal is used in sodium thiosulphate titrations in the form of potassium iodide. At the end of the titrations, a solution containing iodide ions, together with sodium, potassium, tetrathionate and maybe other ions as well as starch and other impurities, remains.

Collect the residues in a communal container which will hold several litres, and carefully add 15% (2M) sodium hypochlorite solution, with stirring. Since an excess of hypochlorite will oxidize iodine, it is important to watch the reaction carefully.

When the iodine has been precipitated, allow it to settle and then pour off the supernatant liquid. Wash with water by decantation. Depending on how much starch is present, the washings will be muddy brown (no starch) or blue (starch).

Now filter off the iodine at the pump, using a large Buchner funnel. Wash thoroughly and allow to dry at room temperature.

The crude iodine still contains insoluble impurities, and is best purified by Soxhlet extraction. Sublimation on a large scale is not recommended.

Set up a 250 cm³ Soxhlet extractor with a litre flask and a double-surface condenser. Use either a thimble or a ceramic disc and glass wool to prevent the iodine blocking the siphon tube. Do not use metal. See p. 12 for details of Soxhlet extraction.

Extract the iodine with tetrachloromethane. As the solution in the flask becomes saturated, iodine crystals form. The iodine is appreciably vaporized with the solvent, but this does not affect the efficiency of the extraction. If the crude iodine will not all fit in the extractor at once, stop the extraction to refill it until no crude iodine is left.

When the extraction has been completed, cool the flask and filter off the iodine at the pump. Allow to dry at room temperature. The filtered crystalline iodine is pure.

Keep the tetrachloromethane solution for use in the next iodine recovery.

Suggested uses for recovered iodine

Since most of the recovered iodine was purchased as potassium iodide, there may be more recovered than is needed for ordinary laboratory use. Unfortunately it is not a simple process to convert it back to potassium iodide. Two suggested uses are:

(*a*) The preparation of organic iodides, as a class exercise, by one of the methods described on pp. 63 and 64. It is suggested that different members of the class make different iodides.

(*b*) The preparation of a solution which can be used in place of potassium iodide in titrations. To make up a 0·5M solution of sodium iodide (which will contain tetrathionate ions, which will not interfere), add 63·5 g of iodine to 124 g of sodium thiosulphate pentahydrate crystals and dilute nearly to 1 litre with water. Stir the solution until all the solid has dissolved. If there is an excess of iodine, add 0·1M sodium thiosulphate dropwise to remove it. If there is no excess of iodine, add 0·1M iodine solution until a faint colour persists and then decolorize with 0·1M sodium thiosulphate. Finally, dilute to exactly 1 litre with water and shake thoroughly.

Solvents

Large volumes of certain solvents are used occasionally in class experiments, such as extraction and chromatography, while smaller volumes are being used regularly for recrystallization and other small-scale uses.

Keep a number of residue bottles clearly

labelled 'Benzene Residues', 'Ether Residues', 'Propanone Residues', etc. Winchesters are a convenient size for solvents which are being used regularly in small quantities. Make sure that they are kept in a place which is well known, and inculcate the habit of putting the residues in them. It is wise to put solid benzene-1,4-diol in ether residue bottles to reduce any peroxides that may form, which would otherwise be dangerous when the ether was distilled.

When sufficient solvent has been collected for recovery, dry it over anhydrous sodium sulphate and filter the dry liquid into a large flask, preferably two-necked, and fit it with a dropping funnel and a fractionating column and water condenser for fractional distillation. Put any of the liquid which will not fit into the flask into the funnel, and distil, using a heating mantle to minimize danger of fire if the flask should crack. Collect the fraction boiling at the correct boiling-point.

Alumina for chromatography

It is difficult to purify alumina completely, but it is possible to clean it sufficiently for re-use in elementary exercises.

Boil the alumina with water for a while, filter it at the pump, then transfer it to a large evaporating basin and heat it until it is thoroughly dry. Transfer it in a suitable ceramic vessel to a muffle-furnace, or an electric pottery kiln, and heat it long enough to burn out or volatilize organic impurities.

Finally, pass it through a sieve of suitable mesh.

Pretreatment and recovery of ion-exchange resins

In these notes a 500 g sample of ion-exchange resin is considered.

Resin, as supplied from the manufacturers, is moist and reasonably pure. For the experiments on p. 108 dry hydrogen-form resin is needed and, if purchased in hydrogen form, the only pretreatment required is to slurry the resin twice with 500 cm^3 of pure water and decant off the cloudy supernatant suspension of 'fines'. The resin can be 'conditioned' with 2 bed-volumes of 2M hydrochloric acid (see below), but this may be omitted and the resin dried at less than 100°C in a large evaporating basin. When dry, the resin is free-running and spreads widely if spilt.

The recovery of the resin after use involves the regeneration, in a column, of the hydrogen form. Resin residues can be collected in 6M hydrochloric acid until sufficient is available to make regeneration worth while.

Set up a glass tube about 50 cm $\times$ 5 cm with a rubber bung, carrying a wide-bore stop-tap, cemented in the lower end with epoxy-resin. Put a filter paper, held down by glass-wool, at the bottom of the tube and pour in 500 cm^3 of 2M hydrochloric acid. Wash the resin residues twice by decantation, using 2M hydrochloric acid, then pour the slurry into the glass tube to form a column of resin, free from trapped air. Regenerate the resin by passing through 4 litres of 2M hydrochloric acid (i.e. 7 bed-volumes) at the rate of 50 cm^3 per minute. Finally, rinse with pure water until the effluent is neutral, dry as above, and the resin is ready for re-use.

III Inorganic Chemistry

GROUP I Li, Na, K, Rb, Cs, Fr

GROUP I consists of highly electropositive metals usually called the *alkali metals* because of the highly alkaline hydroxides they form. Francium does not exist as a stable isotope; tracer techniques show that it behaves similarly to caesium. Lithium forms a much smaller cation than any of the other metals in the group, and as a result shows certain anomalies; its behaviour should be compared with that of magnesium.

Investigation of properties

Cations in solution

Use a (0·1M) solution of the chloride; if lithium chloride is unavailable add 2M hydrochloric acid to lithium carbonate so as just to dissolve it, then dilute with water. Add the following reagents dropwise, and look for signs of precipitates. Try to explain your observations, if any.

(*a*) 2M ammonia followed by 2M ammonium carbonate.
(*b*) 2M ammonia followed by 0·1M ammonium fluoride.
(*c*) 2M ammonia followed by 0·1M disodium hydrogen phosphate.
(*d*) A fresh solution of sodium hexanitrocobaltate (III), $Na_3[Co(NO_2)_6]$.
(*e*) Potassium antimonate, $K[Sb(OH)_6]$, solution.
(*f*) Uranyl zinc acetate solution.

Flame colours

Use the chlorides of the metals if possible. A convenient alternative to the conventional use of a platinum or nichrome wire is to take a clean test tube, half fill it with water and then dip it into a little of the solid to be investigated, moistened with hydrochloric acid. Use a non-luminous Bunsen burner flame, and hold the wire or tube in the edge of the flame.

If possible, use a direct vision spectroscope rather than merely observing the colour.

All the metals in the group give characteristic lines; try to explain them.

Stability of salts

Heat, in hard-glass test tubes, samples of the carbonates of lithium, sodium, potassium. Test the gases evolved by bubbling into fresh barium hydroxide solution. Compare the amount of precipitate obtained under the same conditions from the three carbonates.

Repeat the experiment using the nitrates. Test for the evolution of oxygen, and look for brown nitrogen dioxide. To the residues add 2M hydrochloric acid.

Degree of hydration of crystals

Obtain fresh crystalline specimens of the sulphates of lithium, sodium, potassium. Weigh a sample accurately and heat to constant weight in a weighed crucible.

For details of the technique see p. 182.

Work out x in the formula $M_2SO_4.xH_2O$.

Solubility in organic solvents

Compare the solubilities of lithium chloride and sodium chloride in ethanol and in propanone. Try to explain your results.

The hydrides

Lithium hydride (LiH) is the most stable of the alkali metal hydrides. Investigate the reaction of a small piece of it with water.

GROUP II Be, Mg, Ca, Sr, Ba, Ra

The elements in Group II, the alkaline earth metals, are all electropositive and have only one oxidation state in their compounds, +2. The elements of Group IIB (Zn, Cd, Hg) which show some similarities, especially to magnesium, are considered under 'd-Block elements', p. 44.

Beryllium, like lithium, shows certain anomalies as a result of its very small cation. Its behaviour should be compared with that of aluminium. *Beryllium and its compounds are very toxic.*

Radium is too radioactive to be handled in a normal laboratory. In most respects it resembles barium closely.

Investigation of properties

Cations in solution

Use a soluble salt in 0·1M solution. Add the following reagents dropwise and observe any changes. Try to explain them.

(*a*) 2M sodium hydroxide.
(*b*) 2M ammonia.
(*c*) 2M ammonia, followed by 2M ammonium carbonate.
(*d*) 0·1M ammonium fluoride.
(*e*) 2M sulphuric acid.
(*f*) Saturated calcium sulphate solution (about 0·015M).
(*g*) 0·1M ammonium oxalate, then 2M acetic acid.
(*h*) 0·1M potassium chromate, then 2M acetic acid.
(*i*) 2M ammonium chloride, 2M ammonia and 0·1M disodium hydrogen phosphate.
(*j*) 2 drops of magneson I (*p*-nitrobenzene-azo-resorcinol) and 2M sodium hydroxide.

Flame colours

Carry out the flame test using the chlorides. Omit beryllium and magnesium, which require a higher temperature than a Bunsen flame to give emission spectra. Use a direct vision spectroscope to observe the flame colours if possible.

Stability of salts

Heat, in hard-glass test tubes, samples of the carbonates and nitrates, looking for evolved gases. The beryllium salts, which may not be readily available, both decompose below 200°C.

Degree of hydration of crystalline salts

Gravimetric determination of the loss of water of crystallization from magnesium and calcium chlorides will not be very accurate because these salts are deliquescent; nevertheless, the experiment will give a guide to the result. Carry out gravimetric determinations on the chlorides and the sulphates of the metals. (See p. 182.)

Organometallic compounds

Dry a little magnesium in an oven, and add it to a dry solution of iodoethane in ether. If nothing happens, add a small crystal of iodine and warm very gently on a water-bath. Compare with what happens using calcium, and barium if possible. (See p. 69.)

Nitrides

Magnesium turnings burn readily in air. Investigate the product obtained when magnesium is burnt in a nickel crucible, with a lid, to see if nitride is formed as well as oxide: add cold water to the product and look for the formation of ammonia gas.

Try to burn some magnesium in a stream of nitrogen from a cylinder. The main problem will be found to be the cooling effect of the gas on the heated metal.

Repeat the experiment with some calcium.

Double salts

Magnesium sulphate is one of a number of sulphates of bivalent metals which will form double salts (or lattice compounds) having the general formula $M_2SO_4.M'SO_4.6H_2O$ (the schönites). M^+ is usually K^+ or NH_4^+, M'^{2+} is one of a large range of cations of roughly similar ionic radius (including, as well as magnesium in this group, all the transition and associated metals from vanadium to zinc).

The double salt frequently has a sufficiently low solubility to cause it to be precipitated when saturated solutions of the single salts are mixed. As an example, mix together saturated solutions of ammonium sulphate and copper (II) sulphate. Distinguish clearly between the double salt formed here, and the complex salt whose preparation is described on p. 51.

The usual method of making the double salts is to mix equimolar solutions of the single salts, and then crystallize the double salt.

Potassium magnesium sulphate,

$K_2SO_4.MgSO_4.6H_2O$. Dissolve 8·7 g (0·05 mole) of potassium sulphate in 100 cm³ of water, and add it to a solution of 12·3 g (0·05 mole) of magnesium sulphate heptahydrate in 50 cm³ of water. Evaporate the solution to half its volume and leave it to crystallize, preferably in an ice-bath or refrigerator.

Ammonium nickel sulphate,

$(NH_4)_2SO_4.NiSO_4.6H_2O$. Place equal volumes (20 or 25 cm³) of 2M sulphuric acid in two beakers. Neutralize one with 2M ammonia, using an external indicator, and the other, which needs warming, by adding solid nickel carbonate until an excess is present. Filter the nickel sulphate solution into the ammonium sulphate. Crystals of the double salt should form rapidly, though scratching may be needed.

Ammonium iron (II) sulphate,

$(NH_4)_2SO_4,FeSO_4,6H_2O$, Plate 20 cm³ of 2M sulphuric acid in a 100 cm³ conical flask, and fit it with a stopper carrying a Bunsen valve (see Fig. 6). Add an excess of iron wire and warm gently. Meanwhile neutralize 20 cm³ of 2M sulphuric acid in a beaker with 2M ammonia, using an external indicator. When the reaction in the flask is over and no more hydrogen is evolved, filter the iron (II) sulphate solution into the ammonium sulphate solution. Make sure the mixture is acidic by adding 1 cm³ of concentrated sulphuric acid to the mixture, then leave to crystallize, preferably in a refrigerator.

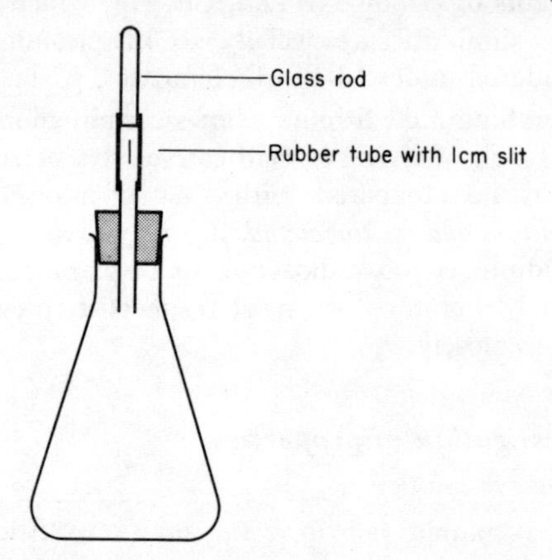

Fig. 6. Conical flask fitted with a Bunsen valve.

Compare the crystal shapes of double salts prepared as above with each other, and with those of the alums. (See p. 28.)

GROUP III B, Al, Ga, In, Tl

Group III contains a non-metal, boron; a familiar metal, aluminium; and three less familiar ones, gallium, indium and thallium. Gallium and indium show similarities to aluminium, but thallium is more profitably compared with lead and silver. The group valency of three is shown by all the elements, but for thallium valency one is more stable, and there are similarities to the alkali metals (for example, TlOH is a soluble alkali). It is useful also to compare boron with silicon, and aluminium with beryllium.

Boron compounds

Orthoboric acid, H_3BO_3. Borax, sodium tetraborate decahydrate $Na_2B_4O_5(OH)_4.8H_2O$, occurs naturally and is the chief source of boron compounds. It is the salt of a weak acid, and can be titrated with a strong acid (cf. sodium carbonate). It may be used as a primary standard.

To prepare orthoboric acid make a solution of borax in hot water (10 g in 100 cm^3) and add concentrated hydrochloric acid until strongly acid. Orthoboric acid is much less soluble in cold water than hot, and crystallizes on cooling in a refrigerator or an ice-bath.

Boron oxide, $(B_2O_3)_n$. Put about 2 g of orthoboric acid in a crucible, and heat to constant weight. Boron oxide, a glassy solid, is formed. Note the much lower melting-point than that of aluminium oxide (why?). Add some cold water to the cooled boron oxide, and note what occurs.

Ethyl orthoborate, $(C_2H_5)_3BO_3$. Volatile esters of orthoboric acid are readily formed. Make a suspension of a little borax in ethanol in a test tube, add a little concentrated sulphuric acid, fit a delivery tube and warm gently. Light the vapour produced (mainly ethanol) and note the colour of the flame caused by the ethyl orthoborate.

Aluminium compounds

Oxide and hydroxide. These are amphoteric, whereas those of boron are acidic, reflecting the much greater ease with which Al^{3+} is formed.

To a solution of an aluminium salt add 2M sodium hydroxide slowly, with agitation, until an excess is present. The aluminate ions formed are probably $Al(OH)_4^-$ (aq).

Repeat the experiment, stopping this time when the gelatinous precipitate of hydrated oxide is present. Filter the solution, wash the precipitate, dry by suction and transfer to a crucible. Heat to constant weight and compare the results with those for boron. Try the effect of acid and alkali on the heated oxide.

Chloride. Anhydrous aluminium chloride, Al_2Cl_6, is an important compound in organic chemistry. It is readily prepared by the action of dry chlorine on heated aluminium. (For practical details see p. 40). To a little anhydrous aluminium chloride add some water and look for evolution of gas. Test the solution with litmus paper.

Compare the action of heat of aluminium chloride hexahydrate with that on anhydrous aluminium chloride.

Complex salts

Potassium trioxalatoaluminate, $K_3[Al(C_2O_4)_3].3H_2O$. A few complex salts of aluminium can be prepared. A well-known example is potassium trioxalatoaluminate, similar to the oxalate complexes formed by Fe^{3+}, Cr^{3+} and Co^{3+}.

$$Al(OH)_3 + 3KHC_2O_4 \longrightarrow K_3[Al(C_2O_4)_3].3H_2O$$

Dissolve 6·7 g (0·01 mole) of aluminium sulphate octadecahydrate in 20 cm^3 of water, and add 30 cm^3 of 2M sodium hydroxide. Filter off the precipitated aluminium hydroxide, wash, and boil with a solution of 7·7 g (0·06 mole) potassium hydrogen oxalate in 80 cm^3 of water. Filter off the excess of aluminium hydroxide, and evaporate the solution to crystallization. Calculate the percentage yield.

Reference: Bailar, J. C. and Jones, E. M. In *Inorganic Syntheses*, 1, 36. McGraw-Hill, New York and Maidenhead, 1939.

Double salts

Aluminium sulphate is one of a number of sulphates of tervalent metals which will form double salts of the general formula $M_2SO_4.M'_2(SO_4)_3.24H_2O$

(or $MM'(SO_4)_2.12H_2O$), generally called alums, which form octahedral crystals.

M^+ can be any of K^+, Rb^+, Tl^+ or Cs^+, or NH_4^+, and metals forming M'^{3+} of suitable radius are Al, Ga, In, Ti, V, Cr, Mn, Fe, Co. The lanthanide metal ions are too large for them to form alums. Methods of preparation of these double salts are similar to those used for the ones described under Group II. Examples are given below. See p. 45 for the preparation of vanadium alum.

Potassium aluminium sulphate,
$KAl(SO_4)_2.12H_2O$. K_2SO_4 is anhydrous and has $M_r = 174$. $Al_2(SO_4)_3.18H_2O$ has $M_r = 666$.

A convenient quantity to use would be 0·02 mole, hence weigh out 3·5 g of potassium sulphate and dissolve in 40 cm³ of water, and add to a solution of hydrated aluminium sulphate (13·3 g in 30 cm³ water). Leave to crystallize. The solutions may be warmed to dissolve the solids more quickly.

Ammonium iron (III) sulphate,
$NH_4Fe(SO_4)_2.12H_2O$, is usually prepared from iron, which is first dissolved in sulphuric acid and then oxidized with nitric acid. To obtain the correct ratio of one mole NH_4^+ to one mole of Fe^{3+}, start with 25 cm³ of 2M sulphuric acid and neutralize with 2M ammonia to form the ammonium sulphate solution.

Now put 50 cm³ of 2M sulphuric acid in a 150 cm³ conical flask, add 6 g (an excess) of iron wire, and fit the flask with a Bunsen valve to allow hydrogen to escape and prevent entry of air. Warm gently, and when there is no more effervescence filter off the excess of iron. Add a further 25 cm³ of 2M sulphuric acid and then concentrated nitric acid until the solution, which goes first to brown nitrosoiron (II) ion, turns yellow.

Mix the iron (III) sulphate solution and the ammonium sulphate solution, evaporate to about half the volume and leave to crystallize. It is an advantage if at this stage, after cooling, 2 cm³ of concentrated sulphuric acid is added.

Potassium chromium (III) sulphate,
$KCr(SO_4)_2.12H_2O$, is most conveniently prepared by the reduction of potassium dichromate in sulphuric acid solution. Suitable reducing agents are sulphur dioxide, ethanol and propanol. Above 60°C, chromium (III) sulphate forms a rather soluble green modification which does not readily give the double salt on cooling, so care should be taken not to exceed this temperature during the preparation.

It is convenient to use 0·02 mole, hence take 5·9 g of potassium dichromate and dissolve in 50 cm³ of water. Add to the cooled solution 5 cm³ of concentrated sulphuric acid and, slowly with stirring, 5 cm³ of ethanol (a slight excess). Keep the temperature below 60°C.

Crystals of the double salt will form on cooling with iced water or in a refrigerator.

Crystals of the alums may be grown on top of other alum crystals. As an example take a well-formed crystal of potassium chromium (III) sulphate and hang it in a saturated potassium aluminium sulphate solution.

GROUP IV C, Si, Ge, Sn, Pb

The elements of Group IV show a steady change in character from a non-metal, carbon, to a metal, lead. Four, the group valency, is shown by all the five elements, but an increasingly stable valency of two is shown by germanium, tin and lead. For lead, two is, in fact, the more stable valency and lead (IV) compounds are strongly oxidizing.

So many and varied are the compounds of carbon that they are dealt with separately as 'organic chemistry'; in this section only those compounds which show the relationship of carbon with the other elements in the group are considered.

Although carbon and silicon both show only one valency and form entirely covalent compounds, the similarity between them is not as great as might be expected. Carbon chemistry is dominated by the very large variety of hydrides and their derivatives, but it is the oxygen derivatives which are more important in silicon chemistry. The relative stabilities of the bonds formed by C and Si with H and O are shown by the bond energies:

C–H 413, C–O 356, Si–H 314, Si–O 451 kJ mol^{-1}

Tin and lead both form ionic and covalent compounds. Tin (II) compounds are readily oxidized but lead (II) compounds are stable. Although germanium itself is now of considerable commercial importance, its compounds are still comparatively unfamiliar. Their properties are intermediate between those of corresponding compounds of silicon and tin.

Hydrides

Methane, CH_4, is readily prepared in an impure form either by hydrolysis of aluminium carbide, Al_4C_3, or by decarboxylation of an acetate. To carry out the latter, grind together in a mortar anhydrous sodium acetate and an excess of soda-lime, put the mixture into a hard-glass test tube fitted with a delivery tube, and heat. The methane, which is insoluble, can be collected over water.

$$CH_3CO_2Na + NaOH \longrightarrow Na_2CO_3 + CH_4$$

Natural gas is largely methane.

Silane, SiH_4, is usually prepared in an impure form by hydrolysing magnesium silicide with hydrochloric acid. The product, which is spontaneously inflammable, contains higher silanes as well as the simplest one.

A general method of preparation of covalent hydrides which can be applied to silane is the reduction of a chloride with lithium aluminium hydride in dry ethereal solution. Since the product is so inflammable, the reaction should be carried out in an inert atmosphere. One mole of lithium aluminium hydride will reduce one mole of silicon chloride:

$$SiCl_4 + LiAlH_4 \longrightarrow LiCl + AlCl_3 + SiH_4$$

Hence 1 g ($\frac{1}{38}$ mole) of $LiAlH_4$ will reduce 44·7 g of $SiCl_4$. *It is essential that conditions are completely anhydrous.*

Oxides

All the elements of the group form dioxides. All have macromolecular structures except CO_2. Carbon dioxide is readily prepared by the action of an acid on a carbonate, and when required for preparations it is usually prepared in a Kipp's gas generator. It is dried by passing through sulphuric acid.

Silicon dioxide, SiO_2, is found widely in nature, often coloured yellow by iron compounds. It can be purified by conversion to the fluoride followed by hydrolysis:

$$SiO_2 + 2CaF_2 + 2H_2SO_4 \longrightarrow 2CaSO_4 + 2H_2O + SiF_4$$

$$SiF_4 + 2H_2O \longrightarrow 4HF + SiO_2$$

Into an old 100 cm^3 round-bottomed flask (whose glass will be attacked by the hydrogen fluoride produced in the experiment), carefully dried and fitted with a dropping funnel and delivery tube, put a mixture of 5 g of dried sand (a slight excess) and 12 g (0·08 mole) of calcium fluoride.

Set up the flask on a sand-bath and arrange the delivery tube to dip below mercury in a 25 cm^3 beaker standing in a 250 cm^3 beaker (see Fig. 7). Fill the 250 cm^3 beaker with water (the mercury prevents suck-back of water, and also the possibility of blocking the delivery tube with precipitated silica), and put about 20 cm^3 of concentrated sulphuric acid in the tap funnel.

Add the sulphuric acid, and warm the mixture with a Bunsen flame. Silicon tetrafluoride gas is evolved, and as it bubbles into the water it is rapidly hydrolysed to a gelatinous precipitate of hydrated silica. This is filtered off at the pump, washed thoroughly to remove hydrogen fluoride solution and finally heated strongly in a crucible.

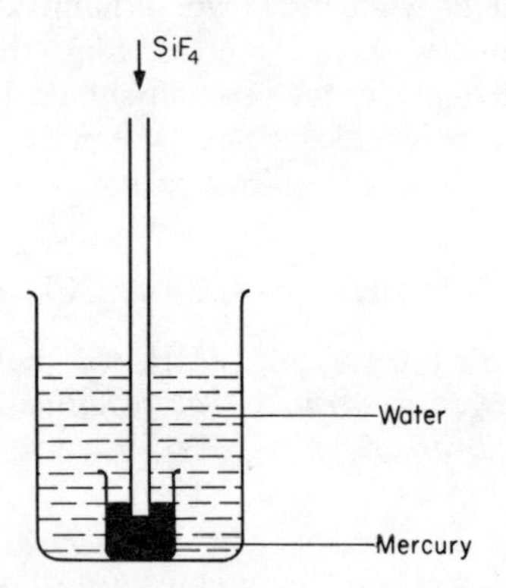

Fig. 7. Hydrolysis of silicon tetrafluoride to form pure silicon dioxide.

Tin (IV) oxide, SnO_2, is most readily prepared by oxidizing tin with concentrated nitric acid. Oxides of nitrogen are evolved and a white precipitate of the hydrated tin (IV) oxide is formed. It is filtered off, washed and heated to give tin (IV) oxide.

When prepared this way, the oxide is almost insoluble in acids and alkalis.

Lead (IV) oxide, PbO_2. Lead is oxidized only as far as Pb^{2+} ions by nitric acid, so it is necessary to use a stronger oxidizing agent to give lead (IV) oxide. A convenient one is sodium hypochlorite, in alkaline solution.

Add 5 g of lead, freshly granulated, to a mixture of 10 cm^3 of concentrated nitric acid and 10 cm^3 of water in a 250 cm^3 conical flask, and heat gently to accelerate the reaction. If the reaction is slow, add water to replace that lost by evaporation. When all the lead has dissolved, cool to room temperature, and add water if necessary to dissolve any lead (II) nitrate that has been precipitated. Now add 2M sodium hydroxide to neutralize excess of nitric acid; add it slowly until a faint white precipitate is just present.

To bring about oxidation to lead (IV) oxide, add a further 50 cm^3 of 2M sodium hydroxide and 30 cm^3 of 2M (commercial 15%) sodium hypochlorite solution, and bring carefully to the boil. The colour of the precipitate changes gradually to dark brown, and at this stage stop heating and allow the precipitate to settle.

Decant carefully as much as possible of the supernatant solution, add 100 cm^3 of water, shake up, allow to settle and decant again. Add 20 cm^3 of 2M nitric acid to dissolve any remaining lead (II) hydroxide, then 50 cm^3 of water. Filter off the precipitate at the pump, wash thoroughly and allow to dry.

Calculate the percentage yield based on the lead used.

Investigate the action of heat on lead (IV) oxide, and its reactions with concentrated hydrochloric acid and 2M sodium hydroxide. Compare its behaviour with the corresponding oxides of silicon and tin. The sample can be analysed by the method given on p. 168.

Halides

The halides, $XHal_4$, are known for all the elements except lead (IV) bromide and iodide. The chlorides illustrate the change in properties from tetrachloromethane, stable to air, water and thermal decomposition; through silicon tetrachloride, stable thermally but rapidly hydrolysed; to lead (IV) chloride, unstable and strongly oxidizing. All these halides, except possibly lead (IV) fluoride, are covalent in character.

Germanium, tin and lead form bivalent halides with increasing stability and ionic character. Those of lead are insoluble in cold water, and are readily prepared by precipitation.

Silicon tetrafluoride, SiF_4. The preparation of this compound is described above, under silicon oxide. It is stable thermally but very readily hydrolysed.

Tin (IV) chloride, $SnCl_4$, is readily prepared from its elements, which combine at room temperature. It is a colourless liquid, boiling at 114°.

A chlorine generator is required; for details see p. 40.

Lead a supply of *dry* chlorine by a gas inlet tube into a 50 cm^3 pear-shaped flask fitted for distillation with vent (see Figs. G, H).

The whole experiment should be conducted *in a fume cupboard*. Place 3 g (0·025 mole) of granulated tin into the flask, assemble the apparatus and turn on the supply of chlorine. The reaction is exothermic, but not sufficiently so for the tin (IV) chloride to be distilled over without additional heating.

When the reaction is complete, distil the product into the receiver, and then redistil. Calculate the percentage yield based on tin used.

Compare the reaction of water with the chlorides of carbon, silicon and tin.

Carbon tetrabromide, CBr_4, is usually prepared by the bromination of propanone. At first tribromomethane is produced in a reaction analogous to those used in the preparation of trichloromethane and tri-iodomethane.

$$CH_3COCH_3 + 3Br_2 + 4NaOH \longrightarrow CHBr_3 + 3NaBr + CH_3CO_2Na + 3H_2O$$

Excess hypobromite then readily replaces the final hydrogen in $CHBr_3$ with bromine:

$$CHBr_3 + Br_2 + NaOH \longrightarrow CBr_4 + NaBr + H_2O$$

The reaction is carried out in dilute solution with vigorous shaking, since tribromomethane, a dense liquid immiscible with water, would otherwise react only very slowly.

Dilute 100 cm^3 of 2M sodium hydroxide to 300 cm^3 with pure water in a 500 cm^3 conical flask fitted with a rubber or glass stopper. Add 2·5 cm^3 (0·05 mole) of bromine and shake the mixture until the colour goes.

Dilute 5 cm^3 of propanone to 50 cm^3 with pure water, and add 9 cm^3 of this solution ($\frac{1}{80}$ mole of propanone), with shaking, to the hypobromite solution. Stopper the flask and shake vigorously. After a little while droplets of tribromomethane appear. Do not stop shaking at this stage but continue until the solid bromide separates out. When the cloudiness due to the tribromomethane droplets has cleared, leave the flask to stand for a while, and then filter off the bromide at the pump.

Now return the filtrate to the conical flask and repeat the process with a second lot of bromine and propanone, adding the precipitate to the first. Wash the precipitate well with pure water and dry it over sulphuric acid in a desiccator. The colourless crystals melt at 94°C, with slight decomposition. Calculate the percentage yield, based on $\frac{1}{40}$ mole of propanone. Yields are not usually better than 50%.

Reference: Palmer, W. G. *Experimental Inorganic Chemistry*, p. 243. University Press, Cambridge, 1954.

Tin (IV) bromide, $SnBr_4$, is prepared by direct combination. If this is carried out with the elements the reaction is vigorous, with inflaming, and it is better to use an inert solvent, such as tetrachloromethane, in which the product is soluble.

Into a 50 cm^3 pear-shaped flask put 20 cm^3 tetrachloromethane and 2·5 cm^3 (0·05 mole) of bromine. Fit a reflux condenser. Weigh out 4 g (an excess) of small pieces of granulated tin and add a little at a time down the condenser, cooling the flask if necessary with a beaker of water. When all the tin has been added, warm gently until all the bromine has reacted (i.e. the colour has gone).

Now fit the flask for distillation with a 250°C thermometer. Distil off the tetrachloromethane (B.P. 77°C) and then the tin (IV) bromide (B.P. 203°C). Redistil the bromide into a weighed receiver, where it will solidify to colourless crystals, M.P. 33°C. Calculate the percentage yield based on the bromine taken.

Carbon tetraiodide, CI_4, is best prepared by halogen exchange between tetrachloromethane and iodoethane, using anhydrous aluminium chloride as catalyst:

$$CCl_4 + 4CH_3CH_2I \longrightarrow CI_4 + 4CH_3CH_2Cl$$

All the reagents must be completely water-free; the chloroethane formed is gaseous at room temperature.

Fit a 150 cm^3 conical flask (preferably with a ground-glass joint) with a calcium chloride tube. Dry about 30 cm^3 of tetrachloromethane and 30 cm^3 of iodoethane over anhydrous calcium sulphate and redistil.

Weigh as quickly as possible into the flask 5 g ($\frac{1}{30}$ mole) of the dry tetrachloromethane and 20 g ($\frac{4}{30}$ mole) of the dry iodoethane and add 1 g anhydrous aluminium chloride. Put the stopper with the calcium chloride tube in place as rapidly as possible.

If all the reagents are completely dry, the reaction will start at once and the reaction mixture will turn red, while chloroethane is produced with effervescence. Shake gently to mix the reagents; when the reaction is complete the mixture is almost solid. It will take at least half an hour to reach this stage.

Now filter off the red crystals at the pump and wash first with ice-cold water to remove aluminium chloride, and then with ethanol to remove any excess of reagents. The red crystalline carbon tetraiodide is dried over sulphuric acid in a desiccator. Calculate the percentage yield based on the organic materials taken. The crystals melt with decomposition at about 170°C; they are also unstable in sunlight.

Reference: McArthur, R. E. and Simons, J. H. In *Inorganic Synthesis*, **3**, 37. McGraw-Hill, New York and Maidenhead, 1950.

Silicon tetraiodide, SiI_4, can be prepared by direct combination of the elements. Iodine vapour in a stream of carbon dioxide is passed over red-hot silicon. It is a colourless solid, M.P. 121°C.

Germanium (IV) iodide, GeI_4, can be prepared by the action of constant-boiling hydriodic acid (containing 57% HI, B.P. 126°C) on germanium (IV) oxide. It is best carried out in an inert atmosphere, nitrogen or carbon dioxide:

$$GeO_2 + 4HI \longrightarrow GeI_4 + 2H_2O$$

Fit a 50 cm³ pear-shaped flask with a gas inlet tube and still-head and condenser. To the gas inlet tube attach a supply of nitrogen or carbon dioxide. In the flask put 0·5 g (0·005 mole) germanium (IV) oxide and 8 cm³ (an excess) of freshly distilled constant-boiling hydriodic acid. Warm gently so that the mixture refluxes until the white oxide has been replaced by orange-red crystals of iodide. This will take up to ten minutes.

Now raise the temperature to distil off the water until the product is nearly dry, then cool and filter off at the pump. Dry in a desiccator, and recrystallize if desired from trichloromethane. Germanium (IV) iodide crystals are reddish-orange and melt at 146°C. Calculate the percentage yield based on germanium (IV) oxide used.

Reference: Foster, L. S. and Williston, A. F. In *Inorganic Syntheses*, **2**, 112. McGraw-Hill, New York and Maidenhead, 1946.

Tin (IV) iodide, SnI_4, is prepared, like the bromide, by direct combination of the elements in an inert solvent (*Caution: toxic*).

Add 2 g (an excess) of granulated tin to a solution of 6·35 g (0·025 mole) of iodine in 25 cm³ of tetrachloromethane in a 100 cm³ flask fitted with a reflux condenser.

Now heat the mixture so that it refluxes gently. Once the reaction has started it is sufficiently exothermic to make heating unnecessary until the reaction is near completion. The iodine colour is not quite completely discharged, but violet iodine vapour is no longer present when the reaction is complete.

Now filter the hot solution through a pre-heated funnel to remove excess of tin. Wash the residue with 10 cm³ of hot tetrachloromethane and add the washings to the filtrate.

Cool the filtrate in a refrigerator or an ice-bath to precipitate the orange crystals of tin (IV) iodide, which are more soluble in hot than cold tetrachloromethane.

Filter off the crystals, and evaporate and cool the filtrate to obtain a further crop. The crystals melt at 144°C. Calculate the percentage yield, based on the iodine used.

Reference: Moeller, T. and Edwards, D. C. In *Inorganic Syntheses*, **4**, 119. McGraw-Hill, New York and Maidenhead, 1953.

GROUP V N, P, As, Sb, Bi

The elements in Group V show the same sort of change in character as those in Group IV from a non-metal, nitrogen, to a metal, bismuth. Nitrogen shows a valency of three only; the other elements all show valencies of three and five, though five-valent bismuth compounds are strong oxidizing agents. The behaviour of nitrogen is, in most respects, anomalous because it is not able to expand its outer shell of electrons beyond eight. In many ways the other elements, phosphorus, arsenic, antimony and bismuth show a steady change; for example, the trioxides vary from entirely acidic, through amphoteric, to entirely basic.

Hydrides of nitrogen

Ammonia, NH_3, is by far the most important of the hydrides. It is made commercially by direct combination (Haber process), but in the laboratory it is more convenient to obtain it when required by other ways: an ammonium salt is warmed with an alkali (e.g. 2M sodium hydroxide), or the concentrated solution of the gas ('0·880' ammonia) may be heated to boiling (at which temperature the solubility is only slight) and the gas dried by passing it through a tower containing fresh calcium oxide. The more usual drying agents, sulphuric acid and calcium chloride, are unsuitable because they both react with ammonia.

Hydrazine, NH_2NH_2, and **hydroxylamine, NH_2OH,** two ammonia derivatives, are both basic, though less so than ammonia itself, and strongly reducing in character. In the laboratory hydrazine hydrate ($N_2H_4.H_2O$), a liquid, is a useful reducing agent, being oxidized chiefly to nitrogen. Investigate its action by adding a drop to a solution of, for example, potassium manganate (VII), potassium dichromate, silver nitrate, chloroauric acid and Fehling's solution. See also p. 168.

Hydroxyammonium chloride ($NH_3OH^+Cl^-$) is the most familiar salt of hydroxylamine. For examples of its use see the preparation of dinitrogen monoxide and the preparation of oximes from aldehydes and ketones on p. 67.

Oxides of nitrogen

Three common oxides of nitrogen, dinitrogen oxide (N_2O), nitrogen monoxide (NO) and nitrogen dioxide (NO_2 in equilibrium with N_2O_4), are all quite readily prepared.

Dinitrogen monoxide, N_2O, is best prepared in a simple gas generator. A solution of sodium nitrite is put in the flask and a solution of hydroxyammonium chloride run in slowly from the dropping funnel. The reaction is exothermic, but warming may be necessary at first to initiate it:

$$NH_3OH^+ + NO_2^- \longrightarrow N_2O + 2H_2O$$

The gas is fairly soluble in water, but may be collected satisfactorily over water if required.

Nitrogen monoxide, NO, is best prepared by adding 2M sulphuric acid from a dropping funnel to a solution containing iron (II) sulphate and sodium nitrite in equimolar proportions and warming the mixture:

$$2NO_2^- + 2Fe^{2+} + 4H_3O^+ \longrightarrow 2Fe^{3+} + 6H_2O + 2NO$$

The gas, which reacts at once with oxygen to give nitrogen dioxide, may be collected over water. The dark brown colour formed in this preparation is the same as that in the 'brown ring test' for nitrates, $Fe[NO(H_2O)_5]^{2+}$. It decomposes on warming.

Nitrogen dioxide, NO_2, is formed when concentrated nitric acid is reduced by many substances. It is usually prepared in the laboratory by the action of heat on an anhydrous nitrate of one of the less reactive metals, commonly lead. Oxygen is produced as well in the reaction and so the nitrogen dioxide is collected by condensation to a liquid (B.P. 22°C).

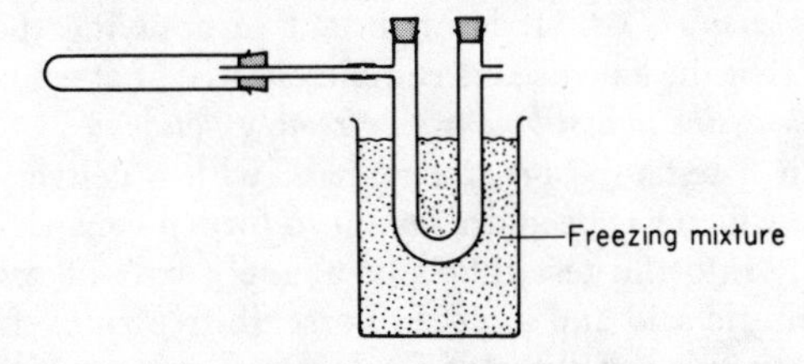

Fig. 8. Condensation of nitrogen dioxide.

Grind up crystals of lead nitrate and put them in a *hard-glass* test tube fitted with a delivery tube leading to a U-tube surrounded by an ice-salt freezing mixture (see Fig. 8). Heat the tube with a Bunsen flame, and note the oxygen emerging from the U-tube. The liquid in the tube is commonly coloured greenish because some of the nitrogen dioxide has been decomposed in the test tube to nitrogen monoxide and oxygen, and the presence of nitrogen monoxide is responsible for the colour (an equimolar mixture of NO and NO_2 condenses to a blue liquid).

Hydrides of the other elements

The thermal stability of the hydrides decreases with increasing atomic number. Bismuth hydride (BiH_3) has only been detected in trace quantities.

Phosphine, PH_3, is often prepared by the action of an alkali on white phosphorus in an inert atmosphere. As prepared this way it is contaminated with P_2H_4 which causes it to be spontaneously inflammable. A similar disadvantage is found when phosphine is obtained by the action of heat on a solid phosphite or hypophosphite. Pure phosphine is best prepared by the reduction of dry phosphorus trichloride by a solution of lithium aluminium hydride in ether:

$$3LiAlH_4 + 4PCl_3 \longrightarrow 3LiCl + 3AlCl_3 + 4PH_3$$

Hence 1 g $LiAlH_4$ will reduce 4·8 g PCl_3. *It is essential that conditions are completely anhydrous.* The gas may be collected over water.

An alternative method which gives the pure gas is analogous to the preparation of ammonia; phosphonium iodide is warmed with a solution of an alkali. Carry out these preparations *in a fume cupboard.*

Arsine, AsH_3, and **Stibine, SbH_3,** are formed by the reduction of soluble arsenic or antimony compounds by zinc in acid solution. This is the basis of *Marsh's Test.* It is important in applying the test that the zinc used is completely free of arsenic. *Both arsine and stibine are extremely poisonous.*

Fit a test tube or 25 cm^3 flask with a delivery tube which has been drawn out to form a constriction. Into the test tube put about 5 cm^3 of 2M sulphuric acid and a piece of arsenic-free zinc. If hydrogen is not liberated freely on warming, add a drop of copper (II) sulphate. Warm the delivery tube near the constriction with a microburner flame.

Now add a trace of an arsenic compound to the test tube, and repeat the experiment. Arsine is readily decomposed to its elements by heat.

Using fresh reagents and clean apparatus, repeat the experiment with an antimony compound.

Investigate the action of oxidizing agents (e.g. a hypochlorite) on any deposits formed in the tube.

Use the experiment to test for the presence of arsenic or antimony in commercial zinc, and in organic materials (e.g. hair).

Oxides of the other elements

Phosphorus (V) oxide, P_4O_{10}, is the most stable oxide of phosphorus. It is formed when any allotrope of phosphorus is burnt in an excess of air or oxygen.

It is important that all the apparatus used for the preparation is thoroughly dried by leaving in an oven for an hour or two. A possible apparatus is shown in Fig. 9.

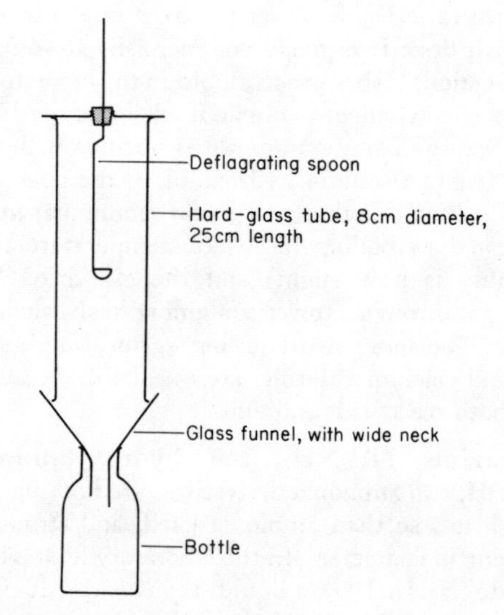

Fig. 9. Preparation of phosphorus (V) oxide

Dry a small piece (about 0·5 g, handle with tongs) of white phosphorus on filter paper and put it into the deflagrating spoon. Touch it with a hot glass rod and quickly put the spoon into position. The reaction will go better if the wide tube is filled with oxygen from a cylinder, rather than with air.

When the reaction is over, tap the sides of the tube so that the oxide drops into the bottle, which should be securely stoppered.

Investigate the effect of adding a drop of water to some of the white solid phosphorus (V) oxide, and also the effect of adding a little of the solid to hot water. Test the solution with a warm solution of ammonium molybdate in M nitric acid.

Antimony (V) oxide, Sb_2O_5, can be prepared in the same way as tin (IV) oxide by oxidizing the element with nitric acid. Carry out the reaction in the same way (see p. 30), and when it is complete filter off the hydrated oxide, wash it with water, drain thoroughly and transfer it to a crucible. Heat it to constant weight. The yellow oxide left is probably impure.

Bismuth (V) oxide. This compound has not been prepared in a pure state, although the compound called sodium bismuthate probably contains bismuth in a +5 oxidation state.

If the procedure described for the preparation of lead (IV) oxide is followed for bismuth, a similar dark brown to black solid is obtained. Its composition may be investigated by the method described on p. 168.

Oxy-salts: test tube reactions

(*a*) Use a solution of a nitrite, a phosphite, an arsenate (III) and an antimonate (III), and solid bismuth (III) oxide. Investigate the reaction with acidified potassium manganate (VII) solution, and with a solution of iodine in potassium iodide.

(*b*) Use a solution of a nitrate, a phosphate, an arsenate (V) and an antimonate (V), and solid sodium bismuthate. Investigate the reaction with a manganese (II) salt solution, and with acidified potassium iodide solution.

Halides

Phosphorus trichloride, PCl_3, is prepared by direct combination, but an excess of chlorine must be avoided to prevent the formation of the pentachloride. The experiment should be carried out *in a fume cupboard.*

Set up the apparatus for distillation with gas inlet and attach a supply of dry chlorine from a cylinder or gas generator. Make sure all the apparatus is quite dry, and arrange the height of the gas inlet tube so that it does not go too near the bottom of the flask. This helps to avoid the presence of an excess of chlorine near the phosphorus.

Either white or red phosphorus can be used in this preparation. Red is easier to handle, but white reacts more readily. Weigh 1 g of dry phosphorus, put it in the flask, and attach this without delay to the still-head. Start the flow of chlorine at once to flush out the air, then warm the flask gently on a water-bath. Keep the flow of chlorine steady but not too fast. When the reaction is over, the phosphorus trichloride should be redistilled using a 110°C thermometer in place of the gas inlet tube. It boils at 75·5°C. Calculate the yield based on the phosphorus used.

Investigate the action of phosphorus trichloride on water (test the solution formed with litmus paper, and with a little potassium manganate (VII) solution), and on ethanol.

Phosphorus pentachloride, PCl_5, is readily formed at room temperature from phosphorus trichloride. The reaction is readily reversed at higher temperatures:

$$PCl_3 + Cl_2 \rightleftharpoons PCl_5$$

Solid phosphorus pentachloride is ionic, $PCl_4^+PCl_6^-$. The reaction should be carried out in completely dry apparatus *in a fume cupboard.*

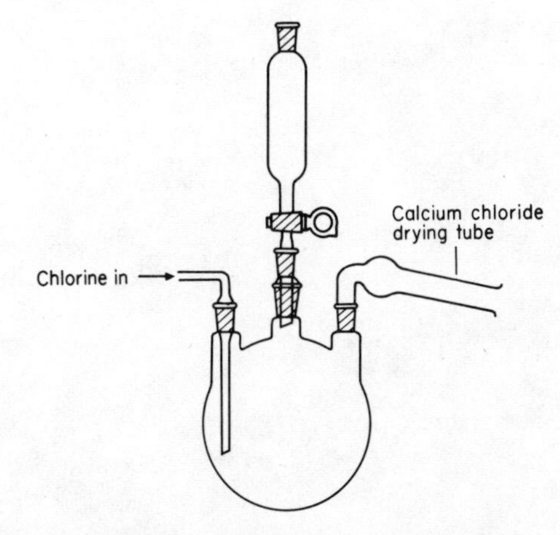

Fig. 10. Preparation of phosphorus pentachloride.

Fit a three-necked, 500 cm³ round-bottomed flask, if available (if not a wide-necked flask with a rubber stopper with three holes can be used instead), with a gas inlet tube, a dropping funnel and a calcium chloride drying tube (see Fig. 10).

Attach to the gas inlet tube a supply of dry chlorine and put 10 cm^3 (15·7 g, 0·12 mole) of phosphorus trichloride in the dropping funnel. Switch on the supply of chlorine, and allow the phosphorus trichloride to drop in at such a rate that it is converted to pentachloride almost at once.

When all the trichloride has been added turn off the gas supply, remove the funnel and tubes and shake out the product (which is a white solid) into a dry container. Weigh it and work out the yield.

Investigate the action of the pentachloride with water, ethanol and glacial acetic acid. Test the result of the reaction with water with warm ammonium molybdate solution in M nitric acid.

Antimony (III) bromide, $SbBr_3$, can be prepared in exactly the same way as tin (IV) bromide (see p. 31). It melts at 96·6°C and boils at 280°C.

Antimony (III) iodide, SbI_3, is prepared in a similar way by direct combination in an inert solvent, in this case toluene.

Add 0·6 g (0·005 mole) of antimony powder to 40 cm^3 of toluene in a flask fitted with a reflux condenser. Heat on a water-bath and, when the toluene is hot, add, in small portions down the condenser, 2 g (an excess) of iodine. When all the iodine has been added, continue heating for twenty minutes, then remove the water-bath and cool the flask. The iodide begins to form as red crystals in the hot liquid, and more are deposited on cooling.

Filter off the crystals at the pump, wash away the excess of iodine with tetrachloromethane, and allow them to dry at room temperature. Weigh the dry crystals and work out the yield.

Reference: Bailar, J. C. and Cundy, P. F. In *Inorganic Syntheses*, **1**, 104. McGraw-Hill, New York and Maidenhead, 1939.

GROUP VI O, S, Se, Te, Po

The only elements in Group VI considered here are oxygen and sulphur. Selenium, tellurium and polonium show the trend to increasing positivity noted in previous groups.

Oxides

Oxygen forms a very large number of compounds with other elements, and these have been used as a basis for classifying elements according to whether the oxides are acidic, basic, amphoteric or neutral.

Many oxides can be prepared by direct combination, but this is rarely a good method of preparing them. Various examples of oxide preparation are summarized below. Practical details will be found under the appropriate group.

(*a*) *Synthesis*

Used for phosphorus (V) oxide.

(*b*) *Oxidation of an element*

Nitric acid is quite commonly used for oxidizing an element to its oxide, usually in hydrated form. Examples of oxides prepared this way are iodine (V) oxide, tin (IV) oxide, antimony (V) oxide. Steam can also be used, for example for magnesium oxide and tri-iron tetroxide.

(*c*) *Displacement of a volatile acidic oxide from a salt*

A less volatile acid or acidic oxide is used. Examples of the use of this method are carbon dioxide, sulphur dioxide and sulphur trioxide.

(*d*) *Displacement of an insoluble oxide from a salt*

This is convenient for the macromolecular acidic oxides, such as those of boron and silicon, and also for many basic oxides, usually precipitated in hydrated form.

(*e*) *Decomposition of an oxy-salt*

This is the method by which basic or amphoteric metal oxides are frequently prepared. Thermal decomposition of carbonates or nitrates is usually used, but for some easily oxidized compounds such as iron (II) oxide or manganese (II) oxide it is necessary to use the oxalate.

(*f*) *Oxidation of a lower oxide*

This can be used for the higher oxides of lead or manganese. The lower oxide is often prepared *in situ* and then oxidized by hydrogen peroxide or sodium hypochlorite. Sulphur trioxide is manufactured in the contact process by a reaction of this type.

(*g*) *Reduction of a higher oxide*

The reaction between carbon dioxide and carbon is an example of this.

Peroxides

The name peroxide is reserved for those oxides containing an oxygen–oxygen single covalent bond which may be regarded as derived from hydrogen peroxide. The two best known are those of sodium and barium which can be prepared by the action of an excess of air or oxygen on the element.

Hydrogen peroxide, H_2O_2. This is prepared commercially by electrolytic oxidation, but in the laboratory it can be made from a metal peroxide. This illustrates its relationship to the metal peroxides.

$$H_2SO_4 + Na_2O_2 \longrightarrow H_2O_2 + Na_2SO_4$$

Cool 25 cm³ of 2M sulphuric acid to near 0°C in an ice-bath. Add, in small portions with stirring, 3·5 g of sodium peroxide. When the addition is complete, keep the mixture cold for about ten minutes, then filter off the sodium sulphate that has formed, using a pre-cooled sintered glass funnel or crucible. It is possible to concentrate the hydrogen peroxide solution by using vacuum distillation.

Reactions of hydrogen peroxide solution

(*a*) Investigate the effect of adding a little 2M hydrogen peroxide to solutions of the following: titanyl (IV) sulphate, ammonium vanadate, potassium dichromate, potassium manganate (VII). Acidify the resulting solution, and try to find out what changes have occurred.

(*b*) Add a little 2M hydrogen peroxide to the following: titanium (III) sulphate, chromium

(III) sulphate, manganese (II) sulphate, vanadium (III) ammonium sulphate; then make alkaline with 2M sodium hydroxide and warm. Try to explain what you see, and correlate the results with those from test (*a*).

(*c*) Investigate the effect of possible catalysts on the rate of decomposition of hydrogen peroxide. Use metals, preferably in finely divided form, various oxides and raw liver.

Sulphur and its compounds

Sulphur differs from oxygen in being able to expand its outer electron shell beyond eight, using 3d orbitals, and this enables it to show oxidation states up to +6.

Conversion of rhombic (α) sulphur to monoclinic (β) sulphur. When sulphur crystallizes from a solution above the transition point (95·5°C), it forms the monoclinic crystals of β-sulphur. A suitable solvent to use is xylene (B.P. *c.* 140°C).

Heat 50 cm^3 of xylene to nearly boiling on a hotplate or in an oil-bath. Add, a little at a time, crushed rhombic (α) sulphur until the solution is saturated, then add a little more xylene and raise to boiling. Now cool the solution slowly by surrounding the flask with cotton-wool or wood-shavings in a box. The long needles of β-sulphur form on cooling.

Leave some β-sulphur to stand for twenty-four hours and observe and explain any change that occurs.

Sulphur trioxide, $(SO_3)_n$. If a sample of this is required in the laboratory, it is most conveniently prepared by dehydrating sulphuric acid:

$$2H_2SO_4 + P_4O_{10} \longrightarrow 4HPO_3 + 2SO_3$$

Set up the apparatus for distillation, avoiding the use of grease on joints, using as receiver an ampoule ready for sealing cooled in an efficient freezing mixture (solid carbon dioxide, or ice and calcium chloride).

Put into the flask 5 g of phosphorus (V) oxide and 2 cm^3 of concentrated sulphuric acid, stopper the apparatus, then warm the flask with a small flame until no more white fumes come over. Seal the neck of the ampoule with a small flame.

After a time it will be observed that the sulphur trioxide has grown into long needles by sublimation. These probably contain chains of SO_4 tetrahedra.

It is also possible to prepare sulphur trioxide on a small scale by passing a mixture of sulphur dioxide and oxygen over a heated catalyst (vanadium (V) oxide or platinized asbestos) in a silica tube.

Sodium thiosulphate, $Na_2S_2O_3.5H_2O$. Boiling sulphite solution will gradually dissolve sulphur:

$$Na_2SO_3 + S \longrightarrow Na_2S_2O_3$$

Set up a 100 cm^3 flask with a reflux condenser and put into it 10 g (0·04 mole) of sodium sulphite heptahydrate, 50 cm^3 of water and 1·5 g (a slight excess) of crushed rhombic (not 'flowers') sulphur. Boil gently for about two hours, or until no more of the sulphur dissolves.

Filter off the excess of sulphur, then evaporate the filtrate to about 10 cm^3. It is unlikely that crystals will form from the cooled solution until one is put in to seed the supersaturated solution. Work out the yield.

Calcium dithionate, CaS_2O_6. Dithionic acid is a strong acid, all the salts of which are soluble in water. It contains a sulphur–sulphur bond, much stronger than the corresponding oxygen bond.

$$MnO_2 + 2SO_2 \longrightarrow MnS_2O_6$$

$$(\text{also } MnO_2 + SO_2 \longrightarrow MnSO_4)$$

$$MnS_2O_6 + Ca(OH)_2 \longrightarrow Mn(OH)_2 + CaS_2O_6$$

Make a suspension of 1 g of manganese (IV) oxide in 50 cm^3 of water and pass through it (*fume cupboard required*) sulphur dioxide from a cylinder until all the manganese (IV) oxide has dissolved.

Add a suspension of 1 g of calcium hydroxide in 10 cm^3 of water and raise the mixture to boiling. Any sulphate ions will be precipitated as calcium sulphate. Filter hot, then evaporate and crystallize out the calcium dithionate. Dry the crystals in a desiccator.

Investigate the action of heat, of acids and of oxidizing and reducing agents on calcium dithionate.

Reference: Pfansteil, R. In *Inorganic Syntheses*, 2, 168. McGraw-Hill, New York and Maidenhead, 1946.

Disulphur dichloride, S_2Cl_2. Sulphur readily forms two chlorides by direct combination: disulphur dichloride (sulphur monochloride) and sulphur dichloride (SCl_2). To obtain the former it is necessary to have an excess of sulphur present.

Set up, *in a fume cupboard*, the apparatus for distillation with gas addition, and attach to the gas inlet tube a supply of dry chlorine from a cylinder or gas generator. Use a second 50 cm³ flask for collecting the product.

Put 3·2 g of sulphur into a 50 cm³ flask, attach it to the still-head and pass chlorine through the apparatus. Heat the sulphur gently until it has all reacted. The yellow-to-red distillate is a mixture of the two chlorides. Attach the receiver in place of the reaction flask, and replace the gas inlet tube with a 200°C thermometer. Add 2 g of sulphur to the chloride mixture, distil and collect the fraction boiling between 135°C and 140°C.

Peroxodisulphates. These are the most familiar of the peroxo-salts, which all contain an oxygen–oxygen bond in the ion. They are convenient strong oxidizing agents. Use potassium peroxodisulphate ('persulphate'), $K_2S_2O_8$, for the following tests:

(*a*) Heat a little of the solid and see if oxygen is evolved.

(*b*) Investigate the action of the following in both acid and alkaline conditions: iron (II) salts, manganese (II) salts (try the effect of adding a drop of silver nitrate to manganese (II) salts before adding the peroxodisulphate), chromium (III) salts, potassium bromide, potassium iodide.

Sulphamic acid, NH_2SO_2OH. This compound, the half-amide of sulphuric acid, is very stable and a strong acid, and has become of importance as a primary standard for volumetric analysis. It is usually prepared by the action of carbamide on 100% sulphuric acid.

Investigate the reaction of a solution of sulphamic acid with:

(*a*) sodium nitrite solution,
(*b*) bromine water,
(*c*) concentrated nitric acid.

Find out from a textbook what happens if you cannot work it out for yourself.

Reference: Palmer, W. G. *Experimental Inorganic Chemistry*, p. 349. University Press, Cambridge, 1954.

GROUP VII F, Cl, Br, I, At

The halogens form the most homogeneous group of non-metals. Though fluorine differs from the rest of the group in a number of ways, including having no other oxidation state in compounds than −1, these differences are not as marked as those between nitrogen and phosphorus, or oxygen and sulphur.

Astatine appears to have no long-lived isotopes, so it has only been studied in trace quantities.

Fluorine

Fluorine chemistry involves practical difficulties because of the highly reactive nature of the element itself, the effect of many of its compounds on glass, and the unpleasant effect of hydrogen fluoride on skin. The last two problems can be overcome more readily since the advent of plastic materials which are inert to hydrogen fluoride.

Reactions involving hydrogen fluoride can be carried out in polythene or polypropylene apparatus (the latter is preferred because of its higher softening point) so long as they do not involve temperatures higher than those reached in a water-bath. Protective clothing (plastic gloves with long gauntlets, and an eye-shield) should be worn to prevent splashes of hydrogen fluoride falling on the skin, since the burns take a long time to heal. It is recommended that only dilute solutions be used.

Using the precautions above, it is a simple matter to prepare the fluorides of metals in the usual way from their carbonates. A polythene funnel should be used for filtration, and the product kept in a polythene specimen tube.

Fluorides

Investigate the solubility of metal fluorides in water by adding 0·1M ammonium fluoride to 0·1M solutions of various metal salts. Note that many complex fluorides are known, so a precipitate may be found to dissolve in excess.

Chlorine

The use of a supply of chlorine is often required, especially for the preparation of anhydrous chlorides, and the following sources are available:

A cylinder of liquid chlorine. Proper precautions should be taken in the storage and use of this.

Preparation from a hypochlorite. This is a simple and cheap method. Set up a gas generator (Fig. K, i or ii) and put in the flask solid bleaching powder (which contains calcium hypochlorite). Add 2M hydrochloric acid from the dropping funnel. No heat is required, but the flask should be agitated occasionally to allow the reagents to mix properly.

An alternative method is to put 2M hydrochloric acid in the flask and to add sodium hypochlorite solution from the funnel.

Preparation by oxidizing hydrochloric acid. The usual oxidizing agents are manganese (IV) oxide, which requires heat, and potassium manganate (VII), which does not. Add concentrated (11M) hydrochloric acid from the dropping funnel to the solid oxidizing agent in the flask.

The preparation from manganese (IV) oxide and hydrochloric acid is easier to control, especially if the oxide is in lump form, and is useful if a steady supply over quite a long time is required.

If the chlorine is required dry, as it usually is, it is passed either through concentrated sulphuric acid in a flask or Drechsel bottle (if potassium manganate (VII) is being used for the preparation, a trap must be present to prevent suck-back into the generator), or through a U-tube containing anhydrous calcium chloride.

Always use a fume cupboard or hood for all experiments involving chlorine.

Preparation of anhydrous chlorides

It is possible to prepare most chlorides by direct combination, the exact method depending on the physical state of the product. Examples of liquid chlorides prepared this way include those of phosphorus (p. 35), sulphur (p. 38) and tin (p. 30). A general method for the preparation of many solid chlorides which sublime readily is given below. The experiment must be carried out *in a fume cupboard* and, since anhydrous chlorides are usually sensitive to water, all the apparatus must be thoroughly dried before use.

Set up the apparatus for heating a solid in a stream of gas (Fig. N), and attach a supply of dry chlorine from a cylinder or a generator. Weigh out about 1 g of the metal to be used, preferably in the form of fine wire or shavings, and put it into the combustion tube with a wad of glass-wool on either side. Connect up the apparatus and allow a stream of chlorine to flush all the air from the tube before starting to heat. Now light the burner and heat the metal strongly to start the reaction. Incandescence may occur, and the chloride gradually sublimes into the receiver. Move the burner about so that the whole of the metal is thoroughly heated.

When the reaction appears to be complete, turn off the chlorine supply, and stop the heating once all the chloride has sublimed. Allow the apparatus to cool, then remove the product to a dry, weighed specimen tube. Weigh it and calculate the yield based on the metal used.

Investigate the action of water on a small amount of the chloride.

Metals whose chlorides can be prepared this way include iron, aluminium, chromium, cobalt, nickel and zinc.

Action of water on chlorides

Nearly all the elements form chlorides, which may be ionic, covalent or intermediate in character. The more electropositive the element forming the chloride, the more likely it is to be ionic (i.e. a salt).

Some of the less electropositive metals form covalent liquid chlorides similar to those of non-metals (compare $SiCl_4$, $SnCl_4$ and $TiCl_4$). The melting- and boiling-points form quite a good guide to the degree of covalent character in the bonding, while further information is furnished by the action of water on the chloride.

Simple ionic chlorides will usually dissolve with no chemical action, but hydrolysis to a greater or lesser extent occurs with most of the others. Evolution of hydrogen chloride usually accompanies hydrolysis. To investigate the action of water, add it dropwise to a small amount of the chloride in a test tube. Look for fumes of hydrogen chloride, and test both the gas and the solution with litmus. Note whether a precipitate forms, and test the solution for possible ions (e.g. for phosphate from a phosphorus chloride). Note colour changes which may indicate the formation of a complex (e.g. with iron (III) chloride).

Some covalent liquid chlorides (e.g. tetrachloromethane) are immiscible with water and react only slowly, if at all. Shake with a solution of silver nitrate at intervals over a period of hours, and look for a white precipitate of silver chloride.

See p. 125 for a quantitative study of hydrolysis.

Interhalogen compounds

Iodine trichloride, ICl_3, is one of the interhalogen compounds. It is a rather unstable volatile yellow solid.

Set up a 500 cm^3 round-bottomed flask fitted with a gas inlet tube and a calcium chloride drying tube. Attach a supply of dry chlorine to the gas inlet tube, and put into the flask 0·5 g of dry iodine crystals. Carry out the experiment *in a fume cupboard.*

Allow a small amount of chlorine to enter the flask and vaporize some of the iodine by warming gently. What changes do you observe? Now add more chlorine until the yellow solid, iodine trichloride, is formed. Detach the chlorine supply from the flask, and allow any excess chlorine to escape by inverting it. What further changes occur? Are they reversed by adding more chlorine?

Investigate the effect of (*a*) heat and (*b*) water on samples of iodine trichloride.

Oxides

The halogen oxides as a class are unstable and strongly oxidizing. The only one that is sufficiently stable to keep is iodine (V) oxide, the anhydride of iodic acid.

Iodic acid, HIO_3, and iodine (V) oxide, I_2O_5. Iodine is quite readily oxidized. The usual oxidizing agent is fuming nitric acid near its boiling-point:

$$I_2 + 10HNO_3 \longrightarrow 2HIO_3 + 10NO_2 + 4H_2O$$

Iodine is lost as vapour if the reaction is carried out in an open vessel. A long reflux air condenser, in which the refluxing nitric acid washes the iodine back into the reaction mixture, helps to overcome this loss.

Put 0·5 g of iodine into a small flask fitted with a long air condenser, and add 15 cm^3 of fuming nitric acid. Heat on a water-bath at 85–90°C, *in a fume cupboard*, until no more brown fumes are evolved and there is no iodine left in the flask. This will take quite a long time, at least an hour.

Transfer the solution from the flask to an

evaporating basin and evaporate to dryness on a steam-bath. The iodic acid crystals can be re-crystallized from 50% nitric acid.

To convert the iodic acid to iodine (V) oxide, heat it to constant weight on a liquid-bath maintained between 200°C and 250°C.

$$2HIO_3 \longrightarrow I_2O_5 + H_2O$$

Investigate the effect of heating a sample of the oxide more strongly.

Barium iodate, $Ba(IO_3)_2.H_2O$. A chlorate in solution will oxidize iodine to iodate:

$$2NaClO_3 + I_2 \longrightarrow 2NaIO_3 + Cl_2$$

Add 1·3 g (0·005 mole) of iodine to a solution of 1·1 g (0·01 mole) of sodium chlorate in 50 cm³ of M nitric acid, and warm the mixture until all the iodine has dissolved.

Barium iodate is sparingly soluble, and can be precipitated by adding a solution of barium nitrate:

$$Ba(NO_3)_2 + 2NaIO_3 \longrightarrow Ba(IO_3)_2 + 2NaNO_3$$

To the solution of sodium iodate add a solution of 2·7 g of barium nitrate in 30 cm³ of water, stir, and cool in a refrigerator or ice-bath. Filter off the crystals of barium iodate monohydrate which are deposited, wash with ice-cold water and dry on filter paper. Work out the yield in terms of the iodine used.

Investigate the action of heat on a sample of the crystals. Also investigate the action of various reducing agents on a solution of an iodate. For the use of iodates in volumetric analysis see p. 165.

Hydrides

Hydrogen chloride, HCl. If a supply of dry hydrogen chloride is required for an experiment, the simplest way of preparing it is to add concentrated sulphuric acid from a dropping funnel to solid sodium chloride or ammonium chloride in a 250 cm³ flask (see Fig. K, i or ii).

Hydrogen bromide, HBr. It is not possible to prepare hydrogen bromide gas in the same way as hydrogen chloride, because it is oxidized to bromine to some extent by the sulphuric acid. However, if 90% phosphoric acid is used instead, the gas is evolved from a bromide on heating.

Hydrogen iodide, HI, is too readily oxidized and dissociated to be prepared by either of the methods above. It is normally made by hydrolysing phosphorus tri-iodide prepared *in situ*.

Set up a gas generator, and attach to the delivery tube a U-tube containing glass-wool smeared with moist red phosphorus (to remove excess of iodine) and a calcium chloride drying tube.

Put into the flask 4 g of iodine and 0·5 g of red phosphorus. Warm gently till the mixture is fused, then drop water slowly from the dropping funnel.

Constant-boiling hydrobromic acid is a useful reagent for certain organic reactions, for example the preparation of bromoalkanes. It boils at 126°C (at 760 mm Hg) and contains 47·8% of hydrogen bromide by weight.

Although the action of sulphuric acid on a bromide results in the production of some bromine, this is avoided if the bromide is in solution and the temperature controlled:

$$KBr + H_2SO_4 \longrightarrow KHSO_4 + HBr$$

Add 30 g of potassium bromide to 50 cm³ of water in a beaker. Surround the beaker with cold water and add slowly, with stirring, 22 cm³ of concentrated sulphuric acid, keeping the temperature below 70°C. When all the acid has been added, cool the solution to room temperature, or below, and filter off the precipitated potassium hydrogen sulphate crystals at the pump.

Transfer the filtrate to a distillation apparatus fitted with a 200°C thermometer. Water will distil over first, then the constant-boiling mixture. Collect the fraction boiling between 124°C and 128°C. The yield should be around 80%.

Reference: Heisig, G. B. and Amdur, E. In *Inorganic Syntheses*, 1, 155. McGraw-Hill, New York and Maidenhead, 1939.

Hydriodic acid cannot be prepared this way. It is usually made by passing hydrogen sulphide into a suspension of iodine in water. The sulphur is filtered off and the filtrate distilled. Constant-boiling hydriodic acid boils at 126°C (at 760 mm Hg) and contains 57% by weight of hydrogen iodide.

GROUP O He, Ne, Ar, Kr, Xe, Rn

The elements in this group are all gaseous at normal temperatures, and until 1962 they were regarded as inert chemically. In that year, however, xenon was found to take part in a number of reactions and to form a solid covalent fluoride. Since then many compounds of xenon, and some of krypton and radon, have been prepared and investigated. So far, only compounds containing bonds with fluorine, oxygen and chlorine have been obtained.

Xenon tetrafluoride, XeF_4, the simplest noble gas compound to prepare, is formed when a mixture of xenon (from a cylinder) and an excess of fluorine (from an electrolytic fluorine cell or a cylinder) is passed through a nickel tube, containing pieces of sheet nickel, heated to dull red heat. The vapour of the fluoride is condensed to white crystals in a freezing mixture of solid carbon dioxide and propanone.

Reference: Holloway, J. H. and Peacock, R. D., *Proc. Chem. Soc.*, 1962, 389.

D-BLOCK ELEMENTS

This term is used to include the elements from scandium to zinc in the first long period, and the corresponding elements in the later periods. Scandium is an unfamiliar element, and has colourless salts in which it has an oxidation state of +3. It will not be considered further. Zinc and the other elements in Group IIb should be compared with the elements of Group IIa (Be–Ra).

The characteristics of the d-block elements which are most familiar are their coloured ions, the readiness with which they form complexes with a large variety of ligands, and their wide range of oxidation states. They and their compounds frequently have a catalytic effect and are used industrially for this purpose; they also have characteristic magnetic properties.

Only a few of the elements in the later periods will be mentioned. In general their higher oxidation states are more stable than those of the lighter elements.

Investigation of properties

1. For this set of tests use 0·1M solutions of salts of the metals in their lower oxidation states (titanium (III) sulphate, ammonium vanadium (III) sulphate, chromium (III) chloride, manganese (II) chloride, iron (II) sulphate, iron (III) chloride, cobalt and nickel sulphates, copper (II) sulphate, silver nitrate, zinc and cadmium chlorides, mercury (I) nitrate and mercury (II) chloride).

(*a*) Add 2M sodium hydroxide dropwise until an excess is present.

(*b*) Add 2M ammonia dropwise until an excess is present.

(*c*) Add 1·5M sodium carbonate.

(*d*) Add 0·1M ammonium thiocyanate until an excess is present.

(*e*) Acidify with 2M hydrochloric acid and pass hydrogen sulphide.

(*f*) If there is no reaction in (*e*), add 2M ammonia until alkaline.

(*g*) Add 2M sodium hypochlorite solution and warm gently.

Look for the precipitation of insoluble compounds and for colour changes. Try to work out what has happened; for example if a precipitate dissolves in an excess of a reagent it is likely that a complex has been formed.

2. Investigate the action of reducing agents on the higher oxidation states. Try sulphur dioxide, sodium dithionite, oxalic acid, tin (II) chloride, hydrazine hydrate, and zinc with 2M hydrochloric acid. An acidic solution is usually needed (why?).

   Use 0·1M solutions of titanyl (IV) sulphate, ammonium vanadate, potassium chromate, potassium manganate (VII), ammonium molybdate (make sure there is no nitric acid present, there often is in the laboratory reagent), sodium tungstate.

3. Investigate the action of the metals on common acids. Use 2M hydrochloric acid, concentrated hydrochloric acid, 2M nitric acid and concentrated nitric acid. Warm if there is no reaction in the cold. Look for the evolution of gases, and try to work out the oxidation state of the salt produced.

Titanium: oxidation states +3 and +4

Titanium (III) salts form violet solutions, while those of titanium (IV), containing the TiO^{2+} ion, are colourless.

Titanium (IV) chloride, a colourless liquid, can be prepared by the method described for tin (IV) chloride (p. 30). It is very easily hydrolysed to the hydrated oxide, a white solid.

Vanadium: oxidation states +2, +3, +4 and +5

All four oxidation states can be obtained in aqueous solution. VO_2^+ (aq) is yellow, VO^{2+} (aq) is blue, V^{3+} (aq) is green, V^{2+} (aq) is violet. The lower oxidation states are obtained by reducing an acidic solution of a vanadate. Either zinc and 2M hydrochloric acid or sodium sulphite and 2M hydrochloric acid can be used.

Try the effect of various oxidizing agents on the lower oxidation states. See p. 56.

+5 to +3: Ammonium vanadium (III) sulphate (vanadium alum), $NH_4V(SO_4)_2.12H_2O$.

This is a good example of a compound prepared by electrolytic reduction. Ammonium vanadate is first reduced to vanadyl (IV) sulphate, using sulphur dioxide, and this is reduced further to vanadium (III) sulphate by electrolysis.

$$2NH_4VO_3 + 2H_2SO_4 + SO_2 \longrightarrow (NH_4)_2SO_4 + 2VOSO_4 + 2H_2O$$

$$2VOSO_4 + H_2SO_4 + 2e^- \longrightarrow V_2(SO_4)_3 + 2OH^-$$

Carefully add 16 cm^3 of concentrated sulphuric acid to 50 cm^3 of water; arrange for a tube from a liquid sulphur dioxide cylinder to be just above the surface of the liquid and giving a slow stream of the gas (carry out the reaction *in a fume cupboard*). Now add, a little at a time, 11·7 g (0·1 mole) of ammonium vanadate (do not add a second portion until the first has dissolved to form a clear blue solution). When the ammonium vanadate has all been added and reduced, boil the solution for a few minutes, and allow it to cool.

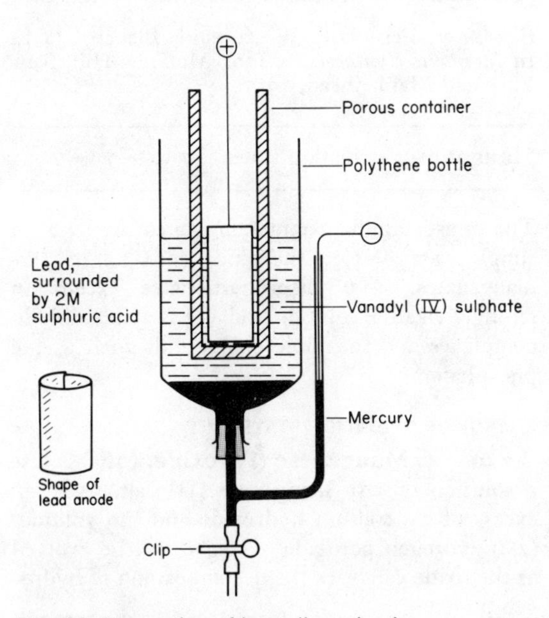

Fig. 11. Preparation of 'vanadium alum'.

Now set up an electrolytic cell with a mercury cathode. A suitable cell, which can be made from a polythene bottle with the bottom cut off, is shown in Fig. 11. Use a 15 V supply with a rheostat in series, work out the surface area of the mercury and arrange for a current density of 0·1 A cm^{-2}. The colour of the solution in the cell will change gradually from blue to green, and the reaction will probably be complete in an hour or two.

When the solution has no trace of blue colour left (test by putting a drop on a filter paper), switch off the current, and run off first the mercury, then the solution. Leave the solution to cool in a refrigerator or ice-bath. If crystals do not form, scratch the inside of the beaker with a glass rod. Filter off the crystals and wash them with ethanol, then with propanone. Work out the yield. Note the shape of the crystals.

Reference: Palmer, W. G. *Experimental Inorganic Chemistry*, p. 320. University Press, Cambridge, 1954.

Chromium: oxidation states +2, +3 and +6

CHANGE OF OXIDATION STATE

+3 to +6: Potassium dichromate, $K_2Cr_2O_7$. In this preparation, chromium (III) is oxidized to chromium (VI) in alkaline solution, using hydrogen peroxide as the oxidizing agent. The chromate is converted to dichromate by acidification after the removal of any excess of hydrogen peroxide (why?).

$$2CrCl_3 + 10KOH + 3H_2O_2 \longrightarrow 2K_2CrO_4 + 6KCl + 8H_2O$$

$$2K_2CrO_4 + 2CH_3CO_2H \longrightarrow K_2Cr_2O_7 + 2CH_3CO_2K + H_2O$$

Dissolve 14 g (0·05 mole) of chromium (III) chloride hexahydrate in 40 cm^3 of water in a 250 cm^3 beaker. Add slowly, with stirring, a solution of 17 g (0·25 mole) of potassium hydroxide in 40 cm^3 of water. The green precipitate of chromium (III) hydroxide formed partially redissolves. Now warm the mixture gently, do not boil, and add slowly, with stirring, 60 cm^3 of '20 volume' (2M) hydrogen peroxide. When it has all been added, bring the solution to the boil for a few minutes to destroy any excess of oxidizing agent, then filter hot through a large (7 cm) Buchner funnel.

Transfer the solution of potassium chromate to an evaporating basin and boil away half the liquid. Add to the hot solution 5 cm^3 (0·09 mole) of glacial acetic acid, and cool the solution in a refrigerator or ice-bath. Filter off the orange crystals formed, wash with a small volume of iced water and dry in a desiccator containing sodium hydroxide. Work out the yield, which will be up to 95%.

Estimate the percentage purity of the crystals by titration (see p. 163).

+6 to +3. See p. 28 for the preparation of 'chrome alum'.

+6 to +3 to +2: Chromium (II) sulphate solution, $CrSO_4$. When a chromate is acidified it forms a dichromate (see above) and this solution is readily reduced in acid solution to the +3 state. Further reduction to chromium (II) can only be accomplished if air is rigorously excluded from the system. Hence all solutions should be boiled to remove dissolved air, and an inert or reducing atmosphere is required in the reaction vessel. This can be achieved by fitting the flask with an efficient Bunsen valve (see Fig. 6, p. 26).

Dissolve 3 g (0·01 mole) of potassium dichromate in 10 cm³ of water, add 10 cm³ of 2M sulphuric acid and 3 g (an excess) of zinc and fit the conical flask with a stopper carrying a Bunsen valve. The solution will change first to green, then eventually to a clear blue. When exposed to the air, the blue chromium (II) ions are rapidly oxidized to green chromium (III).

COMPLEX FORMATION

Potassium trioxalatochromate (III), $K_3[Cr(C_2O_4)_3].3H_2O$. This is isomorphous with the other trioxalato complexes (e.g. those of Al, Fe and Co), and like them shows stereo-isomerism. It is readily prepared by the reduction of potassium dichromate with an excess of oxalic acid:

$$K_2Cr_2O_7 + 7H_2C_2O_4 + 2K_2C_2O_4 \longrightarrow 2K_3[Cr(C_2O_4)_3] + 6CO_2 + 7H_2O$$

Dissolve 3·7 g (0·02 mole) of potassium oxalate monohydrate in 100 cm³ of water, and add 8·8 g (0·07 mole) of oxalic acid dihydrate. When this has completely dissolved, add 2·94 g (0·01 mole) of potassium dichromate, slowly, with stirring.

When the reaction is complete and no more carbon dioxide is evolved, evaporate the solution to about 50 cm³, and add 10 cm³ of ethanol. Leave the solution in a refrigerator or ice-bath for the complex to crystallize. The deep green crystals should be obtained in nearly 100% yield. Wash them with a little water and then with ethanol, and allow them to dry at room temperature.

Reference: Bailar, J. C. and Jones, E. M. In *Inorganic Syntheses*, **1**, 37. McGraw-Hill, New York and Maidenhead, 1939.

Tris(pentan-2,4-dionato) chromium (III), $Cr(CH_3COCHCOCH_3)_3$. This is an example of an un-ionized chelate complex of chromium. It is prepared in a solution made slightly alkaline with ammonia from the hydrolysis of carbamide:

$$CrCl_3 + 3CH_3COCH_2COCH_3 + 3NH_3 \longrightarrow Cr(CH_3COCHCOCH_3)_3 + 3NH_4Cl$$

Dissolve 1·3 g of chromium (III) chloride hexahydrate in 50 cm³ of water and, when it has dissolved, add 10 g of carbamide and 3 g of pentan-2,4-dione. Heat on a steam-bath for about an hour, then filter off the crystals of complex which have formed. They can be recrystallized from a benzene–petroleum spirit mixture. They melt at 216°C.

Reference: Fernelius, W. C. and Blanch, J. E. In *Inorganic Syntheses*, **5**, 130. McGraw-Hill, New York and Maidenhead, 1957.

Manganese: oxidation states +1 to +7

The most familiar oxidation states are +2 in simple salts, +4 in the oxide and +7 in permanganates. +6 disproportionates except in strongly alkaline solution and +3 is stable only in complexes and insoluble compounds such as the phosphate.

CHANGE OF OXIDATION STATE

+2 to +4: Manganese (IV) oxide, MnO_2. Use a solution of any manganese (II) salt. Add an excess of 2M sodium hydroxide and '20 volume' (2M) hydrogen peroxide. Oxygen will be evolved as the oxide catalyses the decomposition of hydrogen peroxide.

Filter off the black precipitate of manganese (IV) oxide in a sintered-glass crucible, and dry it in an oven at about 120°C.

+2 to +7. This is less readily accomplished. The usual oxidizing agent is sodium bismuthate. Warm a solution of manganese (II) nitrate with concentrated nitric acid and add a little solid sodium bismuthate. Note the purple solution containing manganate (VII) ions. Try the effect of other oxidizing agents.

+4 to +7: Potassium manganate (VII), $KMnO_4$. This is a difficult preparation to carry out efficiently except under industrial conditions on a large scale. In the laboratory, potassium chlorate is used to oxidize manganese (IV) oxide to potassium manganate (VI) (K_2MnO_4) in alkaline conditions. The manganate is then oxidized further with chlorine.

$$6KOH + KClO_3 + 3MnO_2 \longrightarrow 3K_2MnO_4 + KCl + 3H_2O$$

$$2K_2MnO_4 + Cl_2 \longrightarrow 2KMnO_4 + 2KCl$$

It is best to approach this as a quantitative experiment to discover what percentage of the manganese (IV) oxide is converted to manganate (VII) ions in solution, without attempting to crystallize the specimen. (Potassium manganate (VII) can be purchased cheaply.)

Mix together and fuse on an iron tray, or in a stainless steel crucible, 6·8 g of potassium hydroxide and 2·5 g of potassium chlorate. Add a little at a time, stirring with a spatula, 5 g of powdered pure manganese (IV) oxide. The mass will solidify and turn green when the reaction is complete.

Now cool the mass, break it up and transfer it to a beaker, add 40 cm^3 of water and boil to dissolve the manganate (VI). Cool the solution and then, *in a fume cupboard*, pass chlorine from a cylinder or generator until the solution has turned from green to purple. Boil the solution to expel any excess of chlorine, filter at the pump through sintered glass (not paper) and cool.

Transfer the solution to a standard 100 cm^3 flask and make up to the mark with water. Use the solution in a burette to titrate a standard solution of an oxalate or an iron (II) salt. The solution should, theoretically, be about 0·6M but will in fact be much less. For details of manganate (VII) titrations, see p. 163.

+7 to +6: Potassium manganate (VI), K_2MnO_4. If potassium manganate (VII) is made strongly alkaline reduction occurs:

$$4KMnO_4 + 4KOH \longrightarrow 4K_2MnO_4 + 2H_2O + O_2$$

To 10 cm^3 of 0·02M potassium manganate (VII) add 100 cm^3 of 2M potassium hydroxide. Note the change of colour. Investigate the effects of heating, dilution and acidification of samples of solution.

COMPLEX FORMATION

Tris(pentan-2,4-dionato) manganese (III), $Mn(CH_3COCHCOCH_3)_3$. Manganese (II) is oxidized to manganese (III), in the presence of a slightly alkaline solution of pentan-2,4-dione, by potassium manganate (VII) which is itself reduced to manganese (III) in these conditions. Bis(pentan-2,4-dionato) manganese (II) is first formed.

$$MnCl_2 + 2CH_3COCH_2COCH_3 + 2CH_3CO_2Na \longrightarrow Mn(CH_3COCHCOCH_3)_2 + 2NaCl + 2CH_3CO_2H$$

$$4Mn(CH_3COCHCOCH_3)_2 + KMnO_4 + 7CH_3COCH_2COCH_3 + CH_3CO_2H \longrightarrow 5Mn(CH_3COCHCOCH_3)_3 + CH_3CO_2K + 4H_2O$$

Dissolve 2·5 g (0·012 mole) of manganese (II) chloride tetrahydrate in 100 cm^3 of water and add 6·6 g of sodium acetate trihydrate to the solution. When this has dissolved add 10 cm^3 of pentan-2,4-dione, mix thoroughly, then make up a solution of 0·5 g of potassium manganate (VII) in 25 cm^3 of water. Add this slowly, with stirring, and follow it with a solution of 6·6 g of sodium acetate trihydrate in 25 cm^3 of water. Heat the mixture on a boiling water-bath for about ten minutes, then cool and leave in a refrigerator or ice-bath. Filter off the crystals and allow them to dry at room temperature. They can be recrystallized from a benzene–petroleum spirit mixture.

Reference: Charles, R. G. In *Inorganic Syntheses*, **7**, 183. McGraw-Hill, New York and Maidenhead, 1963.

Iron: oxidation states +2, +3 and +6

CHANGE OF OXIDATION STATE

Iron (II) ions are slowly oxidized by atmospheric oxygen to iron (III). Common oxidizing agents which are used to bring about this change are manganate (VII), dichromate and nitric acid.

Iron (III) is most readily reduced to iron (II) using zinc in acid solution.

Examples of the preparation of double salts containing iron will be found on pp. 26 and 28.

Anhydrous iron (II) chloride, $FeCl_2$, is prepared by the action of dry hydrogen chloride gas on the metal. Follow the procedure described on p. 40 for the preparation of chlorides but, instead of a chlorine generator, use a hydrogen chloride generator.

Anhydrous iron (III) chloride, Fe_2Cl_6, is readily prepared by direct combination. See p. 40 for details. Compare the appearance, and the action of water on the two chlorides.

Barium ferrate (VI), $BaFeO_4$. Iron (III) ions are oxidized in strongly alkaline solution to ferrate (VI) ions by strong oxidizing agents. The ferrate ions are similar in colour, in solution, to manganate (VII).

It is better to avoid an excess of chloride ions, and iron (III) nitrate is a good starting material. The usual oxidizing agent is sodium hypochlorite:

$$2Fe(NO_3)_3 + 3NaOCl + 10NaOH \longrightarrow 2Na_2FeO_4 + 6NaNO_3 + 3NaCl + 5H_2O$$

Sodium ferrate is very soluble, and the sparingly soluble barium ferrate is formed as a precipitate when barium nitrate solution is added.

Make a solution of 1 g of iron (III) nitrate enneahydrate in 5 cm³ of water and put it into a dropping funnel. Dissolve 2 g of sodium hydroxide in 50 cm³ of 2M sodium hypochlorite solution in a 200 cm³ beaker, and heat to boiling. Allow the iron (III) nitrate to drop slowly (about 1 drop every fifteen seconds) into the stirred solution. When all the solution has been added, boil the solution for a minute or two, then filter the hot solution through sintered glass into a Buchner flask containing a solution of 1 g of barium nitrate in 20 cm³ of water. Filter off the red barium ferrate precipitate, wash with water and dry.

Investigate the action of hydrochloric acid on a sample of the ferrate.

COMPLEX FORMATION

Potassium trioxalatoferrate (III), $K_3[Fe(C_2O_4)_3].3H_2O$. Compare this compound with the similar ones formed by Al, Cr and Co. In this preparation iron (II) oxalate is used as the starting material to avoid the presence of other anions, and the oxidizing agent is hydrogen peroxide:

$$2FeC_2O_4 + H_2C_2O_4 + 3K_2C_2O_4 + H_2O_2 \longrightarrow 2K_3[Fe(C_2O_4)_3] + 2H_2O$$

Suspend 3·6 g (0·02 mole) of iron (II) oxalate dihydrate in a solution of 5·5 g (0·03 mole) of potassium oxalate monohydrate in 20 cm³ of water. Now, keeping the temperature below 50°C, add dropwise 14 cm³ of '20 volume' (2M) hydrogen peroxide from a burette. When it has all been added, boil the solution and add dropwise a solution of 1·3 g (0·01 mole) of oxalic acid dihydrate in 20 cm³ of water. This will dissolve any precipitated iron hydroxide.

Filter the hot solution and add 20 cm³ of ethanol. Cool the solution in a refrigerator or an ice-bath, then filter off the green crystals and wash them with a water–ethanol mixture, then with propanone. Work out the yield; it should be around 75%.

Reference: Palmer, W. G. *Experimental Inorganic Chemistry*, p. 521. University Press, Cambridge, 1954.

Cobalt: oxidation states +2 and +3

+2 is more stable in simple ions, but +3 is more stable in complexes (why is this?). Cobalt is notable for the very wide range of complexes it forms; only a few can be included here.

COMPLEX FORMATION, +2 OXIDATION STATE

Mercury (II) tetrathiocyanatocobaltate (II), $Hg[Co(SCN)_4]$, is an insoluble compound, forming as a blue precipitate from a pink solution.

Dissolve 5·4 g (0·02 mole) of mercury (II) chloride (*poisonous*) in 125 cm³ of water in a large conical flask. Make a solution of 4·7 g (0·02 mole) of cobalt (II) chloride hexahydrate in 50 cm³ of water in a beaker, and add to it a solution of 6·1 g (0·08 mole) of ammonium thiocyanate in 25 cm³ of water. Now add this solution slowly to the mercury (II) chloride solution, and note the blue precipitate forming. When it has all been added, allow it to stand for about ten minutes, then raise to boiling and filter hot. Wash the fine crystals with water and allow to dry. The yield should be near 100%.

COMPLEX FORMATION, +3 OXIDATION STATE

Potassium trioxalatocobaltate (III),
$K_3[Co(C_2O_4)_3].3H_2O$, is another of the trioxalato complexes, similar to those of Al, Cr and Fe. In this preparation, lead (IV) oxide is the oxidizing agent used:

$$2CoCO_3 + 3H_2C_2O_4 + 3K_2C_2O_4 + PbO_2 + 2CH_3CO_2H \longrightarrow 2K_3[Co(C_2O_4)_3] + (CH_3CO_2)_2Pb + 5H_2O + 2CO_2$$

Dissolve 2·4 g (0·02 mole) of cobalt (II) carbonate in a solution of 2·5 g of oxalic acid dihydrate and 7·4 g of potassium oxalate monohydrate in 50 cm³ of water. Warm the solution until a clear purple solution is formed, and then cool to 40°C.

Add slowly, with vigorous stirring (mechanical preferably), 2·4 g of lead (IV) oxide and then, keeping the solution at 40°C, 2·5 cm³ of glacial acetic acid dropwise. The solution will change to a deep green colour.

Filter the solution at the pump, then add to the filtrate 50 cm³ of ethanol. Leave the solution in a refrigerator or an ice-bath and filter off the crystals which are deposited. Wash them with a little water, then with ethanol. The compound should be left to dry in the dark as it is sensitive to light.

Reference: Bailar, J. C. and Jones, E. M. In *Inorganic Syntheses*, 1, 37. McGraw-Hill, New York and Maidenhead, 1939.

Sodium hexanitrocobaltate (III),
$Na_3[Co(NO_2)_6]$. This compound is familiar as a reagent for the identification of potassium ions in solution. Its preparation illustrates the ease with which cobalt is oxidized from +2 to +3 in the presence of ligand ions.

$$Co(NO_3)_2 + 7NaNO_2 + 2CH_3CO_2H \longrightarrow Na_3[Co(NO_2)_6] + 2NaNO_3 + NO + 2CH_3CO_2Na + H_2O$$

Dissolve 8·7 g of sodium nitrite in 10 cm³ of water and, at 50°C, add 2·9 g (0·01 mole) of cobalt (II) nitrate hexahydrate and stir until it has dissolved. Now add dropwise a mixture of 1·7 cm³ of glacial acetic acid and an equal volume of water.

Transfer the solution to a Buchner flask fitted with a tube to enable a current of air to be drawn through the solution (to remove the nitrogen monoxide formed in the reaction), and draw air through for about twenty minutes. Filter the orange-brown solution at the pump, then add to the filtrate 50 cm³ of ethanol to precipitate the sodium salt. Leave to stand until the precipitate has settled, then filter at the pump, wash the precipitate with ethanol and leave to dry.

Investigate the action of a solution of the complex salt with solutions of magnesium, potassium and ammonium salts.

The other two complexes described are both non-electrolytes.

Tris(pentan-2,4-dionato) cobalt (III),
$Co(CH_3COCHCOCH_3)_3$. Pentan-2,4-dione is a weak acid, and it will liberate carbon dioxide from a carbonate. Hydrogen peroxide is used to oxidize the cobalt (II) to cobalt (III):

$$2CoCO_3 + 6CH_3COCH_2COCH_3 + H_2O_2 \longrightarrow 2Co(CH_3COCHCOCH_3)_3 + 4H_2O + 2CO_2$$

Add 2 g (0·017 mole) of cobalt (II) carbonate to 16 cm³ of pentan-2,4-dione and heat to 90–100°C on a water-bath. To the stirred mixture add 40 cm³ of '20 volume' (2M) hydrogen peroxide in a slow stream. The reaction is exothermic, and care is needed to avoid frothing. When the reaction is complete, cool the mixture in a refrigerator or an ice-bath, and filter off the crystals at the pump. They can be recrystallized from a benzene–petroleum spirit mixture. M.P. 213°C.

Reference: Bryant, B. E. and Fernelius, W. C. In *Inorganic Syntheses*, 5, 188. McGraw-Hill, New York and Maidenhead, 1957.

Trinitrotriammine cobalt (III),
$Co[(NO_2)_3(NH_3)_3]$.
Dissolve 6·2 g (0·025 mole) of cobalt (II) acetate tetrahydrate in 50 cm³ of hot water, then cool the mixture in an ice-bath and add a cooled solution of 5·2 g of sodium nitrite in 25 cm³ of '0·880' ammonia. To the mixture, at about 10°C, add 7 cm³ of '20 volume' (2M) hydrogen peroxide slowly, with stirring, and then 1 g of activated charcoal. Keep the mixture cool for about ten minutes, then bring to the boil and keep boiling for half an hour.

Filter the solution while still hot, then cool the

filtrate in a refrigerator or an ice-bath. Filter off the crystals which form, wash with ethanol and leave to dry. A further crop of crystals can be obtained by adding to the filtrate about 0·5 g of charcoal and evaporating to half volume, filtering and cooling as before.

The mustard-yellow crystals are slightly soluble in water. Investigate the conductivity of a solution in pure water using a conductivity bridge (see p. 136).

Reference: Schlessinger, G. In *Inorganic Syntheses*, **6**, 189. McGraw-Hill, New York and Maidenhead, 1960.

Nickel: oxidation states +2 and +4

For nickel the oxidation state +2 is much the most important, although complexes containing nickel (O) (e.g. nickel carbonyl) and an oxide containing nickel (IV) are relatively well known.

COMPLEX FORMATION

Bis(butanedione dioximato) nickel (II), $\mathbf{Ni[(C_4H_7N_2O_2)_2]}$. This compound is almost insoluble in water and, as it is precipitated quantitatively from a solution containing nickel ions in the presence of ammonia, it is used for the estimation of nickel. See p. 183 for practical details.

Tris(ethanediamino)nickel (II) chloride, $\mathbf{[Ni(en)_3]Cl_2}$. 'en' is used as an abbreviation for 1,2-ethanediamine (1,2-diaminoethane), $NH_2CH_2CH_2NH_2$, a well-known bidentate chelate ligand.

Dissolve 12 g (0·05 mole) of nickel chloride hexahydrate in 60 cm³ of water, and add 14 cm³ of 70% ethanediamine hydrate (or an equivalent amount of the pure compound). Filter the solution, then evaporate the filtrate to about 30 cm³ on a steam-bath. Cool, add 2 drops of ethanediamine hydrate and leave in a refrigerator or an ice-bath to crystallize. Filter off the crystals, wash with a little ethanol and leave to dry.

Reference: State, H. M. In *Inorganic Syntheses*, **6**, 200. McGraw-Hill, New York and Maidenhead, 1960.

Bis(ethanediamino) nickel (II) chloride, $\mathbf{[Ni(en)_2]Cl_2}$, can be prepared from the tris-(ethanediamino) complex but is difficult to obtain directly.

Put into a 50 cm³ flask, fitted with a reflux condenser, 9·5 cm³ of methanol and 0·5 cm³ of water. Add 2·4 g of tris(ethanediamino) nickel (II) chloride and 1 g of nickel chloride hexahydrate.

Heat the solution under reflux for about five minutes. The solution will turn to a blue-violet colour. Allow the solution to cool, then put 1 cm³ in a test tube and add propanone, a drop at a time with scratching, until crystals are formed. These crystals will be used to seed the main bulk of the solution.

Now add 20 cm³ of propanone to the rest of the solution, with a few of the seed crystals. When crystals remain in the main bulk of the solution after the propanone is stirred in, wash in the rest of the seed crystals with a few more cm³ of propanone and stir the mixture thoroughly. Filter off the crystals at the pump, wash them with propanone and dry in an oven.

Reference: State, H. M. In *Inorganic Syntheses*, **6**, 198. McGraw-Hill, New York and Maidenhead, 1960.

Copper: oxidation states +1, +2 and +3.

Copper (I) is found only in complexes and insoluble compounds, the simple ion disproportionates in solution. Copper (III) is known in only a few complexes (e.g. $K_3[CuF_6]$).

CHANGE OF OXIDATION STATE, +2 TO +1

Copper (I) chloride, CuCl, is readily prepared by reducing copper (II) ions by sulphite ions in the presence of chloride ions:

$$Na_2SO_3 + 2CuCl_2 + H_2O \longrightarrow Na_2SO_4 + 2CuCl + 2HCl$$

Dissolve 8·5 g (0·025 mole) of copper (II) chloride dihydrate in 10 cm³ of water, and add a solution of 8 g (an excess) of sodium sulphite hexahydrate 40 cm³ of water. A white precipitate slowly settles out from the dark brown solution, whose colour gradually fades. When the reaction appears to be complete, pour the solution into 500 cm³ of water, through which a little sulphur dioxide has been bubbled, and allow the precipitate to settle.

Decant the supernatant liquid, filter off the precipitate at the pump and wash it with water containing dissolved sulphur dioxide. Wash it rapidly with small portions of ethanol, and then propanone, and dry it for a short while in an oven. Put it into a sealed specimen bottle and calculate the yield. After a time the powder will turn greenish. (Why?)

An alternative method of preparation is to use copper metal as the reducing agent. Equimolar quantities of copper (II) oxide and copper are dissolved in an excess of concentrated hydrochloric acid (forming $CuCl_2^-$ ions), and the mixture is poured into water.

Reference: Keller, R. N. and Wycoff, H. D. In *Inorganic Syntheses*, **2**, 1, McGraw-Hill, New York and Maidenhead, 1946.

Copper (I) oxide, Cu_2O. Copper (II) oxide is readily reduced in alkaline solution. The usual reducing agent is glucose, which is oxidized to gluconic acid.

In a 250 cm³ conical flask dissolve 10 g (0·04 mole) of copper (II) sulphate pentahydrate in 100 cm³ of water. Warm the solution and add 4 g of glucose. When this has dissolved, add a solution of 4 g of sodium hydroxide in 50 cm³ of water (or 50 cm³ of 2M sodium hydroxide).

Boil the solution carefully, using a small flame, for about ten minutes, during which time the red-orange copper (I) oxide will be precipitated. Filter off the precipitate from the hot solution, wash it with water, then with ethanol, and dry in an oven. Work out the yield.

Copper (I) iodide, CuI, is particularly insoluble, and is formed when iodide ions are added to copper (II) ions in solution:

$$2Cu^{2+} + 4I^- \longrightarrow 2CuI + I_2$$

This reaction is quantitative, and the iodine can be titrated with standard sodium thiosulphate. The copper (I) iodide can be filtered off at the pump, washed with sodium thiosulphate solution and then water, and dried in an oven.

COMPLEX FORMATION

Tetrammine copper (II) sulphate, $[Cu(NH_3)_4]SO_4.H_2O$. This complex is one of a large number with copper–nitrogen bonds. It is very readily formed and is easily precipitated with ethanol.

Dissolve 5 g (0·02 mole) of copper (II) sulphate pentahydrate in 20 cm³ of water and add slowly, with shaking, 2M ammonia until the precipitate which at first forms is completely redissolved. To the cold solution add 30 cm³ of ethanol and leave in a refrigerator or an ice-bath. Filter off the crystals at the pump, wash with ethanol containing a drop of '0·880' ammonia and dry at room temperature. Investigate the action of heat on the crystals.

Tetrapyridino copper (II) peroxodisulphate, $[Cu(C_5H_5N)_4]S_2O_8$. The pyridine complexes are more stable and less soluble than the ammine complexes.

Dissolve 2·5 g (0·01 mole) of copper (II) sulphate pentahydrate in 50 cm³ of water. Add 4 cm³ (an excess) of pyridine, which will form the deep blue complex solution. To this solution add a solution of 4 g of ammonium peroxodisulphate in 50 cm³ of water. Allow the deep blue crystals to settle, then filter them off at the pump. Wash them with water and allow them to dry.

Bis(ethanediamino) copper (II) tetraiodomercuriate, $[Cu(en)_2]HgI_4$. This is a comparatively insoluble compound containing cation and anion which are both complex. The two complexes are made separately and then mixed.

Dissolve 3·5 g of mercury (II) chloride in 60 cm³ of warm water. Cool, then add 0·5M potassium iodide solution until the red precipitate just redissolves. Now dissolve 2·5 g (0·01 mole) of copper (II) sulphate pentahydrate in 20 cm³ of water, and add 2 cm³ of ethanediamine hydrate. A deep violet solution is formed.

Heat the solution of the mercury complex nearly to boiling, then add, with stirring, the solution of the copper complex. On cooling the solution, fine blue-violet crystals are deposited.

Filter at the pump, wash the crystals with water and then with a little ethanol, and leave them to dry. Calculate the yield.

A similar compound, in which cadmium replaces mercury, can be prepared by replacing the first solution with one containing 2·5 g of cadmium chloride 2½-hydrate and 7 g of potassium iodide in 50 cm³ of water.

Dipyridino dithiocyanato copper (II) **$[Cu(C_5H_5N)_2(SCN)_2]$**. This is one of a number of similar complexes which are sufficiently insoluble to be used in the gravimetric estimation of the metals which form them (Cu, Co, Ni, Mn, Cd, Zn).

Dissolve 2·5 g (0·01 mole) of copper (II) sulphate pentahydrate in 50 cm³ of water and add 2 cm³ of pyridine. To the deep blue solution add slowly, with stirring, a solution of 3 g of ammonium thiocyanate in 50 cm³ of water. The complex is precipitated at once.

Filter off the precipitate at the pump, wash with water and allow to dry at room temperature.

The cobalt, nickel and manganese compounds are of the type $M(C_5H_5N)_4(SCN)_2$. They can be prepared by the addition of 4 cm³ of pyridine to a solution of 0·01 mole of the metal salt, followed by the thiocyanate solution as described above.

Silver: oxidation states +1 and +2

The more important oxidation state of silver is +1, and in this state it usually forms complexes with two molecules of ligand. In the +2 oxidation state it is possible to make some complexes with four molecules of ligand, similar to the copper (II) complexes. An example is given below.

Tetrapyridino silver (II) peroxodisulphate, **$[Ag(C_5H_5N)_4]S_2O_8$**. Dissolve 0·5 g (0·003 mole) of silver nitrate in 10 cm³ of water and add it to a solution of 6 g of potassium peroxodisulphate in 200 cm³ of water. At once add 2 cm³ of pyridine and stir. A yellow colour should appear, followed by the precipitate of the complex. Leave to stand for about ten minutes, then filter at the pump. Wash with a little water, then dry in a desiccator over sodium hydroxide.

Investigate the action of sodium hydroxide on the filtrate and on a small sample of the solid.

Zinc, Cadmium and Mercury: oxidation state +2

This group of elements (Group IIb) shows some characteristic d-block properties, for example in the formation of complexes, but the compounds are colourless and the only oxidation state shown is +2. (Mercury (I) compounds contain a mercury–mercury bond in the Hg_2^{2+} ion.)

Dipyridino dithiocyanato complexes, **$M(C_5H_5N)_2(SCN)_2$**. Insoluble complexes of this type are formed readily by both zinc and cadmium.

Dissolve 2·9 g (0·01 mole) of zinc sulphate heptahydrate in 50 cm³ of water and add 2 cm³ of pyridine. Now add slowly, with stirring, a solution of 3 g of ammonium thiocyanate in 50 cm³ of water. Filter off the precipitated complex at the pump, wash it with water, drain thoroughly and allow it to dry. The yield is quantitative and the precipitation can be used for the estimation of zinc.

To prepare the cadmium compound, use 2·3 g of cadmium chloride 2½-hydrate, and proceed as above.

INVESTIGATION OF THE ELEMENTS IN A PERIOD: Na, Mg, Al, Si, P, S, Cl, Ar

It is instructive to investigate the properties of the elements in one period of the periodic table, looking for trends in the physical and chemical properties of the elements and their compounds. It is useful to express as much of this information as possible in diagrammatic form on graph paper. Numerical information can be obtained from such publications as the *Handbook of Chemistry and Physics* (published by the Chemical Rubber Publishing Co.) or the Nuffield *Advanced Science Book of Data*.

The elements

1. Find out and plot on graph paper against atomic number:

(*a*) Melting-point.
(*b*) Boiling-point.
(*c*) Covalent radius and ionic radius.
(*d*) Density of the solid form.
(*e*) First ionization energy.

2. Investigate the physical properties of the element (state, colour, odour) at room temperature.

3. Investigate the action of the element with the following reagents. Use small quantities. Oxygen is best obtained from a cylinder. Take *special care* in all reactions involving the use of sodium metal which often reacts very vigorously.

(*a*) Oxygen: warm a little of the element in a stream of the gas (not possible for Cl_2 or Ar) (see Fig. 12).

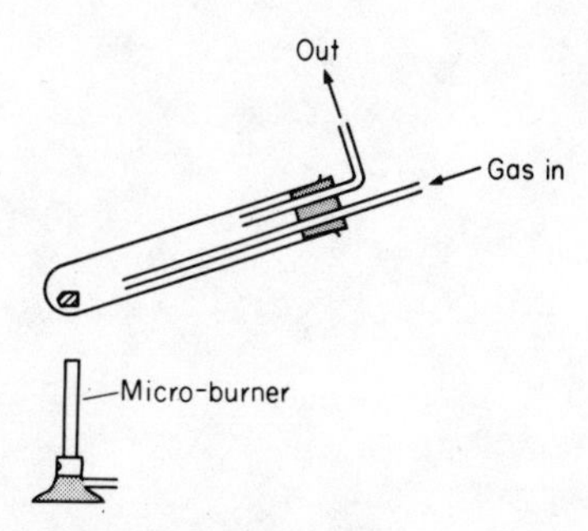

Fig. 12. Heating a small sample of an element in a stream of gas.

(*b*) Chlorine: warm a little of the element in a stream of the gas (not possible for Ar, no point for Cl_2). Use *a fume cupboard.*
(*c*) Water: test the solution with pH paper.
(*d*) 2M hydrochloric acid (omit Na).
(*e*) 2M sodium hydroxide solution (omit Na).
(*f*) Nitric acid: first dilute, then concentrated (omit Na). Test the resultant solution from P and from S for possible oxyacids.

The oxides (note that some elements form more than one oxide)

1. Find out and plot on graph paper against atomic number:

(*a*) Melting-point.
(*b*) Boiling-point (if known).
(*c*) Heat of formation per mole of oxygen atoms.

What information do these figures give us?

2. Investigate the physical properties.

3. Investigate the action with the following:

(*a*) Water: test the solution with pH paper.
(*b*) 2M hydrochloric acid.
(*c*) 2M sodium hydroxide solution.
(*d*) Moist starch–iodide paper (for oxidizing action).
(*e*) Acidified potassium manganate (VII) (for reducing action).

The chlorides (note that some elements form more than one chloride)

1. Find out and plot on graph paper against atomic number:

(*a*) Melting-point.
(*b*) Boiling-point.
(*c*) Heat of formation per mole of chlorine atoms.

(For these purposes treat Cl_2 as chlorine chloride)

2. Investigate the physical properties.

3. Investigate the action with water. Look for signs of hydrolysis: hydrogen chloride evolved, precipitate. Test the solution with pH paper.

The hydrides

Some of these are not readily available; either prepare specimens or else obtain the information from a reference book. Look for signs of whether the hydride is saline (ionic) or volatile (covalent), and investigate its action with water, air, acids and alkalis.

Conclusions

Try to interpret the various diagrams in terms of the types of bonding in the compounds, and relate this to the electronic structure of the element concerned.

See if you can make any generalizations about the way in which the character of the elements in a period changes with increase in atomic number. If any feature stands out as unusual, try to explain it.

REDOX REACTIONS

The concept of oxidation and reduction reactions is a useful but not fundamental subdivision of chemical behaviour which can be expressed:

Redox reactions involve changes in the oxidation numbers of atoms.

The term 'redox' is sometimes preferred as a reduction cannot occur without a balancing oxidation. Thus redox reactions can be separated into two half equations in which reduction is seen as electron acceptance and oxidation as electron donation:

$$2e^- + 2H^+ + H_2O_2 \longrightarrow 2H_2O.$$

$$Fe^{2+} \longrightarrow Fe^{3+} + e^-.$$

The actual mechanism of electron transfer is likely to be complex, but the energy change involved can be measured by a simple procedure if the reaction can be arranged as a suitable voltaic cell. The e.m.f. of the cell is then known as the redox potential of the reaction.

(*a*) *To test for oxidizing agents* add equal volumes of 2M sulphuric acid and potassium iodide solution to a solution of the substance under test. The formation of a dark red solution, a black precipitate or, on warming, a purple vapour suggests the presence of an oxidizing agent. As a more sensitive test for iodine, add starch. Carry out this test with salts of iron (III) and tin (IV); also with sodium nitrate and sodium sulphate.

(*b*) *To test for reducing agents* add equal volumes of 2M sulphuric acid and potassium manganate (VII) solution to a solution of the substance under test. The formation of a brown precipitate, or a change from purple to colourless solution, suggests the presence of a reducing agent. Carry out this test with salts of iron (II) and tin (II); also with sodium nitrite and sodium sulphite.

(*c*) *The relative power of oxidizing agents* can be tested by observing their ability to oxidize bromide ions to bromine. Mix equal volumes of potassium bromide solution, 2M sulphuric acid and add a solution of an oxidizing agent. Note the amount of reaction as judged by colour change or smell of bromine. Warm if necessary.

Test samples of hydrogen peroxide, potassium manganate (VII), potassium dichromate, a salt of iron (III) and potassium hexacyanoferrate (III). Record your results and list the reagents in estimated order of oxidizing power.

(*d*) *The influence of pH* on the reactivity of potassium manganate (VII) can be tested by observing the conditions in which the halide ions can be oxidized.

Mix a *little* potassium manganate (VII) solution with warm solutions of potassium chloride, bromide and iodide. If no reaction is observed in neutral solution (pH 7), add 2M acetic acid to alter the conditions to pH 3. If no reaction is observed, add 2M sulphuric acid to alter the conditions to pH 0. Record your results.

(*e*) *The complex mechanism of potassium manganate (VII) reactions* can be appreciated from the following reactions.

To three separate portions of potassium manganate (VII) solution add 2M sodium hydroxide, 1·5M sodium carbonate and 2M sulphuric acid. Then to each add an equal volume of 0·1M sodium formate solution. Record the immediate result and note any changes after five minutes.

Repeat the experiment using 0·1M sodium oxalate instead of formate, and contrast the results.

(*f*) *In qualitative analysis* redox reactions are used mainly as confirmatory tests. Carry out the tests in Table 5. Describe the results carefully and name the products.

Further, the presence of oxidizing agents will interfere with the proper use of hydrogen sulphide.

Pass hydrogen sulphide gas briefly into acidified solutions of potassium manganate (VII), potassium dichromate and an iron (III) salt. Identify the precipitate.

How can this difficulty be avoided?

Table 5

Cation	*Confirmatory test*
Tin (II)	Add mercury (II) chloride
Tin (IV)	Add a little granulated zinc and 2M hydrochloric acid, followed by mercury (II) chloride
Chromium (III)	Make alkaline with 2M sodium hydroxide and add hydrogen peroxide. Then acidify with cooling
Manganese (II)	Acidify with concentrated nitric acid, then add a *little* sodium bismuthate, then warm

(*g*) *Ink stains* can be removed by oxidation or reduction:

To oxidize, add one drop of blue ink to sodium hypochlorite solution followed by 2M acetic acid.

To reduce, add one drop of blue ink to saturated ammonium oxalate solution followed by '0·880' ammonia.

(*h*) *Oxidation of vanadium (II)*. Dilute 25 cm^3 of 0·1M sodium vanadate in a small conical flask with an equal volume of 2M sulphuric acid and add a small portion of zinc dust. Allow to stand until fully reduced to a violet solution of vanadium (II).

Add the following solutions dropwise to portions of the vanadium (II) solution, when the distinctive colours of the other oxidation states should be produced:

0·1M iron (III) chloride
0·1M iodine
0·1M ammonium chlorostannate (IV)
2M hydrogen peroxide
excess 2M hydrogen peroxide.

Note the colours and relate them to the known oxidation states of vanadium. See p. 44.

IV Organic Chemistry

ALCOHOLS
1. Oxidation of alcohols to carbonyl compounds

THIS is an example of the oxidation of the =CH(OH) group to =CO. It can be carried out on a variety of simple primary and secondary alcohols to give aldehydes and ketones respectively. Longer chain primary alcohols give poor yields producing large amounts of esters, such as hexyl hexanoate from hexanol.

Sodium dichromate is used as oxidant; since it is more soluble than potassium dichromate, the volume of solution required is smaller.

$$(CH_3)_2CHOH + Na_2Cr_2O_7 + H_2SO_4 \longrightarrow (CH_3)_2CO + Na_2SO_4 + Cr_2(SO_4)_3$$

In this experiment the expected yield is 85%.

Requirements

Propan-2-ol, M_r 60·10, 0·785 g cm^{-3}.
Sodium dichromate.

Set up the apparatus for fractionation with addition (see Fig. 13) using a flask of appropriate size.

Place 11·5 cm^3 of propan-2-ol (0·15 mole) in the flask with 10 cm^3 of water. In the dropping funnel place a solution of 15 g of sodium dichromate (0·05 mole) in 100 cm^3 of 2M sulphuric acid (an excess).

Run about 10 cm^3 of sodium dichromate into the flask and gently heat the mixture to boiling. Then add the remainder of the solution at the rate of 3 drops per second. Keep the mixture boiling gently but do not allow condensate to reach the top of the column.

When all the solution has been added, adjust the heating so that distillation occurs at the rate of 2 drops per second. The temperature of distillation must be below 60°C.

Dry the distillate of almost pure propanone by shaking it for five minutes in a corked conical flask with anhydrous sodium sulphate. If drying is complete some of the sodium sulphate should remain powdery.

Meanwhile set up the apparatus for distillation (Fig. D).

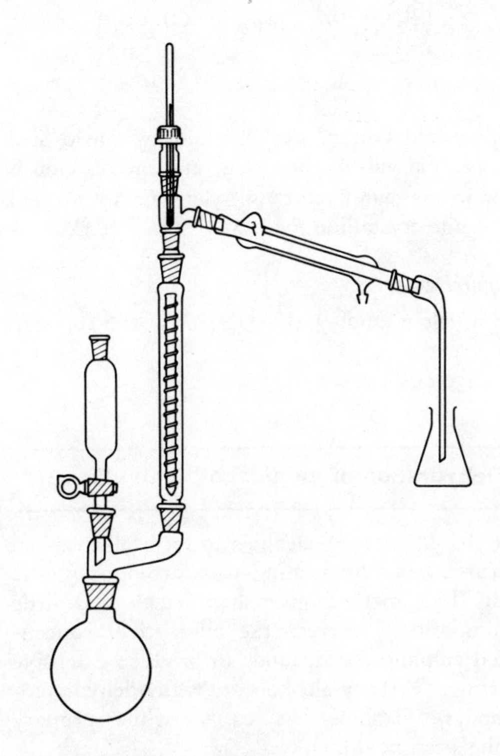

Fig. 13. Oxidation of propanols.

Filter the dried propanone directly into the distilling flask through a small funnel fitted with a plug of cotton-wool.

Redistil the propanone, collecting the fraction boiling between 54°C and 57°C. Calculate your yield based on the alcohol used.

Further preparations

Using the same molar proportions and conditions:
Propanol gives propanal, B.P. 49°C.
Using the same molar proportions but dispensing with the fractionating column:

Pentan-2-ol gives pentan-2-one, B.P. 102°C.
Cyclohexanol gives cyclohexanone, B.P. 156°C.
(the distillate will contain water as well as the ketone).

If potassium dichromate is used for the last two preparations, chrome alum may be obtained from the residual solution by leaving it to evaporate at room temperature.

2. Oxidation of an alcohol to an acid

Alcohols can be oxidized to acids by using sufficient oxidizing agent:

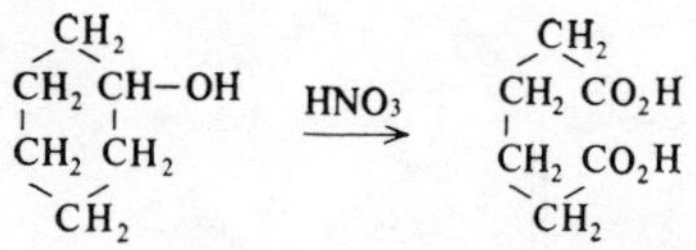

$$\begin{array}{c} CH_2 \\ CH_2\ CH{-}OH \\ CH_2\ CH_2 \\ CH_2 \end{array} \xrightarrow{HNO_3} \begin{array}{c} CH_2 \\ CH_2\ CO_2H \\ CH_2\ CO_2H \\ CH_2 \end{array}$$

The oxidation of cyclohexanol by nitric acid gives adipic acid in good yield, and the reaction is used in the manufacture of nylon-66. Adipic acid is a white crystalline solid, M.P. 152–154°C.

Requirements

Cyclohexanol $C_6H_{11}OH$, M_r 100·16, 0·95 g cm^{-3}.
250 cm^3 beaker.

Place 8 cm^3 of concentrated nitric acid in a 250 cm^3 beaker *in a fume cupboard.* The large beaker is necessary because of the vigour of the reaction. Add 2·1 cm^3 (0·02 mole) of cyclohexanol and withdraw your hand at once. When the reaction subsides, cool the beaker under the tap.

Filter off the crude adipic acid at the pump and recrystallize from hot 2M nitric acid, adding a little activated charcoal to remove coloured impurities. Filter free of charcoal and cool the filtrate in a refrigerator.

Collect the recrystallized product by filtration at the pump, washing free of nitric acid with a little chilled water. Dry on a water-bath.

Determine the melting-point of your product and calculate the yield, based on the cyclohexanol used.

3. Dehydration of an alcohol to an alkene

The dehydration of alcohols to form alkenes can be carried out by heating with orthophosphoric acid. This method gives high yields and little decomposition whereas the alternative, concentrated sulphuric acid, tends to produce extensive charring. Tertiary alcohols are easily dehydrated, secondary alcohols less easily, while primary alcohols can be difficult.

Alkenes with less than five carbon atoms are gases, and so are less convenient to handle.

$$CH_3{\cdot}CH_2{\cdot}C(CH_3)_2{\cdot}OH + H_3PO_4 \longrightarrow CH_3{\cdot}CH{:}C(CH_3)_2$$

The expected yield in this experiment is 85%. 2-methylbut-2-ene is a very volatile liquid with a petrol-like odour.

The product can be converted to its dibromo-derivative by the procedure on p. 64.

Requirements

2-Methylbutan-2-ol, M_r 88·15, 0·81 g cm^{-3}.
Orthophosphoric acid (100%).

Set up the apparatus for fractional distillation (Fig. I).

Place 22 cm^3 (0·2 mole) of 2-methylbutan-2-ol and 6 cm^3 of orthophosphoric acid in the distillation flask.

Connect the flask to the fractionating column and heat with a very small flame. Collect the distillate which boils between 37°C and 39°C and consists of almost pure alkene. After fifteen minutes the reaction should be complete, when the acid will be clear and water will be seen in the column.

Dry the distillate of almost pure alkene by shaking for five minutes in a corked conical flask with granular anhydrous calcium chloride.

Filter the dried alkene directly into the distillation flask through a small funnel fitted with a plug of cotton-wool.

Redistil the alkene, collecting the fraction boiling between 37°C and 38°C. Use a beaker of hot water as a water-bath for heating.

Calculate your yield based on the alcohol used.

Further preparations

Using the same molar amounts, but with 10 cm³ of orthophosphoric acid:

> Cyclohexanol gives cyclohexene, B.P. 84°C (fractionates with water, 70°C to 72°C).
> Octan-2-ol gives oct-1-ene and oct-2-ene, B.P. 123°C (fractionates with water, 90°C to 105°C).

4. Dehydration of an alcohol to an ether

The reaction between ethanol and sulphuric acid can result in the formation of a number of products, including sulphate esters and ethylene, but with an excess of ethanol at about 140°C the main product is ethoxyethane. This is the basis of the industrial production of ether, but it is less efficient on a small scale, partly because of losses by evaporation during purification.

$$2CH_3CH_2OH \longrightarrow (CH_3CH_2)_2O + H_2O$$

Ethoxyethane is a colourless liquid with a characteristic smell, boiling at 35°C. Yields are rarely over 50%.

Requirements

Ethanol, M_r 46·07, 0·79 g cm^{-3}.

Set up the apparatus shown in Fig. 14, for distillation with addition. Make sure that the thermometer (0–200°C) reaches into the liquid, and attach a rubber tube to the vent to make sure that the *highly inflammable* ether vapour is led away to well below bench level. Surround the receiver with iced water.

Place about 1g of clean, dry sand in the flask, with 10 cm³ of ethanol. Cool the flask in water and add 8 cm³ of concentrated sulphuric acid in small portions, with shaking. Attach the flask to the distillation apparatus and make sure that all the joints are fitted well together. Place 9 cm³ (making 0·33 mole in all) of ethanol in the dropping funnel and heat the flask on a gauze with a microburner.

When the temperature reaches about 140°C, the ether should start to distil. Maintain the temperature between 140°C and 145°C, and allow ethanol to drip in at the same rate as the ether distils. When all the ethanol has been added, close the tap in the funnel and maintain the heating until no more ether distils.

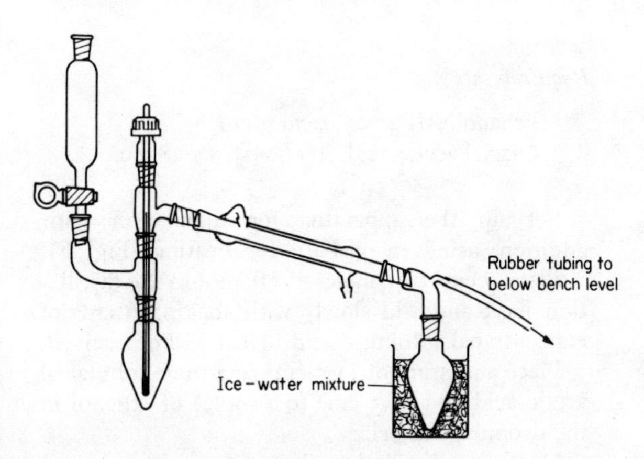

Fig. 14. Preparation of ethoxyethane

Now extinguish the flame, and any others nearby, and transfer the distillate to a separating funnel. Wash with two 5 cm³ portions of 2M sodium hydroxide to remove any dissolved acid vapours, and run the ether into a small conical flask. Add a few pieces of granular anhydrous calcium chloride and cork securely. Allow to stand for fifteen minutes, shaking occasionally. The calcium chloride will absorb any ethanol present as well as water.

Set up the apparatus for distillation and filter the dry ether through a plug of cotton-wool into the flask. Heat the flask with hot water in a beaker (heated elsewhere), and collect in a weighed specimen tube the fraction boiling at 34–39°C.

Calculate the yield based on the ethanol used.

Further preparations

The same technique can be used for other primary alcohols up to butanol. Methoxymethane is a gas, B.P. —23·6°C, so different practical details would be required.

5. Preparation of an ester (method 1)

The formation of esters using the acid and alcohol plus concentrated sulphuric acid is limited to the lower members of the alcohol series. With higher members, and especially branched chain alcohols, extensive charring and alkene formation occurs.

$$CH_3{\cdot}CO_2H + CH_3CH_2OH \longrightarrow CH_3{\cdot}CO_2CH_2CH_3 + H_2O$$

The formation of ethyl acetate using concentrated sulphuric acid has an expected yield of 80%.

Ethyl acetate has a fruity smell which may be familiar as it is frequently used as a solvent for lacquers. Its B.P. is 77°C.

Requirements

Ethanol, M_r 46·07, 0·79 g cm^{-3}.
Glacial acetic acid, M_r 60·05, 1·048 g cm^{-3}.

Set up the apparatus for distillation with addition, using an oil-bath for heating (Fig. F).

Place 6 cm^3 (0·1 mole) of ethanol in the distillation flask and add slowly with shaking 6 cm^3 of concentrated sulphuric acid. Heat will be evolved.

Place a mixture of 11·5 cm^3 (0·2 mole) of glacial acetic acid and 11·5 cm^3 (0·2 mole) of ethanol in the dropping funnel.

Heat the distillation flask by raising the oil temperature to about 155°C and then add the mixture in the funnel at the rate of 1–2 drops per second. Crude ester should soon start distilling over; do not increase the rate of addition otherwise the ester will contain much unreacted ethanol.

When addition of the mixture is complete, raise the oil temperature to about 165°C to complete the distillation.

The distillate of crude ethyl acetate must be cleaned and washed, but use the minimum of aqueous solutions because the ester is appreciably soluble in water (9 g/100 g at 20°C) and forms with it a constant-boiling mixture, B.P. 70·4°C.

Transfer the distillate to a separating funnel and add 5 cm^3 of 1·5M sodium carbonate. Carefully shake the mixture, releasing the pressure of evolved carbon dioxide frequently. Allow the layers to separate and discard the lower aqueous layer. This process removes acidic impurities such as sulphur dioxide from side reactions and acetic acid.

To remove any ethanol from the distillate add to the separating funnel a solution of 5 g calcium chloride in 5 cm^3 water. Shake the mixture, allow the layers to separate and discard the lower aqueous layer.

Transfer the ethyl acetate to a corked conical flask, and dry by shaking for ten minutes or allowing to stand overnight with a little granular calcium chloride. If the ester has been successfully dried, it will be quite clear and the calcium chloride will still appear granular, with no sign of a separate liquid layer.

Set up the cleaned apparatus for distillation and filter the dried ester directly into the distillation flask through a small funnel fitted with a plug of cotton-wool.

Redistil the ester slowly, collecting the fraction boiling between 74°C and 79°C.

Calculate your yield based on the acetic acid used.

Further preparations

Using the molecular proportions and procedure given above, the following esters can be prepared in good yield from the appropriate acids and alcohols:

Ethyl formate, B.P. 54°C.
Ethyl propanoate, B.P. 98°C.
Propyl formate, B.P. 81°C.
Methyl acetate, B.P. 57°C.

6. Preparation of an ester (method 2)

The use of concentrated sulphuric acid in the formation of esters is limited (see the preparation of ethyl acetate, opposite), but the use of an acid anhydride with the alcohol dispenses with the need for a dehydrating agent and gives clean products quickly and in high yield.

This method is most widely used for the preparation of acetates.

$$C_5H_{11}OH + (CH_3CO)_2O \longrightarrow CH_3{\cdot}CO_2C_5H_{11} + CH_3{\cdot}CO_2H$$

In this experiment the expected yield is 90%.

Requirements

3-Methylbutan-1-ol, M_r 88·15 cm³, 0·81 g cm⁻³.
Acetic anhydride.

Set up the apparatus for reflux (Fig. A).

Place 11 cm³ (0·1 mole) of 3-methylbutan-1-ol in the distillation flask and pour down the condenser 12·5 cm³ (a slight excess) of acetic anhydride.

Warm the mixture gently until the reaction starts. When the initial reaction has subsided, heat to boiling and reflux gently for five minutes. Then allow the mixture to cool.

Transfer the cool mixture to a separating funnel, add about double its volume of water and shake the mixture. Allow the two layers to separate and discard the lower aqueous layer.

Add about 10 cm³ of 0·1M sodium bicarbonate to the separating funnel and shake carefully, releasing the pressure of carbon dioxide frequently. Repeat the washing with further amounts of sodium bicarbonate until no more carbon dioxide is produced. Discard the lower aqueous layer each time.

Transfer the ester to a corked conical flask and dry by shaking for five minutes with anhydrous sodium sulphate.

Filter the dried ester directly into the distillation flask through a small funnel fitted with a plug of cotton-wool.

Redistil the ester, collecting the fraction boiling between 135°C and 140°C.

Calculate your yield based on the alcohol taken.

This ester is used in commerce; note the odour.

Further preparations

Using the same molar proportions (i.e. a slight excess of acetic anhydride) the following acetates can be prepared:

Butan-1-ol gives butyl acetate, B.P. 125°C.
Hexan-1-ol gives hexyl acetate, B.P. 169°C.
Cyclohexanol gives cyclohexyl acetate, B.P. 172°C.

PHENOLS

1. Conversion of an amine to a phenol

Phenols are usually prepared from the corresponding amines by diazotization, followed by hydrolysis and decomposition. This method is used widely, and other groups present in the molecule are usually unaffected. The precise details depend on the strength of the amine and on the groups present.

$$2C_6H_5NH_2 + 2H_2SO_4 + 2NaNO_2 \longrightarrow (C_6H_5N_2^+)_2SO_4^{2-} + Na_2SO_4 + 4H_2O$$

$$(C_6H_5N_2^+)_2SO_4^{2-} + 2H_2O \longrightarrow 2C_6H_5OH + 2H_2SO_4$$

Requirements

Aniline, $C_6H_5NH_2$, M_r 93·13, 1·02 g cm^{-3}.
M copper (II) sulphate (250 g l^{-1}).

Mix carefully 3 cm³ of concentrated sulphuric acid with 50 cm³ of water. To the well-stirred mixture add 5 cm³ (0·05 mole) of aniline. Heat the mixture gently until all the precipitated aniline sulphate has redissolved, and then cool the solution to 5–10°C in an ice-bath. Some aniline sulphate may be reprecipitated, but this does not matter.

Dissolve 3·7 g (0·05 mole) of sodium nitrite in 10 cm³ of water and place the solution in a dropping funnel. Stir the aniline sulphate solution with a thermometer, and allow the sodium nitrite solution to run in at such a rate that the temperature does not rise above 10°C. When all the nitrite solution has been added, allow the diazonium solution to stand for five minutes, but no longer.

Transfer the diazonium salt solution to a dropping funnel and, over a period of five to ten minutes, run it into 20 cm³ of M copper (II) sulphate kept between 50°C and 60°C in a 250 cm³ round-bottomed flask. Nitrogen should be evolved and some tarry by-products produced.

When the evolution of nitrogen has stopped, fit the flask for steam distillation (Fig. H). Steam distil until about 150 cm³ of distillate has been collected (this is necessary because phenol is appreciably soluble in hot water, and therefore steam distillation is less effective than for aniline).

Extract the distillate three times with ether, using 20 cm³ portions. Dry the combined ether extracts with anhydrous sodium sulphate, shaking the flask periodically until the solution is completely clear.

Distil off the ether using a water-bath, and finally distil the phenol using a short air condenser. Collect the fraction boiling at 177–183°C. If the phenol starts to crystallize in the condenser, warm it gently to melt it. If the liquid in the receiver fails to crystallize, add a small seed crystal of phenol or scratch the liquid with a glass rod.

Phenol is corrosive and should not be handled. It will turn pink on standing, as it is oxidized by the air. M.P. 41°C, B.P. 182°C.

Calculate your yield, based on the aniline used.

Further preparations

o-Cresol, M.P. 30°C, B.P. 191·5°C
m-Cresol, M.P. 11°C, B.P. 202·8°C
p-Cresol, M.P. 36°C, B.P. 202·5°C
all from the corresponding amines.

Test-tube experiments with phenol

1. Add about 1 cm³ of water to a small crystal of phenol, and then fresh bromine water. The colour of the bromine is discharged at once, and 2,4,6-tribromophenol is precipitated as a white solid with a characteristic smell. Write out the structural formula of the product.

2. Place a few crystals of phenol in a test tube and add about half the bulk of phthalic anhydride. Mix the solids thoroughly, then add 1 or 2 drops of concentrated sulphuric acid and warm very gently for one minute. The mixture will go red. Cool carefully and add water. To a sample of the resulting mixture, which contains phenolphthalein and phthalic acid, add 2M sodium hydroxide. The indicator nature of the

product can be seen by acidification of the alkaline solution. This is called the *Phthalein Test*, and it can be applied to many phenols. Resorcinol will give fluorescein. Investigate what happens with other phenols.

3. Make a solution of one crystal of phenol in a little water, and add 1 drop of iron (III) chloride solution. Note the characteristic colour of the complex formed. Investigate what happens with other phenols.

2. Benzoylation of a phenol

Benzoylation of phenols is carried out chiefly in order to prepare pure crystalline derivatives with sharp melting-points which can be used for identifying the phenols. Benzoyl chloride is hydrolysed much less readily than acetyl chloride, and the reaction is carried out in aqueous alkaline solution, using the Schotten–Baumann technique.

$$C_6H_5COCl + C_6H_5OH + NaOH \longrightarrow C_6H_5{\cdot}CO_2C_6H_5 + NaCl + H_2O$$

Phenyl benzoate is a white crystalline solid, M.P. 68°C. Yield 75%.

Requirements

Phenol, C_6H_5OH, M_r 94·11.
Benzoyl chloride, C_6H_5COCl, M_r 140·57, 1·21 g cm^{-3}.

Carry out this experiment *in a fume cupboard* as far as possible, especially when handling benzoyl chloride.

Dissolve 2 g (0·022 mole) of phenol in 20 cm^3 of 2M sodium hydroxide in a stoppered flask. Add 3 cm^3 (a slight excess) of benzoyl chloride, re-stopper the flask and shake vigorously for several minutes. Heat is evolved, and it is necessary to release the pressure occasionally.

When a solid is formed in the flask, leave it to stand for a few minutes and then filter at the pump. Break up the lumps of solid, wash thoroughly with water and drain dry.

Recrystallize from ethanol, cooling in a refrigerator or ice-bath before finally filtering off the crystals.

Further preparations

Using the same technique, the following can be prepared:

m-Cresol gives *m*-cresyl benzoate, M.P. 54°C.
p-Cresol gives *p*-cresyl benzoate, M.P. 71°C.
2-Naphthol gives 2-naphthyl benzoate, M.P. 107°C.
(*o*-Cresyl benzoate is a liquid, boiling at 307°C).

3. Methylation of a phenol

Aromatic ethers can be made easily from the corresponding phenols by methylation with dimethyl sulphate in alkaline solution. These ethers are more stable and less reactive than the phenols themselves, and they are often used in syntheses where the free phenol would be too reactive.

Dimethyl sulphate is *toxic* and care must be taken not to inhale the vapour, or get the liquid on the skin. Carry out the experiment *in a fume cupboard.*

$$2C_6H_5OH + (CH_3)_2SO_4 + 2NaOH \longrightarrow 2C_6H_5OCH_3 + Na_2SO_4 + 2H_2O$$

Methoxybenzene is a colourless liquid, B.P. 154°C. Yield 80%.

Requirements

Phenol, C_6H_5OH, M_r 94·11.
Dimethyl sulphate, $(CH_3)_2SO_4$, M_r 126·13, 1·32 g cm^{-3}.

Set up the apparatus for reflux (Fig. A), place 4·9 g (0·05 mole) of phenol in the flask and dissolve it in 30 cm^3 of 2M sodium hydroxide. Attach the condenser and add, down the condenser, 5·5 cm^3 (an excess) of dimethyl sulphate. If the reaction mixture shows no signs of getting warm, heat it gently and shake the flask. The reaction should start and continue to reflux without the application of heat. The ether should begin to separate out as an oil on the surface. When the reaction shows signs of subsiding, heat the flask for a few minutes more, then cool.

Transfer the cold mixture to a separating funnel, and run off and discard the lower, aqueous, layer. Wash with a little water, discard the washings and run the rest into a conical flask. Dry with anhydrous sodium sulphate, shaking occasionally until the liquid is clear.

Set up the apparatus for distillation, using an air condenser (Fig. Ei), and distil the methoxybenzene, collecting the fraction boiling at 150–154°C.

Further preparations

This reaction can be applied to a whole range of phenols. Some of the ethers are pleasant-smelling compounds, used in perfumery. Using the same proportions, the following phenols give good results:

o-Cresol gives 2-methylmethoxybenzene, B.P. 171°C.
m-Cresol gives 3-methylmethoxybenzene, B.P. 177°C.
p-Cresol gives 4-methylmethoxybenzene, B.P. 176°C.

If the ether prepared by this reaction is a solid, the procedure is modified: cool the reaction mixture, filter off, wash and dry the ether. Some of the ethers, though solid, have sufficiently low boiling-points to be redistilled after drying. For example:

2-Naphthol gives 2-methoxynaphthalene, M.P. 72°C, B.P. 274°C.
Benzene-1,4-diol gives 1,4-dimethoxybenzene, M.P. 55°C, B.P. 212°C.

In the latter example there are two hydroxyl groups, so twice as much dimethyl sulphate must be used.

Ethyl ethers can be prepared in the same way by using diethyl sulphate. They are usually lower melting, more pleasant smelling compounds.

4. Preparation of 2,4-dichlorophenoxyacetic acid

Chloroacetic acid reacts with a phenol to introduce a carboxymethyl group ($-CH_2 \cdot CO_2H$) in place of a hydrogen. The product is usually a white crystalline solid with a sharp melting-point, which can be used in the identification of the phenol. In addition, the compounds have interesting physiological effects on plants, being used commercially as selective weed-killers. 2,4-Dichlorophenoxyacetic acid is one of the most important of these, ('2,4-D')

$$Cl_2C_6H_3OH + ClCH_2 \cdot CO_2H + 2\,NaOH \longrightarrow Cl_2C_6H_3O \cdot CH_2 \cdot CO_2Na + NaCl + 2\,H_2O$$

2,4-Dichlorophenoxyacetic acid is a white crystalline solid, M.P. 138°C. Yield 70%.

Requirements

2,4-Dichlorophenol, $Cl_2C_6H_3OH$, M_r 163·0.
Chloroacetic acid, $CH_2Cl \cdot CO_2H$, M_r 94·5.

Dissolve 8 g (0·05 mole) of 2,4-dichlorophenol and 5 g (a slight excess) of chloroacetic acid in 60 cm³ of 2M sodium hydroxide, then heat the mixture very gently for ten to fifteen minutes. A solid may be precipitated, but whether it is or not add 40 cm³ of water and 15 cm³ of concentrated hydrochloric acid. A heavy oil should separate out to the bottom of the mixture. Leave to stand overnight, preferably in a refrigerator.

Break up the solid which forms, filter it off at the pump, wash with water and drain thoroughly. Dry the solid in a desiccator.

Recrystallize from benzene. Use 20–30 cm³, under reflux because benzene is very inflammable and has a *toxic vapour*.

Further preparations

Phenol gives phenoxyacetic acid, M.P. 98°C.
o-Cresol gives 2-methylphenoxyacetic acid, M.P. 120°C.
m-Cresol gives 3-methylphenoxyacetic acid, M.P. 154°C.
p-Cresol gives 4-methylphenoxyacetic acid, M.P. 136°C.

Use as a weed-killer

2,4-Dichlorophenoxyacetic acid is the most frequently used of the selective weed-killers. It is harmless to grass, but kills most of the weeds found in lawns. It is usually either the sodium salt or the ethyl ester which is used.

Make a solution of the acid in a slight excess of sodium carbonate solution, at the rate of 2 g of acid to 1 litre of solution needed, and make up the volume required with water. Add a little Teepol to increase the wetting power. Apply to grass at the rate of 1 litre to 10 square metres. Do not cut the grass within four days of applying the solution, and only use it when the grass is growing vigorously. Growth distortion should be noticed in a week, and the weeds should be killed within a few weeks. Carry out a small-scale test on dandelions in a flower-pot or box.

HALOGENOALKANES

1. Preparation of a chloroalkane

Replacement of the hydroxyl group by halogen is fairly easy with tertiary alcohols, but more difficult with secondary and primary alcohols. Also, halogen acid is more easily lost, giving an alkene, from tertiary halides.

Using a tertiary alcohol, the hydroxyl group can be replaced using aqueous hydrochloric acid, but for other alcohols, phosphorus halides or strong dehydrating agents are required.

$$(CH_3)_3COH \xrightarrow[\text{anhydrous } CaCl_2]{\text{concentrated HCl}} (CH_3)_3CCl$$

2-Chloro-2-methylpropane is a volatile liquid with a distinctive odour, B.P. 51°C. In this experiment the expected yield is 85%.

Requirements

2-Methylpropan-2-ol $(CH_3)_3COH$, M_r 74·12, 0·78 g cm^{-3}.
Separating funnel.

Place 20 cm^3 (0·21 mole) of 2-methylpropan-2-ol, roughly 5 g of anhydrous calcium chloride and 70 cm^3 of concentrated hydrochloric acid in a large separating funnel. Shake at intervals until the top layer of product is fully formed. This usually takes about twenty minutes.

Discard the lower aqueous layer. Add about 20 cm^3 of 0·1M sodium hydrogen carbonate to the 2-chloro-2-methylpropane in the funnel. Shake the funnel carefully, releasing the pressure of carbon dioxide frequently. Allow the layers to separate and discard the lower aqueous layer again. The washing with sodium hydrogen carbonate solution is repeated until no more carbon dioxide is formed.

Transfer the upper layer to a small conical flask, add a little anhydrous sodium sulphate and cork securely. Shake occasionally for about five minutes to dry the chloroalkane.

Meanwhile set up the apparatus for distillation (Fig. D). Filter the product directly into the distillation flask through a small funnel fitted with a plug of cotton-wool. Distil the liquid and collect the fraction boiling between 50°C and 52°C.

Weigh your product and calculate the yield based on the 2-methylpropan-2-ol used.

Reactions

(*a*) Shake a little of your product with silver nitrate solution.
(*b*) Boil under reflux for one minute with distilled water, and then add silver nitrate solution.
(*c*) Boil under reflux for one minute with 2M sodium hydroxide solution, then cool. Acidify with 2M nitric acid and add silver nitrate solution.

Account for the different results.

Further preparations

Using the same molar proportions, 2-methylbutan-2-ol gives 2-chloro-2-methylbutane, B.P. 86°C.

2. Preparation of a bromoalkane

A simple method of preparing bromoalkanes is to reflux the alcohol with constant-boiling hydrobromic acid in the presence of sulphuric acid. The method is most suitable for primary alcohols, which are not dehydrated appreciably under these conditions. The preparation is described for 1-bromobutane.

$$CH_3{\cdot}CH_2{\cdot}CH_2{\cdot}CH_2OH + HBr \longrightarrow CH_3{\cdot}CH_2{\cdot}CH_2{\cdot}CH_2Br + H_2O$$

1-Bromobutane is a colourless liquid, B.P. 100°C. In this experiment the expected yield is 80%.

Requirements

Butan-1-ol, C_4H_9OH, M_r 74·12, 0·8 g cm^{-3}.
Constant-boiling hydrobromic acid (see p. 42).

Set up the apparatus for reflux (Fig. A), putting into the flask 9·25 cm^3 (0·1 mole) of butan-1-ol,

20 cm^3 (an excess) of constant-boiling hydrobromic acid and 6 cm^3 of concentrated sulphuric acid. Heat on a gauze and reflux for three-quarters of an hour.

Cool the flask, add 10 cm^3 of water and then replace the reflux condenser with the apparatus for distillation. When the mixture is heated, the 1-bromobutane will distil in the steam. Continue distilling until no more oily drops come over, then stop the heating.

Transfer the distillate to a separating funnel. Decant off the upper, aqueous, layer. Wash the 1-bromobutane with sodium hydrogen carbonate solution to remove dissolved acids, then run the lower, 1-bromobutane, layer into a small conical flask.

Dry with granular anhydrous calcium chloride, shaking occasionally until the liquid is clear.

Redistil, heating on a boiling water-bath. Collect the fraction boiling at 98–102°C.

Further preparations

The following can be prepared by the same method:

1-Bromo-2-methylpropane, B.P. 91°C.
1-Bromopentane, B.P. 129°C.
1-Bromo-3-methylbutane, B.P. 118°C.
1-Bromohexane, B.P. 156°C.

3. Preparation of an iodoalkane (method 1)

The following technique gives better yields for the lower members of the alcohol series than that given below for iodobutane. Even so, it is essential that the reagents are absolutely dry and that the red phosphorus is carefully purified by boiling with a large volume of water, filtering and drying in a warm oven.

$$CH_3CH_2OH \xrightarrow{I_2 \text{ and } P} CH_3CH_2I$$

In this experiment the expected yield is 65%.

Requirements

Ethanol.
Purified red phosphorus.
Iodine.

Set up the apparatus for reflux (Fig. A).

Place 6·5 cm^3 (0·1 mole) of ethanol and 1 g (an excess) of purified red phosphorus in the distillation flask. Add in small quantities 12·7 g (0·05 mole) of well ground iodine. Add about 2 g at a time at one-minute intervals, replacing the condenser between additions. Heat will be evolved; if necessary, cool to prevent actual boiling.

When all the iodine has been added, heat the flask under reflux for about forty minutes on a boiling water-bath.

Now cool the flask and set up for direct distillation. Heat the flask on the water-bath and collect the distillate of crude iodoethane. When distillation becomes slow, heat the flask directly with a small flame.

Transfer the crude iodoethane to a separating funnel and add 5 cm^3 of 2M sodium hydroxide solution. Shake well and, if the lower layer of iodoethane is not colourless, add 2 cm^3 of sodium thiosulphate solution and repeat the shaking.

Transfer the iodoethane to a corked conical flask and dry by shaking for five minutes with granular calcium chloride.

Filter the dried iodoethane directly into the distillation flask through a small funnel fitted with a plug of cotton-wool.

Redistil the iodoethane, collecting the fraction boiling between 68°C and 73°C.

Calculate your yield based on the alcohol used.

Iodoethane darkens on storage but a spiral of bright copper wire will keep it colourless for a longer time.

4. Preparation of an iodoalkane (method 2)

The use of phosphorus halides, or red phosphorus and the halogen, provides a good method of producing organic halides, although the C_1–C_3 alcohols give poor yields by this method. Provided the quantity to be made is small it is possible to add all the reagents at once, thus simplifying the procedure.

$$CH_3{\cdot}CH_2{\cdot}CH_2{\cdot}CH_2OH \xrightarrow{I_2 \text{ and } P} CH_3{\cdot}CH_2{\cdot}CH_2{\cdot}CH_2I$$

In this experiment the expected yield is 85%.

Requirements

Butan-1-ol, M_r 74·12, 0·81 g cm^{-3}.
Red phosphorus.
Iodine.

Set up the apparatus for reflux (Fig. A).

Place 9·25 cm^3 (0·1 mole) of butan-1-ol in the distillation flask plus 1·5 g (an excess) of red phosphorus. Now add 12·7 g (0·05 mole) of iodine and immediately connect to the reflux condenser.

Warm the flask slightly until reaction starts, then remove the flame until the initial reaction subsides. Heat under reflux for about half an hour by which time all the iodine should have reacted, giving a light coloured condensate.

Allow to cool briefly, add 20 cm^3 of water and set up the apparatus for distillation. Distil the reaction mixture, when the product will be collected as an oily layer under a water layer. Continue the distillation until no more oily drops are seen condensing; add more water to the distillation flask if necessary.

Transfer the entire distillate to a separating funnel and, if necessary, add a few cm^3 of sodium thiosulphate solution to remove iodine coloration.

Transfer the lower, iodobutane, layer to a corked conical flask and dry by shaking for five minutes with anhydrous sodium sulphate.

Filter the dried iodobutane directly into the distilling flask through a small funnel fitted with a plug of cotton-wool.

Redistil the iodobutane, collecting the fraction boiling between 128°C and 132°C.

Calculate your yield based on the alcohol used.

Further preparations

Using the same molar proportions, but purifying by carrying out a second steam distillation to avoid the decomposition which occurs in a straight distillation:

Pentan-1-ol gives 1-iodopentane, B.P. 156°C.
Hexan-1-ol gives 1-iodohexane, B.P. 180°C.

5. Bromination of an alkene

This is an example of the addition of a halogen to an alkene. It is a very vigorous reaction and care must be taken not to go too fast at the beginning. Bromine vapour is very irritant and the whole experiment must be carried out *in a fume cupboard.*

$$(CH_3)_2{\cdot}C{:}CH{\cdot}CH_3 + Br_2 \longrightarrow (CH_3)_2{\cdot}CBr{\cdot}CHBr{\cdot}CH_3$$

In this experiment the expected yield is 80%.

2,3-dibromo-2-methylbutane is a heavy, pleasant smelling liquid which darkens a little on keeping. It cannot be distilled directly without decomposition but is steam-distilled successfully.

Requirements

2-Methylbut-2-ene (see p. 58).
Bromine.

Set up, *in a fume cupboard*, the apparatus for reflux with addition (Fig. B).

Place 14 cm^3 (0·2 mole) of 2-methylbut-2-ene and 20 cm^3 of water in the flask. Place 11 cm^3 (a slight excess) of bromine in the funnel and cover with a layer of water. A cork may be placed lightly in the funnel; this will keep down the concentration of bromine vapour escaping, without stopping the bromine from entering the apparatus.

Turn the tap of the funnel *very slowly* until the bromine can just pass through at a rate not greater than 1 drop per second. As the bromine reaches the alkene, which is the upper layer, there is a vigorous reaction to form the dibromide. As the reaction proceeds, the density of the upper layer increases and eventually it will sink to the bottom. At this stage the rate of addition of the bromine may be increased.

Near the end of the reaction, when the water layer becomes coloured with bromine, heat the mixture to boiling. Continue the addition of bromine until there is a permanent yellow colour present. Stop heating and, when the apparatus has cooled, discharge the colour with a few drops of ammonia solution.

Now set up the flask for direct distillation (Fig. D), and distil the mixture of dibromide and water. The distillate consists of water, and a lower layer of the pure product, which merely requires drying.

Transfer the distillate to a separating funnel and run off the lower layer of product into a corked conical flask. Add anhydrous sodium sulphate and shake occasionally for a few minutes.

Filter the product free of drying agent through a small filter funnel fitted with a cotton-wool plug.

Calculate your yield based on the alkene used.

6. Reduction of a halogenoalkane to an alkane

Halogenoalkanes are readily reduced in ethanol by zinc activated by a coating of copper.

$$C_6H_{13}I + 2(H) \longrightarrow C_6H_{14} + HI$$

Hexane is a liquid, B.P. 69°C, with an odour of petrol. The expected yield in this experiment is 60%.

Requirements

1-Iodohexane, $C_6H_{13}I$, M_r 212·1, 1·44 g cm^{-3} (see p. 68).
Zinc powder.

Set up the apparatus for distillation (Fig. D), with a 0–110°C thermometer.

Prepare the zinc–copper couple by placing 5 g of zinc powder in a beaker and adding about 10 cm^3 of copper (II) sulphate solution. Stir with a glass rod and then filter off the metal powder at the pump. Wash first with water, then with ethanol.

Transfer the product to the distillation flask and add a mixture of 8·5 g (0·04 mole) of iodohexane and 10 cm^3 of ethanol.

Have a 250 cm^3 beaker of cold water available, then start the reaction by gentle heating. The reaction is exothermic and should be controlled by cooling if it becomes too vigorous.

The hexane distils over at about 70°C, and when the distillation temperature rises above about 73°C, stop collecting the distillate.

Purify the distillate of crude hexane by adding anhydrous calcium chloride to absorb water and ethanol. Allow to stand for five minutes in a corked conical flask. Redistil the hexane, through a fractionating column if available, and collect the fraction boiling between 68°C and 70°C.

Calculate the yield based on the 1-iodohexane used.

7. A Grignard reagent

The Grignard reagents are the most familiar organo-metallic compounds. They are formed readily by bromo- and iodoalkanes, and less readily by bromoarenes. They are, however, unstable, but can be treated with carbonyl compounds to yield a wide range of products.

$$C_2H_5I + Mg \longrightarrow C_2H_5{\cdot}MgI$$

$$C_2H_5{\cdot}MgI + H_2O \longrightarrow C_2H_6 + Mg(OH)I$$

It is essential that all materials for this preparation should be quite dry: the apparatus and magnesium should be oven-dried, the iodoethane dried with anhydrous calcium chloride and the ethoxyethane dried with sodium wire.

The ethane produced is a gas.

Requirements

Dry iodoethane, C_2H_5I, M_r 155·97, 1·93 g cm^{-3}.
Dry ethoxyethane.
Dry magnesium turnings.

Set up the apparatus for reflux (Fig. A).

Place in the flask 1·5 g (0·06 mole) of magnesium, 5 cm^3 (0·06 mole) of iodoethane and 20 cm^3 (an excess) of ethoxyethane. Warm the mixture gently on a water-bath to start the reaction. If it does not start after a minute or two, add a very small crystal of iodine. The reaction is exothermic, and is complete when the magnesium has all dissolved. Warm occasionally if necessary.

Now convert the apparatus to gas evolution (Fig. Ki), with collection over water.

Place 20 cm^3 of water in the dropping funnel and add *dropwise* to the ethyl magnesium iodide solution. Collect the ethane evolved in gas jars; the theoretical yield is about 1·5 litres.

Test separate gas jars with a lighted splint and with bromine water. How do the results compare with ethene and ethyne?

Consult a textbook about the structure of Grignard reagents and make a list of the different types of product that can be obtained by their use. This preparation can also be carried with bromoethane, using the same molar quantities.

CARBONYL COMPOUNDS
1. Derivatives of carbonyl compounds

Carbonyl compounds form many crystalline solid derivatives, often by simple condensation reactions resulting in carbon–nitrogen double bonds. Not all are formed by both aldehydes and ketones. Tables of melting-points of these derivatives can be found in textbooks of organic analysis.

Aldehyde-ammonias. The simpler aldehydes form crystalline compounds with ammonia. They are not very stable, and usually decompose on heating.

Put a few drops of acetaldehyde in a test tube and blow on to it ammonia gas, prepared by heating '0·880' ammonia in a test tube fitted with a delivery tube. Note the crystals which form. Investigate the action of gentle heat on them.

Bisulphite addition compounds. These compounds also decompose readily in acidic or alkaline conditions, and are used in the purification of carbonyl compounds.

Shake 1 cm^3 of the carbonyl compound with 5 cm^3 of freshly-prepared, saturated sodium bisulphite solution. Note the crystals which form. Decant off the aqueous solution, divide the crystals into two parts, and warm one with 2M hydrochloric acid and the other with 2M sodium hydroxide.

Oximes. The oximes are prepared by the action of hydroxylamine. They are sometimes difficult to obtain as crystals, but cyclohexanone forms a readily crystallized oxime:

$$(CH_2)_5C{=}O + H_2NOH \longrightarrow (CH_2)_5C{=}NOH + H_2O$$

Dissolve 1 g of hydroxyammonium chloride in 5 cm^3 of water in a small conical flask and add 1·5 g of sodium acetate trihydrate. Add 1 cm^3 of cyclohexanone, cork the flask and shake it vigorously for a few minutes. The oxime should be deposited as white crystals.

Cool the flask in iced water, filter off the crystals at the pump using a Hirsch funnel, wash with a little water and drain. Dry on a filter paper.

Recrystallize from aqueous ethanol.

Determine the melting-point, and compare with that quoted in a table of melting-points.

Semicarbazones. The semicarbazones crystallize more readily than the oximes, but are prepared in a similar way.

$$=CO + NH_2{\cdot}NH{\cdot}CO{\cdot}NH_2 \longrightarrow =C{:}N{\cdot}NH{\cdot}CO{\cdot}NH_2 + H_2O$$

Dissolve 0·5 g of semicarbazide hydrochloride in 5 cm^3 of water and add 0·8 g of sodium acetate trihydrate.

Add 0·5 cm^3 (5 drops) of the carbonyl compound and shake. If crystals do not form in a few minutes, try scratching the inside of the tube with a glass rod. If they still do not form, warm gently, then cool (with scratching) in iced water.

Filter off the crystals at the pump using a Hirsch funnel, wash with a little water and drain. Dry on a filter paper.

Recrystallize from aqueous ethanol.

Determine the melting-point, and compare with that quoted in a table of melting-points.

Phenylhydrazones and 2,4-dinitrophenylhydrazones. Phenylhydrazine forms derivatives readily with aromatic aldehydes (see p. 68), but in general 2,4-dinitrophenylhydrazine is preferred because its derivatives are higher-melting and less soluble.

$$=CO + NH_2{\cdot}NH{-}C_6H_3(NO_2)_2 \longrightarrow =C{:}N{\cdot}NH{-}C_6H_3(NO_2)_2 + H_2O$$

The 2,4-dinitrophenylhydrazones are yellow or orange crystals.

Make up a solution of the reagent used in this reaction as follows:

Dissolve 2 g of 2,4-dinitrophenylhydrazine in

4 cm^3 of concentrated sulphuric acid and add carefully, with cooling, 30 cm^3 of methanol. Warm gently to dissolve any undissolved solid and then add 10 cm^3 of water. This solution is called *Brady's Reagent*, and is frequently used to test for the presence of a carbonyl group in a compound: a yellow or orange precipitate is formed.

To prepare the derivative, add 5 cm^3 of the solution to 2 or 3 drops of the carbonyl compound (or a solution in methanol, if it is a solid). The crystals should form at once. If they do not, add a little 2M sulphuric acid. If they still do not form, warm the mixture gently, then cool (with scratching) in iced water.

Filter off the crystals at the pump using a Hirsch funnel, wash with a little water and drain. Dry on a filter paper.

Recrystallize from ethanol or aqueous ethanol.

Determine the melting-point, and compare with that quoted in a table of melting-points.

2. Derivatives of aromatic aldehydes

Aromatic aldehydes are more easily condensed with substances containing the amine or the methylene groups than are aliphatic aldehydes. These condensations are frequently very rapid and simple, and give nicely crystalline products which can be used for identification of the aldehydes.

The following are some examples. These are not 'preparations' in the sense that a maximum yield is required, but sufficient product is made for recrystallization and melting-point determination. The important point is that the materials should be in about the right molecular proportions to conform with the equation, and there should be a slight excess of the reagent: e.g.

$$C_6H_5CHO + C_6H_5{\cdot}NH{\cdot}NH_2 \longrightarrow C_6H_5{\cdot}CH{:}N{\cdot}NH{\cdot}C_6H_5 + H_2O$$

$$2C_6H_5CHO + NH_2NH_2 \longrightarrow C_6H_5{\cdot}CH{:}N{\cdot}N{:}CH{\cdot}C_6H_5 + 2H_2O$$

$$2C_6H_5CHO + (CH_3)_2CO \longrightarrow C_6H_5{\cdot}CH{:}CH{\cdot}CO{\cdot}CH{:}CH{\cdot}C_6H_5 + 2H_2O$$

$$2C_6H_5CHO + \underbrace{CH{\cdot}(CH_2)_4{\cdot}CO}_{} \longrightarrow C_6H_5{\cdot}CH{:}\underbrace{C{\cdot}(CH_2)_3{\cdot}C}_{CO}{:}CH{\cdot}C_6H_5 + 2H_2O$$

The number of common aromatic aldehydes which may be used in the last two reactions are limited, because sodium hydroxide is used for the condensation. Salicylaldehyde, with its free hydroxyl group, forms a salt, but if the phenolic group is first methylated, a cyclohexanone derivative of the methyl ether can be prepared.

The preparation of cinnamic acid from benzaldehyde, by condensation either with acetic anhydride (Perkin's reaction) or with malonic acid, is a further example of the ease with which aromatic aldehydes can be made to condense with methylene groups.

Requirements

Phenylhydrazine, $C_6H_5{\cdot}NH{\cdot}NH_2$, M_r 108·15, 1·1 g cm^{-3}.
Hydrazine hydrate, $NH_2NH_2{\cdot}H_2O$, M_r 50·06, 1·03 g cm^{-3}.
Propanone $(CH_3)_2CO$, M_r 58·08, 0·79 g cm^{-3}.
Cyclohexanone $\underbrace{CH_2{\cdot}(CH_2)_4{\cdot}CO}_{}$, M_r 98·15, 0·95 g cm^{-3}.
Aromatic aldehydes.

Phenylhydrazones. Dissolve 1 cm^3 of phenylhydrazine in 1 cm^3 of glacial acetic acid diluted with 5 cm^3 of water. Add 1 cm^3 of an aldehyde, shake the mixture and allow to stand for a few minutes.

Collect the product by filtration at the pump, wash well with water, drain dry and recrystallize from ethanol.

Determine the melting-point of your product.

Azines. Dilute 1 cm^3 of hydrazine hydrate with 5 cm^3 of water. Add a little over 1 cm^3 of an aldehyde, shake the mixture and allow to stand for a few minutes.

Collect the product by filtration at the pump, wash well with water, drain dry and recrystallize from ethanol.

Determine the melting-point of your product.

Table 6

Aldehyde	*Phenylhydrazone M.P./°C*	*Azine M.P./°C*
Benzaldehyde	159	93
Salicylaldehyde	142	213
Furfural	98	113
Cinnamaldehyde	168	

1,5-diphenylpenta-1,4-dien-3-one. Dissolve 1 cm^3 of propanone and just over 2 cm^3 of benzaldehyde in 20 cm^3 of ethanol in a corked conical flask. Add 3 cm^3 of 2M sodium hydroxide and 20 cm^3 of water and shake the mixture at intervals for ten to fifteen minutes. A pale yellow crystalline solid is slowly precipitated.

Allow the mixture to stand for a further five minutes, then collect the product by filtration at the pump, wash with water and drain dry.

The product can be recrystallized from ethanol, giving pale yellow crystals, M.P. 112°C.

Cyclohexanone derivatives. A condensation occurs between aromatic aldehydes and cyclohexanone very similar to that with propanone, but the reaction is a little more rapid and the products are a brighter yellow.

Dissolve 1 cm^3 of cyclohexanone in about 5 cm^3 of ethanol. Add 2 cm^3 of the aldehyde followed by 2 cm^3 of 2M sodium hydroxide and about 5 cm^3 of water. Stir the mixture and allow to stand, with occasional shaking until the yellow crystalline precipitate is fully formed. Collect the product by filtration at the pump, wash well and drain dry.

It may be recrystallized from ethanol.

3. Oxidation by a hypochlorite

When a hypochlorite reacts with a methyl ketone (i.e. a compound containing the group $CH_3{\cdot}CO^-$), the keto-group is converted to a carboxyl group while the methyl group is eliminated as trichloromethane. The reaction provides a very useful general preparation of acids from methyl ketones, but in this experiment trichloromethane is the product collected.

This preparation is limited by the large volume of hypochlorite required for a small amount of trichloromethane, but the yield is good, being of the order of 60–70%.

$$(1)\quad CH_3{\cdot}CO{\cdot}CH_3 + 3NaOCl \longrightarrow CH_3{\cdot}CO{\cdot}CCl_3 + 3NaOH$$

$$(2)\quad CH_3{\cdot}CO{\cdot}CCl_3 + NaOH \longrightarrow CH_3{\cdot}CO_2Na + CHCl_3$$

The alkali required is present in the hypochlorite solution.

Requirements

Propanone, $(CH_3)_2CO$, M_r 58·08, 0·79 g cm^{-3}.
2M sodium hypochlorite solution (about 15%).

Set up the apparatus for distillation with addition, using a flask of appropriate size (Fig. F).

Run 3 cm^3 (0·04 mole) of propanone into the flask through the funnel, followed by 120 cm^3 of 2M sodium hypochlorite solution. Shake the mixture and allow to stand for a few minutes. The mixture becomes warm and cloudy.

Distil the mixture, collect the distillate until the contents of the flask are clear and no more oily drops of trichloromethane are seen in the condenser.

Now run in a further 30 cm^3 of sodium hypochlorite at once and, if the hypochlorite was under strength (which often occurs), a further small yield of trichloromethane will be obtained on distillation.

Separate the lower heavy layer of trichloromethane in the distillate, and transfer to a small conical flask. Add one or two reasonable sized pieces of granular anhydrous calcium chloride, and cork the flask and shake occasionally for a few minutes.

The trichloromethane, which is reasonably pure, may be decanted from the drying agent and stored. A purer sample is obtained by a further distillation, but the quantity available is very small.

For further purification, decant the dry trichloromethane into a distillation flask and distil slowly from a beaker of hot water for a water-bath, collecting the fraction boiling from 59–62°C.

Calculate your yield based on the propanone used. Note the odour and density of trichloromethane. Consult a textbook about its uses.

Reactions

(*a*) Test your product for chloride *ions* with silver nitrate; then attempt hydrolysis with warm 2M sodium hydroxide or 2M sulphuric acid and repeat the test for chloride ions.

(*b*) Warm some product with Fehling's solution. What explanation can you offer of the result?

4. Preparation of tri-iodomethane

This preparation should be compared with the last one. It is carried out with propanone, but any other compound which contains the same group, or one that can be oxidized to it, could be used instead (i.e. carbonyl compounds containing the CH_3CO- group, or alcohols containing the $CH_3CH(OH)-$ group). The reaction is often used as a test for these compounds.

$$CH_3{\cdot}CO{\cdot}CH_3 + 3KI + 3NaOCl \longrightarrow CH_3{\cdot}CO{\cdot}CI_3 + 3KCl + 3NaOH$$

$$CH_3{\cdot}CO{\cdot}CI_3 + NaOH \longrightarrow CHI_3 + CH_3{\cdot}CO_2Na$$

Tri-iodomethane is a yellow crystalline solid with a characteristic smell, M.P. 119°C.

Requirements

Propanone, $(CH_3)_2CO$, M_r 58·08, 0·79 g cm^{-3}.

Dissolve 10 g (0·06 mole) of potassium iodide in 25 cm^3 of water in a 250 cm^3 conical flask and add 10 cm^3 of 2M sodium hydroxide. Add 1·5 cm^3 (0·02 mole) of propanone, washing out the measuring cylinder with a little water, and adding the washings. Cork the flask securely.

Set up a dropping funnel over the flask, and into it put 35 cm^3 of 2M (15%) sodium hypochlorite solution. Remove the cork from the flask and allow the hypochlorite solution to drop in at a fairly rapid rate, shaking continuously. When it has all been added, replace the cork, shake vigorously and then allow to stand for five minutes.

Filter at the pump, wash the yellow crystals with a little water and drain thoroughly. Dry on a watch-glass on a steam-bath. Recrystallize from ethanol.

Calculate the yield, based on the propanone used.

Alternative method

To carry out this reaction as a test, dissolve or mix 3 drops of the compound with 4 cm^3 of water, add 2 cm^3 of 2M sodium hydroxide and then iodine in potassium iodide until a little colour remains on shaking. Tri-iodomethane will be precipitated in a positive test. Warm to 50°C if nothing happens cold, then cool.

5. Cannizzaro's reaction

When an aldehyde with no hydrogen on the α-carbon atom is treated with a concentrated solution of an alkali, dimerization occurs, forming an ester. This undergoes hydrolysis to an alcohol and the salt of an acid. The final result is the equivalent of the oxidation of one molecule of aldehyde by the other, in other words, disproportionation. For example, with benzaldehyde:

$$2C_6H_5CHO \longrightarrow C_6H_5CO_2C_6H_5 \xrightarrow{NaOH} C_6H_5CH_2OH + C_6H_5CO_2Na$$

The acid is obtained from the salt using an acid:

$$C_6H_5CO_2Na + HCl \longrightarrow C_6H_5CO_2H + NaCl$$

Phenylmethanol is a colourless liquid, B.P. 205°C. Yield 90%.

Benzoic acid is a white crystalline solid, M.P. 121°C. Yield 90%.

Requirements

Benzaldehyde, C_6H_5CHO, M_r 106·13, 1·05 g cm^{-3}.

Dissolve 4 g (an excess) of sodium hydroxide in 4 cm^3 of water in a small conical flask. Cool, and add 5 cm^3 (0·05 mole) of benzaldehyde. Cork the flask and shake vigorously until a stable thick emulsion has formed. Leave to stand for several hours, overnight if possible.

Add 20 cm³ of water to the mixture, warm and stir to dissolve the solid present, and cool thoroughly.

Transfer the solution to a separating funnel and extract with three 5 cm³ portions of ether. Keep the aqueous layer for the benzoic acid preparation (see below). Combine the ether extracts in a small conical flask and dry with anhydrous sodium sulphate, shaking occasionally until the liquid is clear.

Distil off the ether on a water-bath, using the apparatus for distillation with vent (Fig. G). When no more ether distils from boiling water, empty the distillate into an ether residue bottle and replace the receiver. Distil, using a small Bunsen flame, until the receiver reaches 100°C. Now run out the water from the condenser and distil the phenylmethanol, collecting the fraction boiling at 202–208°C (Fig. E, i or ii).

To obtain the benzoic acid from the solution of sodium benzoate remaining from the ether extraction, add slowly, with stirring, 8 cm³ of concentrated hydrochloric acid. Cool thoroughly and filter at the pump. Wash with cold water and drain.

Recrystallize from hot water; 50 cm³ in a 100 cm³ beaker should be adequate.

Calculate your yields based on the benzaldehyde used.

6. Reduction by sodium borohydride

Sodium borohydride is more selective in action than lithium aluminium hydride; thus ketones and aldehydes can be reduced to alcohols while cyanide and ester groups are unaffected. Furthermore, the reagent may be used in aqueous solution:

$$C_6H_5{\cdot}CO{\cdot}CO{\cdot}C_6H_5 \xrightarrow{NaBH_4} C_6H_5{\cdot}CHOH{\cdot}CHOH{\cdot}C_6H_5$$

Diphenylethanedione is reduced in good yield to diphenylethanediol, a white crystalline solid of M.P. 139°C.

Requirements

Diphenylethanedione, $C_6H_5{\cdot}CO{\cdot}CO{\cdot}C_6H_5$, M_r 210·23.
Sodium borohydride, M_r 37·84.

Dissolve 0·7 g ($\frac{1}{300}$ mole) of diphenylethanedione in 7 cm³ of 95% aqueous ethanol by warming in a conical flask. Then cool under the tap to produce a fine suspension of solid.

Add 0·15 g (a large excess) of sodium borohydride. Let the reaction mixture stand for ten minutes, during which time heat should be evolved and the yellow colour disappear.

Add 5 cm³ of water and heat to boiling to destroy excess sodium borohydride; then add more water (about 10 cm³) until solid product just begins to deposit.

Cool well and collect the product by filtration at the pump. Dry on a water-bath.

Determine the melting-point and calculate the yield based on the diphenylethanedione used.

AMINES

1. Reactions of aliphatic amines

Aliphatic amines are usually volatile liquids, although the lowest ones are gaseous at room temperature.

The most convenient ones to use are:

Primary: 1-butylamine, cyclohexylamine
Secondary: diethylamine, diprop-2-ylamine
Tertiary: triethylamine
Quaternary base: tetraethylammonium hydroxide solution

Amine chlorides are readily available as crystalline solids, but many of them are very deliquescent, so it is not advisable to try to keep them for long. To obtain the amine from its chloride, heat it with 2M sodium hydroxide (cf. ammonia preparation). This is the most convenient way of obtaining methyl amines.

Test tube reactions

Use primary, secondary and tertiary amines, and a quaternary base, and compare the results. Describe what you observe, particularly gas evolution, colour change, precipitate formation and smell.

(*a*) Note the smell of the amine.

(*b*) Attempt to ignite the vapour obtained by warming a little of the amine in a test tube fitted with a delivery tube.

(*c*) Investigate the action of the amine with water, and test with a universal indicator.

(*d*) Add a drop or two of concentrated hydrochloric acid to a little of the amine.

(*e*) Add the amine dropwise to solutions of copper (II) sulphate, iron (II) sulphate, iron (III) chloride and other metal salts.

(*f*) To 2 drops of the amine add 2M hydrochloric acid until just acid to litmus, then add a few ml of M sodium nitrite solution. If no reaction occurs when cold, warm gently.

(*g*) To 2 drops of the amine add a few cm³ of '20 volume' (2M) hydrogen peroxide and shake for a few minutes.

(*h*) To 2 drops of the amine add 3 drops of benzoyl chloride and 4 cm³ of 2M sodium hydroxide. Cork the tube and shake vigorously for a few minutes.

Try to interpret your observations, using a textbook if necessary. Compare the reactions with those of aromatic amines by repeating the tests with aniline, N-methylaniline and NN-dimethylaniline.

2. Reduction of a nitrocompound to an amine

The reduction of a nitro-group is readily accomplished by tin in acid solutions. Under these conditions, the chief product is an amino-group. The reaction is of particular importance as a stage in the conversion of benzene to a very large range of compounds which can be prepared from aniline.

$$2C_6H_5NO_2 + 3Sn + 14HCl \longrightarrow (C_6H_5NH_3^+)_2SnCl_6^{2-} + 4H_2O + 2SnCl_4$$

Aniline is a colourless liquid, readily oxidized in the air to brown, with a characteristic smell. B.P. 184°C. Yield 80%.

Requirements

Nitrobenzene, M_r 123·11, 1·20 g cm^{-3}.
Granulated tin.

Place 5 cm³ (0·05 mole) of nitrobenzene and 13 g (an excess) of tin in a 250 cm³ round-bottomed flask, and fit it for reflux with a water condenser. Pour down the condenser 5 cm³ of concentrated hydrochloric acid and shake the flask. If the heat of the reaction causes the mixture to boil very rapidly, cool the flask. When the reaction slackens, pour a further 5 cm³ of concentrated hydrochloric acid down the condenser, cooling again if necessary.

Continue this way until 25 cm^3 of acid has been added.

Now heat the flask gently for ten minutes on a gauze over a Bunsen flame and then remove the condenser, transfer the flask to a boiling water-bath and heat for fifteen minutes. This will complete the reduction, and any residual nitrobenzene will be volatilized in the steam.

Cool the flask and add, with shaking, a solution of 20 g of sodium hydroxide in 30 cm^3 of water, keeping the flask cool. Now add 2M sodium hydroxide in sufficient quantity to redissolve all the precipitated tin salts and make the solution strongly alkaline.

Equip the flask for steam distillation (Fig. H), using a steam generator, and steam distil until 100 cm^3 of distillate has been collected. Add 10 g of sodium chloride to the distillate and shake until it has all dissolved. Transfer to a separating funnel and extract three times with ether, using 10 cm^3 portions. Dry the combined ether extracts with anhydrous sodium sulphate, shaking the flask periodically until the solution is clear.

Distil off the ether, from a water-bath, using the apparatus for distillation with addition (Fig. G), with a receiver with vent. When no more ether distils, empty the distillate (into an ether residues bottle) and replace the receiver. Distil using a small Bunsen flame until the temperature reaches 100°C. Now run out the water from the condenser and, with a 250°C thermometer, distil the aniline using a Bunsen flame. Collect the fraction boiling at 180–186°C. To get a clean specimen, it helps to add a little zinc dust before the final distillation.

Calculate the yield, based on the nitrobenzene used.

Test tube experiments with aniline

1. In three test tubes prepare three solutions: 1 cm^3 of aniline in 5 cm^3 2M hydrochloric acid, 0·8 g of sodium nitrite in 3 cm^3 of water, and 1·4 g of 2-naphthol in 6 cm^3 of 2M sodium hydroxide. Add the sodium nitrite solution to the cold phenylammonium chloride solution, and after about twenty seconds add the mixture to the 2-naphthol solution. Note the red dye formed. What is its structure?
2. Add 1 drop of aniline to 1 drop of trichloromethane. Add a small pellet of potassium hydroxide and a few cm^3 of ethanol. Warm, and smell *cautiously*. Note the obnoxious smell of phenyl isocyanide. Do not throw the mixture down the sink, but add enough concentrated hydrochloric acid to fill the tube, and leave for at least ten minutes for hydrolysis to occur before discarding. *The product is toxic.*
3. Add a drop of aniline to a suspension of bleaching powder in water. Repeat the experiment with sodium hypochlorite.

3. Reduction of *m*-dinitrobenzene to *m*-nitroaniline

m-Nitroaniline cannot be prepared by direct substitution from aniline or acetanilide because the groups already present in the ring are *o,p*-directing. It is necessary to start with a compound with a group already in the *meta* position. It has been found that certain reducing agents reduce one of the two nitro-groups in a dinitro-compound, in this case *m*-dinitrobenzene, before the other is attacked at all. The reducing agent used here is sodium disulphide, which is oxidized mainly to sodium thiosulphate.

$$C_6H_4(NO_2)_2 \xrightarrow{Na_2S_2} NO_2C_6H_4NH_2$$

m-Nitroaniline is a yellow crystalline solid, M.P. 114°C. Yield 75%.

Requirements

m-Dinitrobenzene, $C_6H_4(NO_2)_2$, M_r 168·11.
Sodium sulphide.

Prepare a solution of sodium disulphide by adding 2·1 g of sulphur ('flowers') to a solution of 8 g of sodium sulphide enneahydrate in 30 cm^3 of water, and boiling the mixture until a clear orange solution is obtained. Transfer the cooled solution to a dropping funnel.

Into a suitable flask put 5 g (0·03 mole) of

m-dinitrobenzene, 100 cm^3 of water and 50 cm^3 of ethanol. Heat the solution to boiling, and allow the sodium disulphide solution to drop in slowly. The reaction is exothermic, but a little heating may be needed to keep the solution boiling. The reaction mixture should eventually go an orange-brown colour. When all the sodium disulphide solution has been added, boil the solution for a further five minutes.

Filter the hot solution through a hot Buchner funnel, then cool the filtrate to room temperature and finally in a refrigerator or an ice-bath. Filter off the crystals, wash with water and drain dry.

Recrystallize the solid from aqueous ethanol.

4. Acetylation of an amine

The acetyl group is very easily introduced into amines, sometimes for the purpose of protection (see *p*-nitroaniline) and also to produce a crystalline solid derivative, easily purified, which, having a sharp melting-point, can be used to identify an amine.

The commonest acetylating agent is a mixture of acetic acid and acetic anhydride.

$$C_6H_5NH_2 + CH_3CO_2H \longrightarrow C_6H_5NH{\cdot}CO{\cdot}CH_3 + H_2O$$

Acetanilide is a white crystalline solid, M.P. 113°C. Yield 90%.

Requirements

Aniline, $C_6H_5NH_2$, M_r 93·13, 1·02 g cm^{-3}.
Acetic anhydride, $(CH_3CO)_2O$, M_r 102·09, 1·08 g cm^{-3}.
Acetic acid, CH_3CO_2H, M_r 60·05, 1·05 g cm^{-3}.

Aniline is readily oxidized in the air. To obtain a pure specimen for this preparation, put about 5 cm^3 in a flask with a little zinc dust and distil using an air conditioner (Fig. Ei).

Set up the apparatus for reflux, with a water condenser (Fig. A), and put into the flask 2 cm^3 (0·022 mole) of aniline, 3 cm^3 (an excess) of acetic anhydride and 2 cm^3 of glacial acetic acid. Heat the mixture with a small flame and reflux for ten minutes.

Cool the mixture, then pour it into about 25 cm^3 of water in a beaker, with stirring. The acetanilide should solidify into white lumps. Cool the mixture, filter off the solid at the pump and wash it with water.

Recrystallize the acetanilide from hot water. Calculate the yield, based on the aniline used.

Test tube preparation of acetyl derivatives

If only a small sample is required, for example for a melting-point determination, carry out the experiment as follows: add 1 cm^3 of acetic anhydride to 0·5 cm^3 of aniline, bring to the boil and pour the mixture into about 5 cm^3 of water.

The experiment can be carried out using a large range of aromatic amines.

o-Toluidine gives *o*-acetotoluide, M.P. 110·4°C.
m-Toluidine gives *m*-acetotoluide, M.P. 65·5°C.
p-Toluidine gives *p*-acetotoluide, M.P. 146°C.

5. Benzoylation of an amine

This reaction should be compared with the acetylation of an amine. It is carried out for the same purposes, but the conditions it requires are different. Benzoyl chloride is hydrolysed much less readily than acetyl chloride, and the reaction is carried out in aqueous alkaline solution, using the Schotten–Baumann technique (see also p. 86).

$$C_6H_5NH_2 + C_6H_5COCl + NaOH \longrightarrow C_6H_5NH{\cdot}CO{\cdot}C_6H_5 + NaCl + H_2O$$

Benzanilide is a white crystalline solid, M.P. 163°C. Yield 80%.

Requirements

Aniline, $C_6H_5NH_2$, M_r 93·13, 1·02 g cm^{-3}.

Benzoyl chloride, C_6H_5COCl, M_r 140·57, 1·21 g cm^{-3}. (*Lachrimatory*)

Shake together, in a stoppered conical flask, 2 cm^3 (0·022 mole) of redistilled aniline and 20 cm^3 of 2M sodium hydroxide, so that they form a fine mixture. Aniline is not soluble in sodium hydroxide solution and the shaking is necessary to prevent the formation of two layers. Now add 3 cm^3 (a slight excess) of benzoyl chloride, restopper the flask and shake vigorously for several minutes. Heat is evolved and it is necessary to release the pressure occasionally.

When a solid is formed in the flask, leave it to stand for a few minutes, and then filter at the pump, wash the solid thoroughly with water and drain dry.

Recrystallize the benzanilide from ethanol. Calculate the yield, based on the aniline used.

Further preparations

Using the same technique, the following can be prepared:

o-Toluidine gives *o*-benzotoluide, M.P. 143°C.
m-Toluidine gives *m*-benzotoluide, M.P. 125°C.
p-Toluidine gives *p*-benzotoluide, M.P. 158°C.

6. Diazotization and coupling to form a dye

This is an example of the formation of an azo-dye and entails the formation of a diazonium salt. These salts are unstable, and it is necessary to carry out the reaction below 10°C. The diazonium salt cannot be kept so the second half of the process, the coupling with 2-naphthol in alkaline solution, has to be carried out at once.

NH_2 —HCl, $NaNO_2$→ $N_2^+ Cl^-$
2-naphthol, NaOH
HO
N=N

Benzene-azo-2-naphthol is a red crystalline solid, M.P. 131°C. Yields vary widely.

Requirements

Aniline, $C_6H_5NH_2$, M_r 93·13, 1·02 g cm^{-3}.
2-Naphthol, $C_{10}H_7OH$, M_r 144·17.

Put 0·9 cm^3 (0·01 mole) of aniline into a small beaker and add 15 cm^3 of 2M hydrochloric acid. Put the beaker into an ice-bath and cool the solution to 5°C. Meanwhile, in a larger beaker, dissolve 2 g (an excess) of 2-naphthol in 18 cm^3 of 2M sodium hydroxide and dilute the solution to 100 cm^3.

Dissolve 1 g (a slight excess) of sodium nitrite in 10 cm^3 of water in a test tube or a small conical flask.

The solutions are now all ready. Do not start the next stage of the reaction unless you have twenty minutes available in order to complete it.

Add to the cold phenylammonium chloride solution, while stirring with a thermometer, small portions of the sodium nitrite solution. Keep the temperature just below 10°C; if it rises rapidly at any stage, add a small piece of ice to the reaction mixture. When the nitrite solution has all been added, allow the solution of the diazonium salt to stand in the ice-bath for a few minutes. Yellow crystals may start to be deposited.

Now pour the diazonium salt solution slowly, with stirring, into the alkaline 2-naphthol solution. A red precipitate should form.

Allow the mixture to stand for a few minutes, then filter at the pump, wash thoroughly with water and drain dry.

The solid can be recrystallized from glacial acetic acid. Keep back some of the crude product and purify it by the chromatographic method described on p. 115.

Further preparations

Using the same proportions, but smaller quantities, prepare samples of other azo-dyes. For example:

diazotized aniline and alkaline phenol: soluble, orange;

diazotized *m*-nitroaniline and alkaline *m*-cresol: insoluble, yellow.

Test for primary aromatic amines

To 1 drop of the amine dissolved in 1 cm³ of 2M hydrochloric acid, add two or three crystals of sodium nitrite. Add the mixture to a solution of a little 2-naphthol in 2M sodium hydroxide solution. The formation of a red precipitate confirms the presence of a primary aromatic amino-group. The test is very sensitive, and it can also be applied to testing for a *nitro-group* as follows: to 1 drop of the suspected nitro-compound add 0·5 cm³ of concentrated hydrochloric acid and a small piece of tin. After a few seconds, decant the solution, dilute with 1 cm³ of water, add a few crystals of sodium nitrite and pour into a solution of 2-naphthol in 2M sodium hydroxide. A red precipitate (or maybe only a colour) of an azo-dye confirms nitro-compound (so long as it is already known that there is no amino-group present).

7. Preparation of an iodoarene from an amine

In this reaction an amino-group is replaced, via the diazonium salt, with a halogen atom. For chlorine and bromine, complex copper (I) ions are used, but for iodine the simple iodide ion gives high yields.

$$C_6H_5NH_2 \xrightarrow[NaNO_2]{HCl} C_6H_5N_2^+Cl^- \xrightarrow{KI} C_6H_5I$$

Iodobenzene is a pale yellow liquid, with a characteristic smell. B.P. 188°C. Yield 80%.

Requirements

Aniline, $C_6H_5NH_2$, M_r 93·13, 1·02 g cm^{-3}.

Dissolve 4·6 cm³ (0·05 mole) of aniline in a mixture of 12 cm³ of concentrated hydrochloric acid and 10 cm³ of water in a beaker, and put it in an ice-bath. Make up separate solutions of 3·5 g (a slight excess) of sodium nitrite in 15 cm³ of water, and 9 g (a slight excess) of potassium iodide in 20 cm³ of water.

Once the reaction is started, about twenty minutes will be needed to complete it.

Cool the phenylammonium chloride solution to about 5°C by stirring with a thermometer, then add the sodium nitrite solution in small portions, keeping the temperature just below 10°C. If the temperature starts to rise rapidly, add a small piece of ice to the reaction mixture.

When the nitrite solution has all been added, leave the mixture to stand in the ice-bath for a few minutes and then transfer it from the beaker to a 250 cm³ round-bottomed flask, which will be used for steam-distillation at a later stage.

Now add the potassium iodide solution, all at once. A brisk reaction should occur, with evolution of nitrogen. When it has subsided, heat the flask on a water-bath for ten minutes, during which time the crude iodobenzene should separate out as a dark, oily lower layer.

Decant the upper, aqueous, layer, and wash the lower layer twice by decantation with water, and once with a little sodium thiosulphate solution to remove free iodine. Finally, wash once more with water.

Add 50 cm³ of water and set up the apparatus for distillation (Fig. D). The iodobenzene distils in the steam and forms a pale yellow lower layer in the distillate. When no more oily drops distil, transfer the distillate to a separating funnel, run off the lower layer into a small conical flask and dry with anhydrous sodium sulphate, shaking occasionally until the liquid is clear.

Calculate the yield based on the aniline used.

Further preparations

Using the same molar proportions and conditions:

o-Toluidine gives 2-iodotoluene, B.P. 211°C.
m-Toluidine gives 3-iodotoluene, B.P. 204°C.
p-Toluidine gives 4-iodotoluene, M.P. 35°C, B.P. 211°C.

4-Iodotoluene distils in steam as oily drops which solidify in the condenser into small buttons that float out in the condensed water. If there is a tendency to block the condenser, stop the water flow in the jacket for a while to enable it to warm up.

ACIDS AND DERIVATIVES

1. Hydrolysis of an ester

Most esters can be effectively hydrolysed by refluxing with a slight excess of aqueous sodium hydroxide. The alcohol formed is isolated directly, and the acid is recovered from its sodium salt. The precise nature of the isolation will depend on the physical properties of the acid and the alcohol. If the alcohol is soluble, it is distilled out; if insoluble, it can be steam distilled or extracted with ether.

$$C_6H_5CO_2CH_2CH_3 + NaOH \longrightarrow C_6H_5CO_2Na + CH_3CH_2OH$$

In this experiment the ethanol produced is distilled out, and the benzoic acid is precipitated by a strong acid. Benzoic acid is a white crystalline solid, M.P. 121°C. Yield 80%.

Requirements

Ethyl benzoate, $C_6H_5CO_2C_2H_5$, M_r 150·18, 1·05 g cm^{-3}.

Set up the apparatus for reflux (Fig. A).

Place 5 cm^3 (0·03 mole) of ethyl benzoate in the flask and add 30 cm^3 (an excess) of 2M sodium hydroxide. Heat the mixture to boiling and reflux for about fifteen minutes. At the end of this time the oily drops of ester should have disappeared.

When no more drops of ester are visible, fit the flask for fractionation and distil the mixture slowly. About 1 cm^3 of ethanol is obtained as distillate, at a temperature of 78°C. Stop the distillation when the temperature rises above this figure.

The alcohol fraction can be tested in various ways:

(*a*) Burn a little on the end of a clean wire or metal spatula.

(*b*) Warm a few drops with a little 2M sulphuric acid and potassium dichromate solution. Note the distinctive odour of acetaldehyde. Why does the solution go green?

(*c*) Using a delivery tube, distil out a few drops from the reaction mixture of test (*b*). This will contain a little acetaldehyde. To this distillate add a little 0·5M potassium iodide and then 2M sodium hypochlorite. A yellow precipitate of tri-iodomethane is produced which has a distinctive smell.

The residual solution in the flask after fractionation contains sodium benzoate. To precipitate the benzoic acid as a white solid from this solution slowly add 20 cm^3 (a large excess) of concentrated hydrochloric acid, stirring continuously. Allow to cool to room temperature, then filter off the benzoic acid at the pump.

Recrystallize the benzoic acid from the minimum of hot water, filter off and dry. Weigh the dry product and calculate your yield. To test for purity take the melting-point of a sample.

Hydrolysis of methyl *o*-hydroxybenzoate, 'Oil of Wintergreen'. Use the same molar amount of the ester, with 60 cm^3 of 2M sodium hydroxide. The yield of *o*-hydroxybenzoic acid is 88%.

Hydrolysis of diethyl oxalate. Use the same molar amount of the ester, with 60 cm^3 of 2M sodium hydroxide and 70 cm^3 of water since sodium oxalate is not very soluble in water. Distil off the alcohol from the residual solution, cool and add 1 drop of phenolphthalein, then neutralize with acetic acid. Isolate the oxalic acid as its calcium salt by adding a solution of 5 g of anhydrous calcium chloride in 20 cm^3 water. Filter and dry. The yield of calcium oxalate is quantitative.

Hydrolysis of phenylmethyl acetate. Use the same molar proportions as for ethyl benzoate. Phenylmethanol is isolated by steam distillation. Acidification of the residual solution will free the acetic acid, which can be detected by its odour, but it is difficult to isolate.

2. Conversion of an acyl chloride to an amide

A wide range of amides can be made quickly, and in good yield, by the action of the acid chloride on ammonia. The reaction can be very vigorous, especially with the lower aliphatic acid chlorides. Aromatic acid chlorides can be used more easily with ammonia solution, but considerable heat is evolved:

$$C_6H_5COCl \xrightarrow{NH_3} C_6H_5CONH_2$$

Benzoyl chloride has an unpleasant lachrimatory vapour.

The expected yield of benzamide in this experiment is 75%.

Requirements

Benzoyl chloride, C_6H_5COCl, M_r 140·57, 1·2 g cm^{-3}.

Carry out this experiment *in a fume cupboard* as far as possible, especially when handling benzoyl chloride.

Add 12 cm^3 (0·05 mole) of benzoyl chloride to 40 cm^3 of water and 10 cm^3 of '0·880' ammonia (an excess) in a corked conical flask. Shake at intervals for ten minutes, releasing any pressure developed as the solution warms up.

After ten minutes cool under the tap, then filter off the crude benzamide at the pump. Retain the filtrate.

Recrystallize the crude benzamide from about 30 cm^3 of hot water, adding more water if necessary. To avoid hydrolysis of the product, do not prolong the heating.

Collect the benzamide by filtration at the pump and dry on a water-bath.

Determine the melting-point of your product and calculate the yield on the basis of the benzoyl chloride used.

The loss in yield is partly due to hydrolysis of the benzoyl chloride to ammonium benzoate. The amount can be determined by precipitation of the benzoic acid: add a small excess of concentrated hydrochloric acid to the filtrate and chill to precipitate the benzoic acid. Filter through a pre-weighed sintered-glass crucible and dry in an oven at about 80°C. Dry to constant weight as anhydrous benzoic acid and calculate the percentage of benzoyl chloride hydrolysed to benzoic acid.

3. Conversion of an ester to an amide

Some amides can be made by the action of ammonia on the ester. This reaction does not work on many esters, but ethyl oxalate is a successful example:

$$(CO_2C_2H_5)_2 \xrightarrow{NH_3} (CONH_2)_2$$

Oxamide is an odd substance; it is insoluble in most solvents and extremely high melting, about 420°C. The yield in this experiment is rather variable but usually about 80%.

Requirements

Diethyl oxalate $(CO_2C_2H_5)_2$, M_r 146·15, 1·08 g cm^{-3}.

Add 7 cm^3 (0·05 mole) of diethyl oxalate to 70 cm^3 (an excess) of 2M ammonia in a corked conical flask. Shake at intervals for fifteen minutes, cooling occasionally if necessary. Oxamide should begin to form at once as a white precipitate.

Collect the oxamide by filtration at the pump, wash with a little water and drain as dry as possible. Retain the filtrate. Dry in an oven at about 110°C.

Calculate the yield based on the diethyl oxalate used.

The loss in yield is partly due to the hydrolysis of ester to ammonium oxalate. The amount can be determined by precipitation as calcium oxalate: neutralize the filtrate to litmus with 2M hydrochloric acid, then add an excess of 0·1M calcium chloride. Filter through a pre-weighed sintered-glass crucible, wash with ethanol and dry in a desiccator. Dry to constant weight as calcium oxalate dihydrate and calculate the percentage of ester hydrolysed to oxalate.

4. Conversion of an acid to an amide

Certain ammonium salts can be dehydrated to form the corresponding amides; ammonium acetate is a familiar example. The preparation has to be carried out slowly, since the yield of acetamide depends very largely on the degree of separation of the amide from other products formed.

To suppress the dissociation of ammonium acetate into acetic acid and ammonia, an excess of glacial acetic acid has to be present. Separation of the amide from this acetic acid requires very slow fractionation.

$$CH_3CO_2NH_4 \longrightarrow CH_3CONH_2 + H_2O$$

Acetamide is a colourless crystalline solid, M.P. 82°C, which usually has a mouse-like odour.

Requirements

Ammonium acetate, $CH_3CO_2NH_4$, M_r 77·09.
Glacial acetic acid.

Set up the apparatus for fractional distillation (Fig. I).

Place 7·7 g (0·1 mole) of ammonium acetate and 12 cm³ (an excess) of glacial acetate acid in the flask, and heat so that the mixture refluxes for fifteen minutes. During this stage the ammonium acetate is dehydrated to form acetamide.

Now fractionate the mixture. The flame must be controlled so that not more than 1 drop in two seconds distils over. The temperature gradually rises, fractionation being continued until the temperature reaches 170°C.

The distillate contains mainly water and acetic acid, but a little acetamide will be present. The residue in the flask will be the bulk of the acetamide.

Allow the flask to cool, then convert to direct distillation through an air condenser (Fig. Ei). Collect the fraction boiling between 215°C and 225°C. Leave the distillate of acetamide to solidify; if necessary, overnight in a refrigerator. If the product is wet or has not solidified, it should be redistilled.

Use your product for the following tests:

(*a*) To a small quantity of acetamide add about 1 cm³ of 2M sodium hydroxide solution. Allow to stand for a few minutes, occasionally testing for the formation of ammonia, by smell and by litmus paper. Unlike the reaction with ammonium salts, no ammonia is produced under these conditions.

(*b*) Boil the mixture from test (*a*) and notice the immediate production of ammonia.

Amides are hydrolysed to ammonia by boiling with sodium hydroxide. They are not hydrolysed in the cold.

Write the equation of the reaction.

5. The 'nylon rope trick' (preparation of a polyamide)

An impressive demonstration of the formation of a polyamide polymer is easily carried out by allowing the constituent molecules to react at the interface between two solvents.

This experiment was originated by P. W. Morgan and S. L. Kwolek of E. I. du Pont de Nemours and Co. Inc. and published in *J. Chem. Educ.*, 1959, **36**, 182 and 530.

Requirements

Sebacoyl chloride, $ClCO{\cdot}(CH_2)_8.COCl$, M_r 239·15, 1·12 g cm⁻³.
1,6-Diaminohexane, $NH_2{\cdot}(CH_2)_6{\cdot}NH_2$, M_r 116·21.

Prepare a solution of 1·5 cm³ of sebacoyl chloride in 50 cm³ tetrachloromethane in a 200 cm³ beaker and, separately, a solution of 2·2 g of 1,6-diaminohexane (***caution:*** *this is caustic*) and 4 g of sodium carbonate in 50 cm³ of water.

Clamp the beaker, and above it clamp the roller system (made from tubing) as shown in Fig. 15. Allow about two metres drop from the roller to the receiver.

Now pour the aqueous solution carefully on to the tetrachloromethane solution and, using crucible tongs, pull the interfacial film out, over the rollers and down towards the receiver. When a long enough rope has formed the process will go

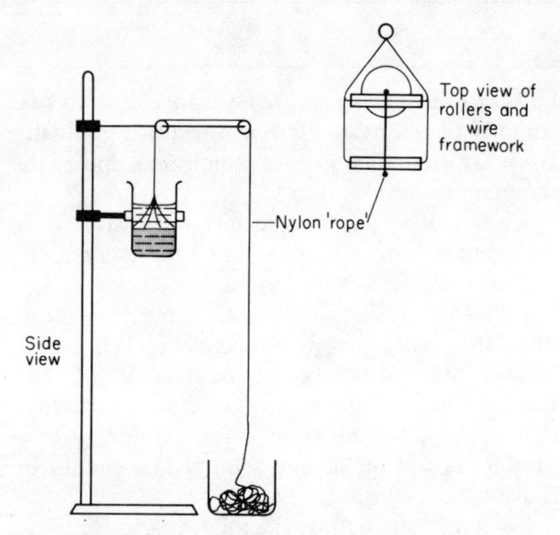

Fig. 15. The 'nylon rope trick' (The nylon rope could be wound up by hand on a roller instead of being allowed to fall).

on of its own accord until the reagents are used up.

To obtain a dry specimen of the nylon polymer, wash it thoroughly in 50% aqueous ethanol, then in water until litmus is not turned blue by the washings. Dry in an oven at about 110°C.

What is the formula of nylon, and what is its average molecular weight? Consult a textbook.

The procedure for the hydrolysis and chromatography of nylon is given on p. 113.

6. Depolymerization of a polymer

Some polymers can be decomposed to the monomers by the action of heat alone. A simple example is polymethyl methacrylate, known commercially as Perspex or Plexiglass:

$$[CH_2{\cdot}C(CH_3){\cdot}CO_2CH_3]_n \longrightarrow nCH_2{:}C(CH_3){\cdot}CO_2CH_3$$

Methyl methacrylate (methyl 2-methylpropenoate) is a colourless liquid with a characteristic smell, B.P. 100°C. (*Caution: toxic.*)

Set up the apparatus for distillation (Fig. D) and fit an air-bath, made from an old tin can, round the flask.

Cut up pieces of Perspex so that they will fit in the flask, filling it about three-quarters full.

Heat the air-bath with a Bunsen flame. The monomer should distil over readily. When no more liquid distils, redistil the methyl methacrylate using a Bunsen flame.

If the monomer is left in a vessel on its own, it is likely to repolymerize. To prevent this, add a few crystals of benzene-1,4-diol.

Further preparations

Try the effect of carrying out the same procedure on other plastics available. Remember that many monomers are gaseous and often inflammable, so take sensible precautions.

PROPERTIES OF AROMATIC RINGS

1. Nitration of benzene to nitrobenzene

Nitration is the process by which a nitro-group is introduced into a molecule. Nitration of aromatic compounds is particularly valuable because the nitro-group can be readily converted to an amino-group, which in turn gives rise to many other groups.

The ease of nitration of the benzene ring is determined by the groups already present. This preparation, and the three which follow, show the different conditions necessary to nitrate the ring in the presence of different groups.

$$C_6H_6 + NO_2^+ \longrightarrow C_6H_5NO_2 + H^+$$

Nitrobenzene (B.P. 210°C) is a pale yellow liquid, with a characteristic smell.

Requirements

Benzene, C_6H_6, M_r 78·11, 0·88 g cm^{-3}.

Place 8 cm^3 of concentrated nitric acid in a 50 cm^3 flask, and add 9 cm^3 of concentrated sulphuric acid, mixing it in a little at a time. Now, stirring with a thermometer, add 7 cm^3 (0·075 mole) of benzene in portions of about 2 cm^3. Make sure the temperature does not rise above 60°C, but do not cool the mixture too much or the reaction may not start at once. If the temperature does rise above 60°C, it can become uncontrollable later.

When all the benzene has been added, attach a reflux water condenser and heat the mixture on a water-bath at 60°C for thirty minutes. Shake the mixture occasionally during the refluxing. At the end of the refluxing, cool the flask.

Now pour the mixture, with stirring, into 100 cm^3 of water in a beaker. Decant off the upper aqueous layer as far as possible, then transfer the residual liquid to a separating funnel and add 20 cm^3 of 1·5M sodium carbonate. Shake gently, frequently releasing the pressure. Allow to stand, then decant the upper, aqueous layer.

Repeat the process until no carbon dioxide is formed on adding the carbonate, then transfer the lower, nitrobenzene, layer to a small conical flask and dry it with anhydrous sodium sulphate, shaking occasionally until the solution is clear.

Set up the apparatus for distillation using an air condenser (Fig. Ei), and distil the nitrobenzene, collecting the fraction boiling at 207–211°C. Calculate the yield, based on the benzene used.

2. Nitration of nitrobenzene to *m*-dinitrobenzene

The presence of one nitro-group deactivates the ring to electrophilic substitution, so that rather more vigorous conditions are needed for the introduction of a second nitro-group. Although the main product is *m*-dinitrobenzene, smaller amounts of other nitro-products are formed as well, so that the yield is unlikely to be above 60%.

$$C_6H_5NO_2 + NO_2^+ \longrightarrow C_6H_4(NO_2)_2 + H^+$$

m-Dinitrobenzene is a pale yellow solid, M.P. 90°C.

Requirements

Nitrobenzene, $C_6H_5NO_2$, M_r 123·11, 1·20 g cm^{-3}.

Place in a 50 cm^3 flask 4 g of potassium nitrate and 8 cm^3 of concentrated sulphuric acid. Add 2·5 cm^3 (0·024 mole) of nitrobenzene and shake gently. Fit the flask with a reflux air condenser and set up the apparatus *in a fume cupboard*.

Heat the mixture gently to boiling, reflux on a sand-bath for about fifteen minutes, with occasional shaking. If there is any sign of the mixture darkening, raise the flask from the sand-bath and remove the flame for a short while.

Now allow the flask to cool to about 50°C (the temperature at which it can be handled without discomfort), remove the reflux condenser and pour the mixture, with stirring, into 50 cm^3 of water in a beaker. The crude dinitrobenzene will solidify.

Filter off the solid at the pump, and wash thoroughly with water to remove all traces of acid.

Recrystallize the solid from ethanol. Leave to

crystallize in an ice-bath or a refrigerator because *m*-dinitrobenzene is quite soluble in ethanol at room temperature. Filter off the crystals at the pump and dry at room temperature.

A further crop of less pure crystals can be obtained from the filtrate by adding water until a precipitate is formed. Add enough water so that the precipitate just dissolves at boiling-point. Cool again in an ice-bath or a refrigerator, and filter and dry as before.

Calculate the yield, based on the nitrobenzene used.

3. Nitration of a phenol

The hydroxyl group activates the benzene ring, and phenol is considerably more reactive than benzene and is sensitive to oxidation. Precautions have to be taken when nitrating to avoid extensive oxidation by nitric acid. The orientation rules lead us to expect a mixture of *o*- and *p*-nitrophenols, and they are formed in a ratio of about 3 : 1. It is easy to separate them, because *o*-nitrophenol is volatile in steam, while the *p*-isomer is not. (Why is this?)

OH OH NO_2 OH

2 dil. nitric acid → +

NO_2

o-Nitrophenol is a yellow crystalline solid, M.P. 46°C. Yield 40%.

p-Nitrophenol is a colourless crystalline solid, M.P. 114°C.

Requirements

Phenol, C_6H_5OH, M_r 94·11.

Make an emulsion of 4·7 g (0·05 mole) of phenol and 3 cm^3 of water by warming them together, with slight shaking. Add the cooled emulsion to a solution of 5 cm^3 of concentrated nitric acid in 15 cm^3 of water in a flask which can be equipped for distillation. Stir with a thermometer, and ensure that the temperature does not rise above 30°C.

When all the emulsion has been added, continue to stir the mixture for a while, cooling it if the temperature rises above 50°C.

Now leave the mixture to stand overnight.

Next day, add 25 cm^3 of water, shake thoroughly, allow the lower layer to settle and decant off the water. Repeat with several lots of water until the lower layer solidifies into lumps.

Add 30 cm^3 of water, set up the apparatus for distillation (Fig. D) and distil over the mixture of water and *o*-nitrophenol until only water is coming over. If the yellow *o*-nitrophenol crystallizes in the condenser, stop the water flow so that the jacket warms up for a while.

Cool the distillate thoroughly and filter at the pump. Allow to dry at room temperature.

Recrystallize the *o*-nitrophenol as follows: add a little ethanol to the crystals in a small conical flask, warm it gently so that some of the crystals dissolve, while the rest melt. Now add more ethanol, a little at a time, until the crystals are all dissolved at about 40°C. Now add drops of water until a faint permanent precipitate forms, then scratch the inside of the flask with a glass rod and cool the flask while stirring continuously.

Finally, cool the flask in a refrigerator or an ice-bath and filter off the cold crystals. Dry at room temperature.

The *p*-nitrophenol remains in the flask after the distillation process. It is difficult to extract it successfully.

Further preparation

If the procedure above is followed for *p*-cresol, a good yield of 4-methyl-2-nitrophenol, M.P. 36·5°C, is obtained.

4. Nitration of an amine

The amino-group activates a benzene ring to a greater extent than the hydroxyl group, and aniline is very readily oxidized by nitric acid, so that direct nitration is impossible. It is therefore necessary to protect the amino-group by acetylating it, and then carry out the nitration. The acetyl group does not change the positions to which the substituent is directed but, since it is a fairly large group, the yield is nearly all *p*-nitroacetanilide.

The acetyl group is removed by alkaline hydrolysis.

$$\left(C_6H_5NH_2 \xrightarrow[+CH_3CO_2H]{(CH_3CO)_2O} \right) C_6H_5NH{\cdot}CO{\cdot}CH_3 \xrightarrow{HNO_3 \,|\, H_2SO_4} p\text{-}NO_2C_6H_4NH{\cdot}CO{\cdot}CH_3 \xrightarrow{NaOH} p\text{-}NO_2C_6H_4NH_2$$

p-Nitroaniline is a bright yellow crystalline solid, M.P. 148°C. Yield 90%.

Requirements

Acetanilide, $C_6H_5{\cdot}NH{\cdot}CO{\cdot}CH_3$, M_r 135·16.

Add 2·7 g (0·02 mole) of acetanilide to 2·5 cm^3 of glacial acetic acid, stir well, and add 5 cm^3 of concentrated sulphuric acid. Cool the hot, clear solution in a freezing mixture of ice and sodium chloride, stirring with a thermometer until the temperature is between 0°C and 5°C. Now add dropwise, with stirring, 1·5 cm^3 of fuming nitric acid (1·51 g cm^{-3}), keeping the temperature below 25°C.

When all the acid has been added and the temperature starts to fall, remove the solution from the freezing mixture and allow to stand for ten minutes. If the temperature starts to rise rapidly, return the solution to the freezing mixture for a while.

Now pour the solution into 50 cm^3 of water, when the *p*-nitroacetanilide should be precipitated as a yellow crystalline powder. Filter at the pump, wash with water and drain thoroughly.

To hydrolyse off the acetyl group, transfer the crystals to a flask fitted for reflux, and add 8 cm^3 of ethanol and a solution of 1 g of sodium hydroxide in 5 cm^3 of water. Heat under reflux for about ten minutes, after which time there should be a clear yellow solution. Cool the flask to crystallize the *p*-nitroaniline.

Filter at the pump, using the filtrate to wash out out the last part of the solid from the flask, and wash with a little water. Drain thoroughly and recrystallize from aqueous ethanol.

Calculate the yield, based on the acetanilide used.

5. Friedel-Crafts reaction

In the Friedel–Crafts reaction, anhydrous aluminium chloride is used to catalyse the alkylation or acylation of the benzene ring. It is essential that water be excluded rigorously. Carry out the experiment *in a fume cupboard.*

$$CH_3{\cdot}COCl + C_6H_6 \longrightarrow CH_3{\cdot}CO{\cdot}C_6H_5 + HCl$$

Acetophenone is a low-melting, colourless solid, M.P. 20°C and B.P. 201°C.

Requirements

Benzene (dried over sodium), M_r 78·11, 0·88 g cm^{-3}.
Acetyl chloride, M_r 78·5, 1·1 g cm^{-3}.
Anhydrous aluminium chloride.

Set up the apparatus for reflux with addition, with exclusion of water (Fig. C). Place in the 50 cm^3 flask 6 g of anhydrous aluminium chloride (***caution:*** *open with care*) and 18 cm^3 (an excess) of dry benzene. Place 3·5 cm^3 (0·05 mole) of acetyl chloride in the dropping funnel. Add the acetyl chloride slowly to the cooled benzene.

When the acetyl chloride has all been added, stopper the dropping funnel (or replace it with a stopper), and heat the mixture on a water-bath at 50°C for one hour.

Now cool the flask, and pour the contents into 50 cm^3 of water in a separating funnel. Shake the stoppered funnel, leave to stand until the layers separate, then run off and discard the lower, aqueous layer. Add 20 cm^3 of 1·5M sodium carbonate, shake with care, releasing the pressure occasionally, allow to stand, and then run off and discard the lower, aqueous, layer. Run the upper layer, acetophenone and excess of benzene, into a small flask, and dry with anhydrous sodium sulphate, shaking occasionally until the solution is clear.

Set up the apparatus for distillation with vent (Fig. G). Distil off the benzene (B.P. 80°C) using a water-bath, then, with a 250°C thermometer and with the water run out from the condenser, distil off the acetophenone, collecting the fraction boiling at 198–203°C.

Calculate the yield, based on acetyl chloride used.

6. Formation of a heterocyclic ring: 2-hydroxy-4-methylquinoline

This is a simple example of a ring-closing reaction, forming a heterocyclic system. Since the product is very stable, concentrated sulphuric acid can be used as the dehydrating agent.

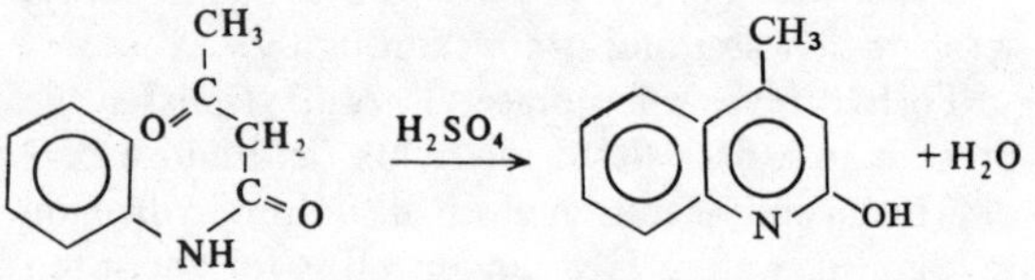

2-Hydroxy-4-methylquinoline is a white crystalline solid, M.P. 220°C.

Requirements

Acetoacetanilide
$C_6H_5 \cdot NH \cdot CO \cdot CH_2 \cdot CO \cdot CH_3$, M_r 177.

Warm 6 cm^3 of concentrated sulphuric acid in a small conical flask to 70°C. Add 4·5 g (0·025 mole) of acetoacetanilide in small portions, keeping the temperature as near 70°C as possible. The solid should dissolve rapidly, and the temperature is liable to rise suddenly.

When all the solid has been added, heat to 90–100°C for a few minutes, then allow to cool to room temperature. Pour the cold solution into 100 cm^3 of water, allow to stand for a short while and then filter off the precipitated white powder.

Wash the product with a little water, drain dry and recrystallize, if desired, from ethanol.

EXTRACTION OF NATURAL PRODUCTS

1. Oils by steam-distillation

Many plants contain pleasant-smelling oils, sometimes called essential oils. These can frequently be extracted by steam-distilling the plant material.

Since the proportion of oil in plant material is usually small, it is necessary to steam-distil a large quantity of material to obtain a small yield of product. This means that, if possible, a very large flask should be used, ideally a wide-necked 5 litre one. This makes it comparatively easy to introduce and, more important, to remove the plant material, which often becomes rather swollen and soggy during the process.

It is wise to carry out four or five extractions consecutively in the same apparatus in order to obtain a reasonable amount of product. The material should be minced or chopped up immediately before putting it in the flask.

Oil from orange peel. Fit a 5 litre flask for distillation with addition, using a water condenser (Fig. F), and put into it 500 cm^3 of water and a little porous pot. Chop or mince sufficient orange peel to half-fill the flask, and pour it into the flask with gentle shaking to obtain an even distribution. Fill the dropping funnel with water, heat the flask on a sand-bath or in a heating mantle to bring to the boil and collect the distillate until no more drops of oil come over.

Add water from the funnel when the level in the flask gets low.

When the distillation is finished, stop heating, cool the flask, empty out the used peel, refill with a second lot and repeat the heating as before. Repeat the extraction process until sufficient oil has been obtained.

Extract the oil from the distillate with several lots of ether, combine the ether extracts and dry them with anhydrous sodium sulphate. Distil off the ether. The oil, which smells strongly of oranges, can be fractionated if desired.

Further preparations

The method can be tried on any aromatic smelling plant which is available in sufficient quantity; for example: bay, cypress, mint, cloves, thyme, and lemon or grapefruit peel. In most of these extractions, the product is a complex mixture of substances, but fennel will give a fairly pure sample of anethole (4-propenylmethoxybenzene, $CH_3{\cdot}CH{:}CH{\cdot}C_6H_4{\cdot}OCH_3$).

The oils extracted do not always smell the same as the original plant, since decomposition of the less stable ones occurs during the extraction.

2. Pigments by Soxhlet extraction

Red and blue flowers usually contain pigments called *anthocyanins*, while most yellow and orange flowers contain pigments derived from the hydrocarbon carotene, called *carotenoids*. These can be quite complex, and it is very difficult to obtain a pure pigment in crystalline form from a plant material. However, if the extraction is followed by a chromatographic separation, interesting and attractive results can be obtained.

The most convenient way of extracting the pigment is to use a Soxhlet extractor, with the appropriate solvent. It is usually necessary to refill the extractor many times, because there is only a very small amount of pigment even in brightly coloured flowers.

Yellow and orange flowers are extracted with ethyl acetate, in which the carotenoids are soluble, while the anthocyanins are extracted from blue or red flowers with methanol containing a little hydrochloric acid. This ensures that the pigment is present in the salt form, which is generally more stable. The salts are nearly all red, irrespective of the original colour of the flower.

The extraction of pelargonidin from pelargonium ('geranium') petals. Assemble a 250 cm^3 Soxhlet extractor, a litre flask and a double surface condenser (Fig. J). Half fill the flask with methanol (ethanol can be used, but it makes the experiment longer), add 20 cm^3 of 2M hydrochloric acid and a little porous pot. Fill the extractor with fresh scarlet 'geranium' petals, and reflux gently. See p. 12 for the use of the Soxhlet extractor.

When the first lot of petals are white, stop the extraction and replace them with a second batch. At least five batches should be used altogether.

The solution should now be deep red. Distil off most of the solvent, on a water-bath, until the final volume is about 25 cm^3.

Filter the concentrated solution into a small conical flask, add 1 cm^3 of concentrated hydrochloric acid and leave the stoppered flask in a refrigerator for several days.

If no crystals have appeared after several days, evaporate a little of the solvent and cool again. Eventually a fine powder will be deposited, which can be seen under the microscope to be crystalline.

The pigment is rather more soluble in ethanol, and it is difficult to get crystals down.

In the original plant, the pigment has sugar molecules attached and is called pelargonin chloride. The sugar molecules are hydrolysed off by hydrochloric acid, and the product is pelargonidin chloride.

Further preparations

The same method can be applied to a number of other red or blue plants: paeonies, cornflowers, roses, antirrhinums or delphiniums. Copper beech leaves also contain anthocyanin pigments.

Investigate the action of dilute acids, dilute alkalis and ammonia gas on the petals.

Chromatographic separation. Prepare a short, wide alumina column according to the procedure on p. 114. Extract the pigment from one batch of petals and, when cold, run the extract through the column. The result varies from flower to flower, but can be very colourful. In most cases, the red pigment is adsorbed as a blue band and, when it emerges, it is as a green solution. The green solution turns red with acid.

The blue form appears to be a complex on the alumina surface; the green form is the acid-free pigment, the acid having been adsorbed on the alumina. It is likely that other bands of impurities will also be observed.

The extraction of the pigments from dandelion petals. Collect a number of dandelion (or similar) flowers. Cut off the yellow petals and dry them in an oven at a little below 100°C until they are brittle.

Assemble a Soxhlet extractor as in the previous experiment (Fig. J), but use dry ethyl acetate as the solvent. Care is needed to make sure that the petals will not block up the siphon tube.

Reflux gently until the siphoning liquid is pale yellow. The petals never lose their colour entirely. Repeat the extraction several times.

When the extraction is complete, dry the solution with anhydrous sodium sulphate. This will remove traces of water from the petals and will render the chlorophyll pigments insoluble.

Filter the solution, which should be brilliant orange. It contains not only a complex mixture of pigments, but also plant waxes which make it hard to crystallize. The pigments are best separated on an alumina column.

Distil off the solvent on a water-bath, and then leave the flask to stand until there remains no smell of ethyl acetate.

Now, making sure all flames are extinguished, add 25 cm^3 of 60–80°C petroleum spirit, and warm gently in a beaker of warm water until all the solid has dissolved.

Prepare an alumina column not less than 40 cm long, using 60–80°C petroleum spirit, and filter the pigment solution directly on to the column. Allow the solvent to pass through at a rate not greater than 3 drops per second.

A very complex pattern of coloured bands will form, and elution with more solvent will enable some of them to be obtained from the column. When no more bands show signs of coming off, change the solvent to a mixture of 90% petroleum spirit and 10% ether. This should elute the remaining bands.

Collect the different pigments in separate receivers. The purity of an individual fraction can be improved by passing it through a second, shorter, column, but it is difficult to obtain a solid sample from any of the solutions.

3(a). Caffeine from tea or coffee

Caffeine, a heterocyclic base ($C_8H_{10}N_4O_2$, M.P. 235–237°C), is found in tea and coffee, and is also present in cola drinks.

The method of extraction depends on its ready solubility in trichloromethane. The extraction from tea goes more smoothly, since trichloromethane emulsions form less readily and there are less coloured impurities.

Weigh out roughly 50 g of tea or roast ground coffee, or 20 g of 'instant' coffee, and warm it in a beaker with 200 cm³ of water, boiling gently for fifteen minutes. Remove the solids by filtering through muslin at the pump, and wash with a little hot water.

Heat the filtrate, with the washings, to boiling and add 100 cm³ of 0·3M lead acetate to precipitate acids and albumin. Filter a small sample through a cotton-wool plug, and test for complete precipitation with a little lead acetate solution.

Filter at the pump, placing muslin over the filter paper to prevent clogging, and add 2M sulphuric acid to the filtrate to precipitate all lead ions. Filter off the lead sulphate.

Add 2M ammonia to the filtrate until it is neutral to litmus, then evaporate to 100 cm³. Allow to cool a little, add 5 g of decolorizing charcoal and bring to the boil cautiously. Filter free of charcoal at the pump.

Extract the filtrate with two 40 cm³ portions of trichloromethane, inverting frequently rather than shaking vigorously (which may cause emulsification) and combine the extracts. Dry the trichloromethane solution with anhydrous sodium sulphate.

Distil off most of the trichloromethane, then evaporate to dryness in a beaker on a water-bath *in a fume cupboard*. About 1 g of caffeine should remain. Recrystallize from hot water.

The caffeine can be purified by chromatography on an alumina column from benzene (3 parts) and trichloromethane (1 part).

Compare with authentic caffeine by chromatography on a silica gel slide. Develop with trichloromethane (9 parts) and ethanol (1 part). Detect by exposing to iodine vapour.

3(b). Alternative procedure for extraction of caffeine from tea

This involves adsorption from ethanolic solution on a solid oxide. Put 50 g of tea (tea-bags are convenient) into a 100 cm³ Soxhlet extractor and extract with 250 cm³ of ethanol for about two hours. Prepare a suspension of 30 g of magnesium oxide in 150 cm³ of water in an evaporating basin, add the ethanolic extract and stir thoroughly. Put the basin on to a steam bath and heat, with occasional stirring, until dry.

The next stage is desorption from the oxide, using water. Boil the solid residue with 250 cm³ of water in a beaker for about ten minutes, filter at the pump and return the solid to the beaker, retaining the filtrate. Boil the solid with a further 200 cm³ of water and filter again into the same flask. Combine the filtrates, and add sufficient 2M sulphuric acid to make the solution acid to litmus (pH less than 7). Set up apparatus for distillation and distil off water until the volume is about 100 cm³.

Trichloromethane is used to extract the caffeine from the aqueous solution. Put the aqueous solution in a separating funnel and shake with four successive 25 cm³ portions of trichloromethane. If the resulting 100 cm³ is coloured, shake with about 10 cm³ of 2M sodium hydroxide, and then with water. Filter off any solid matter which may be present, transfer the solution to a basin over a steam bath *in a fume cupboard* and evaporate off the trichloromethane.

The caffeine (M.P. 235°C) may be recrystallized from hot water. The yield should be about 1 g.

4. Cystine from hair

L-cystine, $(S{\cdot}CH_2{\cdot}CH(NH_2){\cdot}CO_2H)_2$, M_r 240·31, is soluble in strong acids and alkalis but not in weak acids such as acetic acid, and this is the basis of the method of extraction.

Weigh out roughly 50 g of well-chopped hair into a beaker, wash free of grease with tetrachloromethane or with ethanol and dry.

Set up the apparatus for reflux (Fig. A), using a 250 cm^3 flask, and reflux the cleaned hair with 100 cm^3 of concentrated hydrochloric acid for about six hours. At the conclusion of the refluxing, a biuret test should be negative.

Remove a 10 cm^3 portion of the solution and neutralize the remainder with 10M sodium hydroxide, testing with litmus paper. Now return the 10 cm^3 portion and neutralize to pH 5 by adding saturated sodium acetate solution.

Allow the solution to stand overnight, then collect the brown precipitate by filtration at the pump.

Extract the precipitate twice, by boiling with a mixture of 15 cm^3 of concentrated hydrochloric acid and 15 cm^3 of water.

Add 2 g of decolorizing charcoal to the combined extracts and bring to the boil cautiously. Filter off the charcoal. The filtrate should be colourless.

Neutralize the filtrate to pH 5 with a hot saturated solution of sodium acetate; then allow the solution to cool, when colourless crystals of cystine should be deposited.

Collect the crystals by filtration at the pump, wash with a little cold water and ethanol and drain well. The yield should be about 3 g.

Cystine may be purified by recrystallization from hot water.

Use some of your product for the following tests:

(*a*) Confirm the presence of sulphur by boiling a small portion with 2M sodium hydroxide and lead acetate. Lead sulphide should be precipitated.

(*b*) Confirm the presence of sulphur and nitrogen by Middleton's test as given on p. 194.

(*c*) Confirm the nature and check the purity of your product by paper chromatography according to the procedure on p. 113. Use authentic L-cystine for comparison.

(*d*) Use a polarimeter to investigate the optical activity of your product.

V Physical Chemistry

DETERMINATION OF RELATIVE MOLECULAR MASS
1. Gas syringe method

Using the gas laws relationship

$$PV = nRT$$

where n equals the number of moles of gas (equivalent to m/M_r), the relative molecular mass of a liquid can be found if a known mass is vaporized in known conditions to give a determinable volume of vapour.

100 cm³ gas-tight syringes are very suitable for this experiment. They are used ungreased and care is necessary to avoid damage by abrasive dirt: wash in clean soapy water, rinse well in pure water and ethanol and allow to dry at room temperature.

A hypodermic syringe may be used to inject a known amount of liquid into a heated gas syringe where the volume of vapour produced can be determined. The gas syringe may be heated by steam in a glass tube or in a light-bulb oven. Because of non-uniformity of temperature within the oven the apparatus is calibrated by using a liquid of known relative molecular mass.

Requirements

- Syringe oven.
- 100 cm³ gas syringe.
- Rubber caps (vaccine).
- 1 cm³ syringe with No. 1 needle (no plastic parts).
- Silicone rubber seal.
- Propanone, $CH_3{\cdot}CO{\cdot}CH_3$, M_r 58·1.
- Liquids boiling below 100°C.

Switch on the oven to heat up for ten minutes before use. Fit a rubber cap over the end of the syringe so that 5 cm³ of air remain in it and slide it into the oven.

Fit the needle to the 1 cm³ syringe and draw in about 0·5 cm³ of propanone. Hold upright and eject any air present, then press a silicone seal on to the needle point. This protects the needle and avoids accidental loss of liquid. Weigh accurately the syringe and contents.

Note the volume of air in the hot gas syringe after first rotating the barrel to ensure that it is free running and the air is at atmospheric pressure. Now remove the silicone seal from the 1 cm³ syringe and push the needle through the rubber cap until it reaches into the main barrel of the gas syringe. Slowly inject about 0·2 cm³ of liquid into the gas syringe then pull the needle out and replace the silicone seal on the needle point. Take care not to inject too much liquid and so produce an excessive volume of gas.

Reweigh accurately the syringe and contents and note the volume of gas in the gas syringe.

To repeat the experiment, inject more propanone into the gas syringe after removing the rubber cap and pumping out the first sample. The rubber cap may be used several times before it leaks.

The propanone results are used to calibrate the apparatus which can now be used to determine the relative molecular masses of liquids boiling below 100°C. To calculate the results use the expression

$$M_r = \frac{m}{V} \times \left(\frac{VM_r}{m} \text{ for propanone}\right)$$

Derive this expression from the relationship given in the introduction to this experiment.

Reference: Badman, K. W., *S.S.R.*, 1968, **49**, 904.

2. Gaseous effusion method

The kinetic theory of gases suggests that at a fixed temperature gases have different average molecular velocities which are related to their different relative molecular masses. The relationship can be studied by determining the time taken for gas samples to diffuse in identical conditions.

Requirements

- 100 cm³ gas syringe.
- Effusion tube.
- Three-way stop-tap.
- Stop-clock.
- Dry samples of gases.

The effusion tube consists of a short length of glass tube sealed at one end, except for a pin-hole, with a disc of aluminium foil (Fig. 16). The open end of the tube is fitted with a piece of plastic tube for attachment to the gas syringe. To make an effusion tube, cut with a cork borer a disc of thin aluminium foil and stick it to the glass tube by means of an expoxy-resin such as 'Araldite'. A suitable pin-hole is readily made by pricking the foil with an ordinary pin.

Place two pieces of sticky paper on the barrel of the gas syringe to indicate clearly two marks about 70 cm³ apart. Now connect the syringe to the effusion tube while the piston is drawn out. Clamp

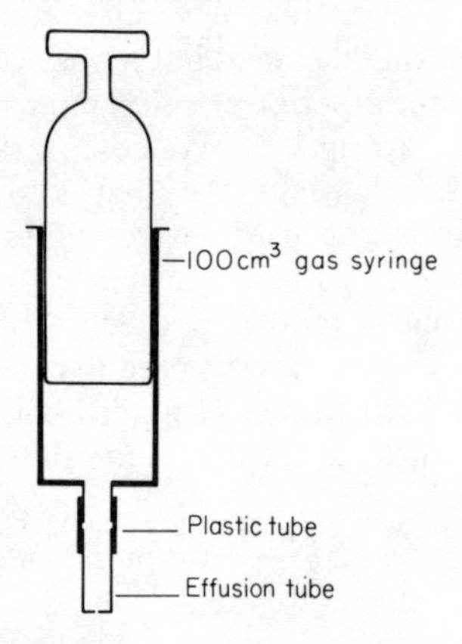

Fig. 16. Gas syringe for gaseous effusion.

the syringe upright and release the piston so that air is forced out by the weight of the piston alone. Note the time taken for the piston to travel between your two marks. If the pin-hole is satisfactory the time should be between one and two minutes.

To repeat the experiment with other gases, the syringe is disconnected from the effusion tube

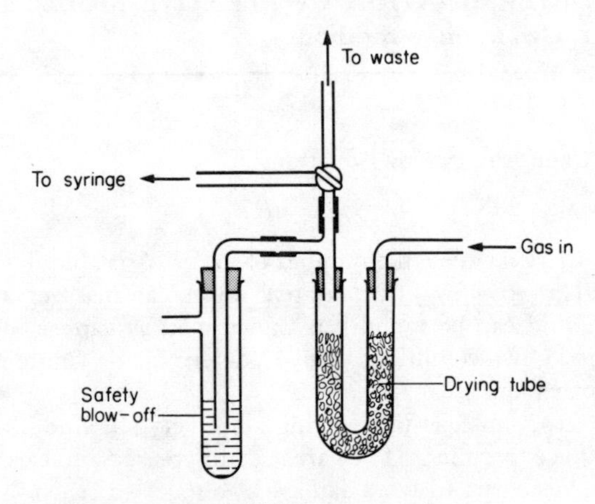

Fig. 17. Apparatus for filling a gas syringe.

and attached to a cylinder of a suitable gas via a three-way stop-tap and drying tube (see Fig. 17). Hydrogen, oxygen and butane ('Camping Gaz') give a good relative molecular mass range, but a wider variety of gases is obviously available if you prepare your own gases by conventional laboratory methods.

Flush out the syringe twice with the chosen gas before filling it for the experiment. Reconnect the full syringe to the effusion tube and note the time taken for the piston to travel between the two marks, as before.

Repeat the experiment with the same gas to obtain consistent results, then carry out further experiments with other gases.

Plot graphs of *effusion time* against *relative molecular mass*, *square of relative molecular mass* and *square root of relative molecular mass*. The correct relationship should give a straight line graph. Derive the correct relationship from the kinetic theory of gases; consult a textbook about Graham's Law of Diffusion.

3. Cryoscopic method

The de-icing of roads in cold weather by the spreading of rock salt is just one example of the lowering of the freezing-point of a solvent when a solute is added.

The phenomenon can be studied quantitatively to investigate the relationship between freezing-point depression and concentration. The technique is due to Beckmann, who developed a thermometer for measuring accurately small temperature changes.

A class could study a range of solutes.

Requirements

Hard-glass test tube (150 mm × 25 mm) with air jacket and stirrer.
Thermometer to −10°C, with 0·1 graduations.
Solvent such as benzene or nitrobenzene.
Solute, such as naphthalene $C_{10}H_8$, M_r 128·17, or diphenylethanedione, $(C_6H_5CO)_2$, M_r 210·23,
or benzoic acid, $C_6H_5CO_2H$, M_r 122·13.
Ice-bath.

Weigh accurately about 20 cm³ of solvent into a hard-glass test tube. Fit the test tube with its stirrer and thermometer and place in the ice-bath (see Fig. 18). The ice-bath should contain water with a little ice.

When crystals of solvent first appear, remove the test tube from the ice-bath, fit the air jacket and return to the ice-bath. Stir well and record the steady temperature which corresponds to the freezing-point of pure solvent. Remove the apparatus from the ice-bath and allow the solvent to melt completely.

Weigh accurately a sample of solute (roughly 0·3 g) into the solvent, and stir to dissolve the sample completely. Determine, as before, the freezing-point of the solution. Supercooling may occur, but when the first crystals appear the temperature will rise to the true freezing-point.

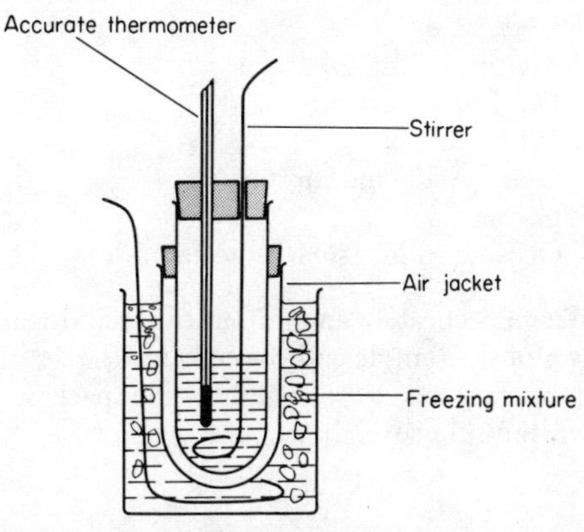

Fig. 18. Apparatus for the determination of freezing-points.

Repeat the determination for at least two further additions of solute. When the experiment is finished, place your solution in a solvent residues bottle, not in a sink.

Plot a graph of *concentration* (in g solute/1000 g solvent) against *freezing-point depression.* Note the depression caused by 1 mole of solute/1000 g solvent; this is known as the cryoscopic constant of the solvent. Would the same result be obtained with other solutes? What would be the influence of association or dissociation of a solute? Consider any benzoic acid results available. Consult a text-book about Raoult's Law.

How could unknown relative molecular masses be determined?

4. Rast's method

If Raoult's Law is obeyed, 1 mole of a substance dissolved in 1000 g of DL-camphor will cause a melting-point depression of 37·7°C. Small quantities will cause a proportionately smaller depression.

Requirements

Melting-point apparatus.
Hard-glass test tube.
Oil-bath.
Glass pestle and mortar.
DL-camphor.
Organic solids (such as acetanilide).

Weigh accurately an ignition tube; add roughly 0·05 g of the sample and reweigh accurately. Now add roughly 0·5 g of camphor and pack it well down the tube; weigh accurately again.

Heat an oil-bath to 200°C, then dip in the ignition tube. When the mixture is molten, shake to mix well. Remove from the oil-bath, wipe off oil and chill at once in cold water. The hot period should be brief to avoid sublimation of camphor, and the chilling rapid to avoid separation of the components on crystallization.

Dig out the mixture and grind in a glass mortar. Fill a melting-point tube with a 1 cm length of sample and seal the tube above the sample.

Determine to 0·1°C the melting-point, which is when the last crystal disappears.

Also determine the melting-point of pure camphor by the standard procedure given on p. 17.

Use this method to determine the relative molecular mass of an organic compound you have prepared, or a compound provided.

5. Ebullioscopic method

The influence of solutes on the boiling-point of solvents can be studied: loss by evaporation must be avoided and superheating occurs readily.

A suitable form of apparatus is due to F. G. Cottrell, in which boiling liquid is pumped over the thermometer bulb. If this apparatus is not available a simple set-up (see Fig. 19a) may be used to obtain reasonable results.

A class could study a range of solutes.

Requirements

Washburn and Read apparatus, with Cottrell pump.
Solvent such as benzene or nitrobenzene.
Solute, such as naphthalene $C_{10}H_8$, M_r 128·17, or diphenylethanedione, $(C_6H_5CO)_2$, M_r 210·23,
or benzoic acid, $C_6H_5CO_2H$, M_r 122·13.
Suitable thermometer, with 0·1 graduations.
Microburner.

Measure an exact volume of the solvent into the Cottrell apparatus (see Fig. 19b) using a safety pipette (20·0 cm³ should be suitable).

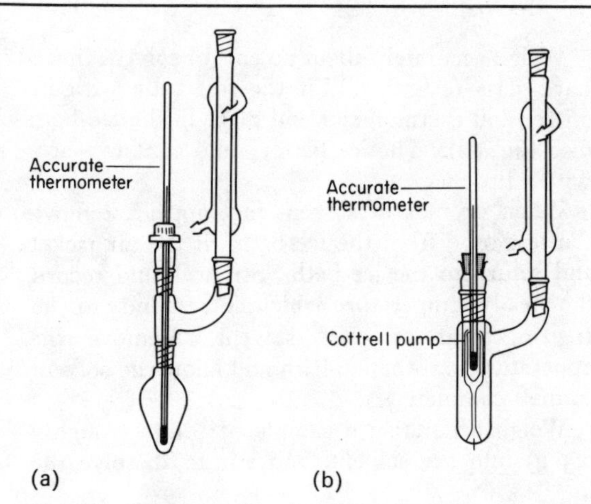

Fig. 19. Apparatus for the determination of boiling-points of solutions.

Set up the apparatus where it will be free from draughts, and heat with a very small flame which should make contact with it. When a steady rate of boiling is observed, as judged by the 'drip rate' from the thermometer, read the thermometer to see if the temperature is steady.

By adjusting the flame size, find the range of drip

rates which correspond to a steady temperature. The best rate of heating for the experiment corresponds to the middle of the range.

Using the optimum rate of heating, record the steady temperature, which will be the boiling-point of the pure solvent.

Allow the apparatus to cool before opening to introduce a sample of solute (***benzene vapour is both inflammable and toxic***). Weigh accurately into the Cottrell apparatus a sample of solute (roughly 0·5 g). Determine the boiling-point of the solution using exactly the same rate of heating as before.

Repeat the determination for two further additions of the same solute. When the experiment is finished, place your solution in a solvent residues bottle, not in a sink.

Plot a graph of *concentration* (in g solute/1000 g solvent) against *boiling-point elevation*. Note the elevation caused by one mole of solute/1000 g solvent; this is known as the ebullioscopic constant of the solvent. Would the same result be obtained with other solutes? What would be the influence of association or dissociation of a solute? Consider any benzoic acid results available. Consult a textbook about Raoult's Law.

How could unknown relative molecular masses be determined?

SOLUBILITY DETERMINATION

1. Titrimetric method

To determine the solubility of salts accurately a suitable analytical procedure is necessary: a neat method is to convert the salt to an equivalent amount of acid by use of a cation exchange resin. The acid is readily determined by titration with standard alkali. Alternatively, if suitable, the cation can be determined by direct titration with EDTA.

Any excess of saturated solution should not be thrown away; after dilution it can be used to restock reagent bottles.

Requirements

Constant-temperature water-bath, set between 20°C and 50°C.
Thermometer, 0–100°C.
Suitable salts, such as metal sulphates.
Titration apparatus and a 250 cm³ standard flask.
Strong cation exchange resin in hydrogen form
Standardized 0·1M sodium hydroxide
} or standard 0·05M EDTA.

Place about 50 cm³ of pure water in a 200 cm³ conical flask and heat to 20°C above the temperature of the constant-temperature water-bath. Weigh out roughly 20 g of the chosen salt and add it to the hot water. Stir the mixture well to dissolve the salt. If necessary, add further 10 g portions of the salt until a permanent undissolved residue remains.

Stopper the flask containing the solution and place overnight in the constant-temperature water-bath to reach equilibrium. Record the temperature of the water-bath.

A portion of the saturated solution is now transferred to a pre-weighed weighing bottle. Shake the flask well (while still held in the water-bath) and allow the excess crystals to settle. Then filter the saturated solution through a dry, fluted filter paper. Reject the first few cm³ of filtrate before collecting about 10 cm³ in the weighing bottle.

Cool the weighing bottle to room temperature and weigh accurately.

Completely transfer the weighed solution, and any crystals which have been deposited, to a standard flask and dilute to the calibration mark with pure water.

Pass 10·0 cm³ aliquots through a cation exchange column to convert to acid for determination by titration with sodium hydroxide; for details see Expt. 3, p. 109.

Alternatively, many metals can be determined by titration with EDTA; for details see Expt. 2, p. 171.

If a 10·0 cm³ aliquot is not suitable change your pipette.

Calculate the solubility of the salt as g of salt/100 g of water at the water-bath temperature. The titration results are used to calculate the weight of salt in the sample taken, while the weight of sample corresponds to the combined weight of salt and water.

2. Conductivity method

Saturated solutions of sparingly soluble salts can be regarded as being at 'infinite dilution' for conductivity measurements. Hence:

$$\Lambda^{\infty} = \kappa(\text{sat.}) \times 1/1000\, M(\text{sat.})$$

where Λ^{∞} is the molar conductivity at infinite dilution in ohm⁻¹ m² mol⁻¹

κ(sat.) is the electrolytic conductivity of the saturated solution in ohm⁻¹ m⁻¹

M(sat.) is the molarity of the saturated solution.

Λ^{∞} is the sum of the conductivities of the separate ions; the conductivity of the saturated solution is

Table 7

Ion	*Λ(ionic) at 25°C /ohm⁻¹ m² mol⁻¹*
Ba^{2+}	0·130
Ca^{2+}	0·119
Pb^{2+}	0·139
F^-	0·0554
SO_4^{2-}	0·160

measured and the molarity of the saturated solution is then readily calculated.

See Electrochemistry, Expts. 4 and 5, pp. 136 and 137.

Requirements

Conductivity bridge, with dip cell of known constant.
Water-bath at 25°C.
250 cm³ polypropylene bottle.
Hard-glass test tube (150 mm × 25 mm).
Sparingly soluble salts ($BaSO_4$, $PbSO_4$, CaF_2).
Conductance water.

Wash a sample of finely powdered salt (roughly 1 g) with conductance water to remove any traces of soluble impurities. Grind the salt in a mortar and remove the water by decantation. Carry out this procedure at least twice.

Shake the purified salt with about 200 cm³ of conductance water in a polypropylene bottle, and leave overnight in a water-bath at 25°C to reach saturation.

When the solution is saturated use three separate small portions to rinse a hard-glass test tube. Then transfer a larger portion by decantation, and clamp the test tube in the water-bath. Hold the dip cell in the saturated solution and take readings every half minute until constant, showing that thermal equilibrium has been established.

Also take a reading for pure conductance water at 25°C.

Calculate the electrolytic conductivity of the saturated solution, (remembering to subtract the conductivity of the conductance water from the readings), using the relationship:

$$\kappa = \text{cell constant} \times 1/\text{resistance}$$

and hence calculate the concentration of the saturated solution using the relationship given in the introduction.

Also calculate the solubility product of the salt.

3. Solubility product of calcium hydroxide

Saturated solutions of sparingly soluble ionic solids should conform to the Equilibrium Law (p. 124) so that, for example:

$$Ca(OH)_2(s) \rightleftharpoons Ca^{2+}(aq) + 2OH^-(aq)$$

gives $$K_{s_0} = [Ca^{2+}] \, . \, [OH^-]^2$$

This can be tested by shaking solid calcium hydroxide with sodium hydroxide solution of various dilutions and determining the total alkali by titration with standard acid.

Requirements

Four corked conical flasks.
0·1M sodium hydroxide (carbonate free).
Calcium hydroxide solid.
200 cm³ measuring cylinder.
Titration apparatus.
Standard 0·05M hydrochloric acid.
Phenolphthalein indicator.

Prepare 200 cm³ portions of 0·05M and 0·025M sodium hydroxide by approximate dilution of the 0·1M solution. Determine the precise concentrations of the three solutions by titration of aliquots with standard 0·05M hydrochloric acid, using phenolphthalein indicator (3 drops).

Place portions of about 100 cm³ each of pure water, 0·1M, 0·05M and 0·025M sodium hydroxide into separate conical flasks. Add roughly 2 g of calcium hydroxide (a large excess) to each conical flask. Cork the flasks, shake well and leave overnight to reach equilibrium. Do not forget to label the flasks.

Check the solutions after reaching equilibrium to see that a solid residue of excess calcium hydroxide remains.

Filter the contents of the conical flasks separately using a dry filter paper and rejecting the first few ml of filtrate in each case. Titrate aliquots of the four filtrates with the standard 0·05M hydrochloric acid using phenolphthalein indicator (3 drops).

Calculate K_{s_0} for calcium hydroxide in each case. Is it reasonably constant, compared to the variation in hydroxide ion concentration? The hydroxide ion concentrations are calculated from the titration of the mixtures; the increases in value over the sodium hydroxide are used to calculate the calcium ion concentrations. Express the concentrations in moles of ion per litre.

PHASE RULE STUDIES
1. Transition temperatures

A change in physical structure is often accompanied by some other change which can be measured. Thus changes in the degree of hydration of a salt are accompanied by changes in heat content (ΔH). This may be observed as an arrest in the cooling curve of a water–salt mixture; the temperature of arrest corresponds to the transition temperature.

Requirements

Hard-glass test tube fitted with an air jacket.
Pestle and mortar.
Water-bath.
Thermometer, 0–100°C, with 0·1 graduations.
Stop-clock.
A suitable pure salt hydrate, such as
sodium carbonate decahydrate
disodium hydrogen phosphate dodecahydrate
sodium sulphate decahydrate
sodium thiosulphate pentahydrate
manganese (II) chloride tetrahydrate
strontium chloride hexahydrate.

Finely powder sufficient of a suitable salt hydrate to half fill a hard-glass test tube. Heat the half-filled tube (without an air jacket) in a water-bath until the salt hydrate is completely molten. This may be difficult in some cases.

In order to make an approximate determination allow the tube to cool rapidly while stirring *carefully* with the thermometer. Note any temperature arrest (there will often be 'supercooling' so a *temperature rise* may be observed immediately before the arrest).

Now reheat the tube to 10°C above the temperature arrest noted, fit the air jacket and allow to cool slowly while stirring. Record the temperature every half minute, and continue the experiment until a steady temperature has been observed for several readings.

Plot a graph of *temperature* against *time* and record the accurate transition temperature. Consult a textbook to discover the corresponding change in hydration. Why must the salt hydrate be pure?

2. The biphenyl–naphthalene system

The compounds biphenyl and naphthalene have similar molecular structures and form an almost ideal solution with each other. Four determinations of melting-point relative to composition should be sufficient to plot the phase diagram, although a class could study a much wider range of compositions.

Requirements

2 hard-glass test tubes (150 mm × 25 mm).
Thermometer, 0–100°C.
Naphthalene, $C_{10}H_8$, M_r 128·17.
Biphenyl, $(C_6H_5)_2$, M_r 154·21.

Weigh accurately a sample of naphthalene (roughly 6 g) into a hard-glass test tube.

Melt the naphthalene by holding the tube in a boiling water-bath. Then determine the freezing-point by allowing the tube to cool in air while stirring *gently* with the thermometer. The temperature arrest corresponds to the freezing-point. Do not remove the thermometer or in any way lose naphthalene.

Weigh accurately a sample of biphenyl (roughly 3 g) and add to the naphthalene in the test tube. Determine the freezing-point of the mixture as before, the temperature being noted when *about a quarter* of the molten mixture has solidified. This procedure allows for any erratic behaviour due to supercooling.

Repeat the experiment with an accurate weight of pure biphenyl (roughly 6 g) in a fresh tube. Then add an accurate weight of naphthalene (roughly 1 g) to the biphenyl and make a final determination of freezing-point.

Plot a graph of *melting-point* (= freezing-point)

against *percentage of naphthalene*. Since the solution is almost ideal the graph consists of two intersecting straight lines. Record the data corresponding to the intersection.

What is the significance of the intersection? What are the molecular structures of biphenyl and naphthalene? Consult a textbook if necessary.

Reference: Washburn, E. W. and Read, J. W., *Chem. Abs.*, 1915, **9**, 1570.

3. The naphthol–naphthalene systems

The systems 1-naphthol with naphthalene[1] and 2-naphthol with naphthalene[2] illustrate the effect of molecular structure on physical behaviour. This experiment is best studied on a class basis, each student contributing one result.

Requirements

Melting-point apparatus.
Pestle and mortar.
Naphthalene, $C_{10}H_8$, M_r 128·17.
1-Naphthol, $C_{10}H_7OH$, M_r 144·17.
2-Naphthol, $C_{10}H_7OH$, M_r 144·17.

For each system it is necessary to prepare four 1 g mixtures of precisely-known composition in which the molar percentage of naphthalene should be about 25%, 55%, 65% and 85%.

In separate containers, weigh roughly the required amounts of the compounds for your assigned mixture. Then place one of the compounds in an accurately pre-weighed test tube and weigh accurately; add the other compound and reweigh accurately. The *exact* composition of the mixture is now known.

Melt the mixture by holding the tube in an oil-bath at about 150°C. Cool quickly by first dipping in boiling water and then holding in cold water. In this way a homogeneous mixture is obtained.

Release the mixture by placing the tube under a cloth and smashing with a sharp tap from a pestle. ***Be cautious***, and throw away the broken glass at once. Powder the mixture by grinding in a mortar.

Determine the melting-points of the pure components and each mixture by the standard method as described on p. 17. As the mixtures are non-ideal, two temperatures must be determined for each mixture: record the temperature at which melting starts (the solidus point), and also record the temperature at which melting is complete (the liquidus point). Slow heating is essential and a second determination is desirable.

Plot a graph of *liquidus* and *solidus points* against *percentage molar composition*. If there appears to be an eutectic point, estimate the corresponding temperature and composition.

From your graph determine the composition of the first crystals to form when liquid mixtures containing 50% naphthalene are cooled.

What is the significance of eutectic points, liquidus and solidus curves? How could the eutectic be determined more accurately? Consult a textbook.

Look up the molecular structures of 1-naphthol, 2-naphthol and naphthalene.

References: [1]Crompton, H. and Whiteley, M. A., *J. Chem. Soc.*, 1895, 327.
[2]Miers, H. A. and Isaac, F., *J. Chem. Soc.*, 1908, 927.

4. The mutual solubility of phenol and water

In organic preparations the mutual solubility of substances is of practical importance: products can be 'wet' with dissolved water, while yields can be reduced by solution in water, e.g. the preparations of ethyl acetate, p. 60, and aniline, p. 76.

Requirements

Two burettes.
Six dry hard-glass test tubes (150 mm × 25 mm).
80% aqueous phenol (100 g phenol mixed with 25 cm³ water).
200 cm³ beaker as a water-bath.
Thermometer, 0–110°C.
Residues bottle.

Place water in one burette and 80% aqueous phenol (***Caution: caustic***) in the other burette. Prepare according to Table 8 exact mixtures, which should be turbid, in the hard-glass test tubes, and calculate the percentage of phenol by weight in each mixture.

Table 8

Mixture number	*Volume of water/1 cm³*	*Volume of 80% aqueous phenol/1 cm³*
1	2	8
2	4	6
3	5	5
4	6	4
5	8	2
6	8·5	1·5

Clamp one of the mixture tubes upright in the water-bath and heat gently. Stir the contents of the mixture tube carefully with the thermometer and record the temperature at which the mixture first becomes clear. Stop heating and allow the water-bath to cool. Record the temperature at which turbidity reappears. The two temperatures should differ only slightly and the average can be taken as the temperature at which miscibility occurs.

Repeat the experiment with the other mixtures.

At the conclusion of the experiment place your mixtures in the residues bottle.

Plot a graph of *temperature* against *percentage of phenol by weight*. What is the temperature above which phenol and water are miscible in all proportions (the consolute point)? What other types of solubility curve are obtained with partially miscible liquids? Consult a physical chemistry textbook and make notes on suitable examples.

5. Clathrates

Clathrates are true crystals formed between two substances, yet there is no new compound formed. It appears that one substance has a crystal structure in which the second substance can be trapped: quinol crystallizes with a cage structure and urea with a hexagonal spiral.

Requirements

Quinol, $C_6H_4(OH)_2$, M_r 110·11.

Carbamide, NH_2CONH_2, M_r 60·06.

The quinol–methanol clathrate. Weigh accurately a sample of quinol (roughly 2 g) and prepare a saturated solution in about 5 cm³ of methanol. Allow the excess methanol to evaporate overnight from a deep beaker and dry the crystalline residue for ten minutes at about 100°C. Cool the crystals and weigh accurately.

Compare the crystalline form of pure quinol and your product. Assuming the weight increase is due to trapped methanol, calculate the molar proportions in your product.

Reference: Sawyer, A. K., *J. Chem. Educ.*, 1964, **41**, 661.

The quinol–hydrogen sulphide clathrate. This experiment must be carried out *in a fume cupboard*. Dissolve roughly 5 g of quinol by warming in about 50 cm³ of water. Maintain an atmosphere of hydrogen sulphide over the solution as it cools (see Fig. 20). Shake the solution regularly. When the solution is cold filter off the crystals at the pump, wash well with water and dry at room temperature.

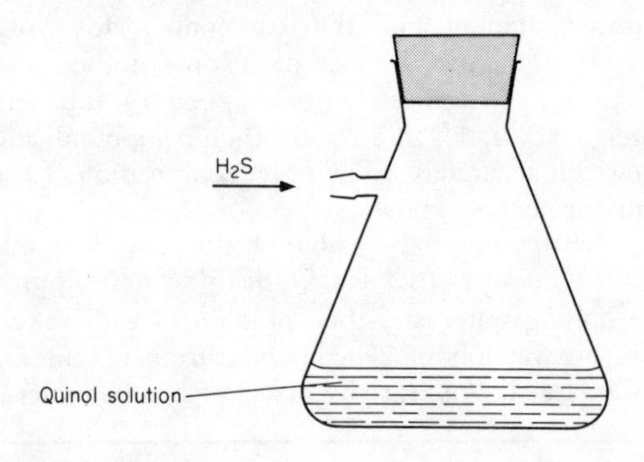

Fig. 20. Adsorption of hydrogen sulphide in the preparation of clathrates.

The filtrate of hydrogen sulphide water should be neutralized with 2M sodium hydroxide and washed away with plenty of water.

Compare the crystalline form of pure quinol and your product. Does the clathrate smell of hydrogen sulphide? Grind some of the crystals and test again for hydrogen sulphide.

Reference: Sawyer, A. K., *J. Chem. Soc.*, 1964, **41**, 661.

The carbamide–propanone adduct. Dissolve roughly 5 g of carbamide in about 25 cm^3 of methanol and add about 1 cm^3 of propanone. On standing, crystals of clathrate should separate. Filter off the crystals at the pump, wash with methanol and dry at room temperature.

Compare the clathrate crystals with pure carbamide. Dissolve some clathrate crystals in water, and test for carbamide and propanone by suitable experiments.

Carbamide will form adducts under similar conditions with butanone, n-octane, octanol, oleic acid and others. If crystallization is slow, a higher concentration of carbamide should help as an equilibrium change is occurring.

Reference: Kobe, K. A. and Reinhart, L. R., *J. Chem. Educ.*, 1959, **36**, 300.

6. Steam-distillation

If two immiscible liquids distil without influence on one another then their vapour pressures in the mixture should be the same as their independent vapour pressures at that temperature. An immiscible pair suitable for study is water and chlorobenzene.

Requirements

Steam-distillation apparatus (250 cm^3 flask).
10 cm^3 measuring cylinders.
Thermometer, 0–100°C.
Chlorobenzene, C_6H_5Cl, M_r 112·6, 1·107 g cm^{-3}.

Set up the apparatus for steam distillation as shown in Fig. 21. The steam inlet tube must go close to the bottom of the distillation flask so that the steam will keep the contents well mixed during distillation. A boiling-stick may be used. Place about 60 cm^3 of chlorobenzene and about 60 cm^3 of pure water in the 250 cm^3 distillation flask. The steam generator should be about one-third full.

Heat the steam generator so that the steam supply maintains a steady, fairly rapid rate of distillation. The flask may also be heated gently but only until distillation commences.

Reject the first portion of distillate, then collect about five fractions of 10 cm^3 each in measuring cylinders.

Record the total volume of a fraction, then transfer to a separating funnel. When the layers have separated, run the lower layer back into the measuring cylinder and record its volume. (Which layer is the chlorobenzene?) This procedure is necessary because of the distorted meniscus between the two liquids.

During the distillation note the temperature from time to time; it should remain steady—if there is a slight variation record the mean temperature.

Also record the barometric pressure during the course of the experiments.

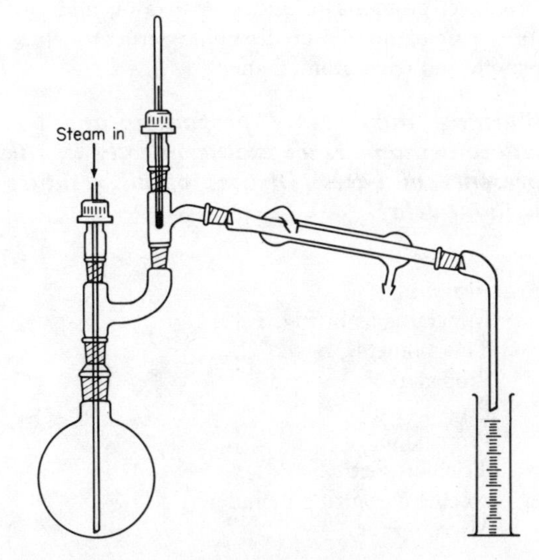

Fig. 21. Apparatus for steam-distillation.

Calculate the number of moles of each liquid in each fraction. Is the ratio (moles of chlorobenzene/moles of water) the same for all the fractions collected? (Consider how accurate the measurements have been.)

Determine the vapour pressure of pure chlorobenzene and pure water at the distillation temperature from graphs plotted from the information in Table 9. Is the vapour pressure ratio (chlorobenzene/water) the same as the distillate mole

Table 9

Temperature /°C	*Water vapour pressure /mm Hg*	*Chlorobenzene vapour pressure /mm Hg*
70	235	100
80	355	145
90	525	210
100	760	290
110	1075	400

ratio (chlorobenzene/water)? What can you deduce?

Is the sum of the vapour pressures approximately equal to the barometric pressure? Why should this be?

Note that the chlorobenzene has been successfully distilled below its normal boiling-point. Look up, and list, a few organic preparations in which steam-distillation is used. Why is it used? See especially p. 64.

7. Miscible liquids: boiling-point changes

If two miscible liquids of boiling-points 50°C and 100°C were mixed in equimolar amounts, a reasonable suggestion for the boiling-point of the mixture would be 75°C. This experiment is designed to test this prediction for a pair of liquids of similar structure, propan-1-ol and propan-2-ol, and also for a pair of liquids of dissimilar structure, propanone and trichloromethane.

Warning: mixtures of propanone and trichloromethane react exothermically in the presence of bases. Dispose of all residues without delay.

Requirements

Apparatus as in Fig. 19(*a*).
Thermometer, 0–100°C.
Propan-1-ol.
Propan-2-ol.
Propanone.
Trichloromethane.
10 cm^3 measuring cylinder.
Microburner.

Set up the apparatus as shown in Fig. 19(*a*) and place in the distillation flask 10·0 cm^3 of propan-1-ol, which should just cover the thermometer bulb. Heat continuously to obtain a gentle reflux. Record the steady temperature, which will be the boiling-point of the pure liquid.

Remove the microburner, and allow the apparatus to cool at least 5°C. Add 2·5 cm^3 of propan-2-ol to the distillation flask **(*Caution: inflammable*)** and determine the boiling-point of the mixture, using the same procedure as before.

Make further additions of 2·5 cm^3 of propan-2-ol determining the boiling-point after each addition until a total of 10·0 cm^3 have been added.

Now empty the distillation flask and repeat the procedure, but start with 10·0 cm^3 of propan-2-ol in the flask and make 2·5 cm^3 additions of propan-1-ol.

Finally repeat the whole procedure using propanone and trichloromethane as the pair of liquids. For this pair make two additions of 1·0 cm^3 followed by 2·0 cm^3 additions to a total of 10·0 cm^3.

Plot graphs of *boiling-point* against *percentage composition by volume*. Are the boiling-points of the mixtures roughly proportional to their molar composition? Plotting the results in terms of molar composition will alter the position of the curves obtained but not their general shape.

Consult a textbook about Raoult's Law.

8. Miscible liquids: fractional distillation

The proper study of the phase equilibria of miscible liquids requires special apparatus, so this experiment is an investigation of one aspect of the propan-1-ol/water system.

Requirements

Fractional distillation apparatus (250 cm^3 flask).
30 cm^3 specific gravity bottles.
Propan-1-ol, C_3H_7OH, M_r 60·1, 0·805 g cm^{-3}.

Set up in a draught-free enclosure the apparatus for fractional distillation, according to Fig. I. Place about 100 cm^3 of propan-1-ol and about 50 cm^3 of pure water in the 250 cm^3 distillation flask. A boiling-stick should also be added.

Heat the flask gently with a naked flame so that distillation proceeds at about 1 drop a second. Take care that the fractionation column does not flood. Reject the distillate until the temperature remains steady; then collect two fractions of about 30 cm^3 each.

Record the temperature of the vapour distilling.

Measure the specific gravity of each fraction; then use the table to calculate the composition of each fraction.

Does the propan-1-ol/water system function as an azeotropic mixture? With maximum or minimum boiling-point? Consult a physical chemistry textbook and record the complete phase diagrams of some typical azeotropic mixtures. As straightforward fractionation will not separate an azeotropic mixture, what techniques can be used?

Table 10

propan-1-ol with water /% by mass	*mass per cm³ at 20°C /g*
100	0·805
90	0·826
80	0·848
70	0·870
60	0·892
50	0·914

9. Miscible liquids: enthalpy changes

Requirements

Tetrachloromethane, CCl_4, M_r 153·84, 1·59 g cm^{-3}.
Trichloromethane, $CHCl_3$, M_r 119·39, 1·44 g cm^{-3}.
Propanone, $(CH_3)_2CO$, M_r 58·08, 1·36 g cm^{-3}.
25 cm^3 measuring cylinder.
Thermometer, 0–100°C.
Six hard-glass test tubes (150 mm × 25 mm).
Solvent residue bottles.

Measure out, as approximately equimolar amounts, two samples of each liquid into separate, labelled hard-glass test tubes ($\frac{1}{5}$ mole of tetrachloromethane is 19·3 cm^3; of trichloromethane is 16·6 cm^3; of propanone is 8·5 cm^3). Stand the samples for a few minutes so that they will all be at room temperature. Record room temperature.

Add a sample of propanone to a sample of trichloromethane and mix by a quick shake. Stir the mixture carefully with the thermometer and record the temperature change of the mixture.

Add a sample of propanone to a sample of trichloromethane and mix by a quick shake. Stir the mixture carefully with the thermometer and record the temperature change of the mixture.

Repeat the experiment with other pairs of samples.

This experiment is not accurately quantitative, but an approximate molar heat of mixing can be calculated.

What explanation can you offer of the results? Consult a textbook about hydrogen bonding.

CHROMATOGRAPHY
1. Partition between solvents

When a substance is shaken with two immiscible solvents, a dynamic equilibrium will be established with the substance unequally divided between the two solvents.

The selective extraction of a particular cation is a valuable analytical method, especially useful in the separation of radioisotopes; 'counter-current distribution' can be used for the extraction and separation of closely related organic compounds.

Partition is also important in chromatography on cellulose.

Introductory test tube experiments. Organic solvents should be placed in residue bottles after use and not poured into a sink.

Requirements

Ether (ethoxyethane).
Trichloromethane.
Butanedione dioxime
50% v/v pentan-2,4-dione in trichloromethane.
2 w/v 8-hydroxyquinoline (oxine) in trichloromethane.
Solvent residue bottles.

(*a*) Shake iron (III) chloride solution with an equal volume of ether (***caution:** have no burners alight*); now add concentrated hydrochloric acid, cool, and shake again. Separate the two layers and test each with ammonium thiocyanate. Record the colour intensity in each phase. How effectively has the iron (III) been extracted into the ether?

(*b*) Repeat (*a*) using uranyl nitrate and nitric acid. This was first developed by Peligot in 1842.

(*c*) Add excess ammonia solution to a nickel salt solution and shake with trichloromethane to which butanedione dioxime has been added. In which layer is the red complex?

(*d*) Shake any iron (III), cobalt (II), copper (II) or nickel salt solution with ether; now add ammonium thiocyanate and shake again. Does the addition of thiocyanate result in extraction into the ether?

(*e*) Shake any iron (III), cobalt (II), copper (II) or nickel salt solution with trichloromethane; now shake a second sample with 2% oxine or 50% pentan-2,4-dione in trichloromethane. Does the addition of a complexing agent result in extraction into the trichloromethane?

Consult advanced textbooks to find the formulae of some of the complex substances which have been prepared and write a general account of these experiments. See Chapter III for the preparation of crystalline specimens.

The partition of acetic acid between water and organic solvents. The partition of acetic acid in the water–butan-1-ol and the water–cyclohexane systems can be studied by different groups of students and the results compared.

After the experiment, a good recovery of butan-1-ol can be made from the collected residues by saturating the aqueous phase with sodium chloride (salting out) before separating the layers. The excess acid should be neutralized to prevent esterification.

Requirements

Three burettes.
100 cm³ separating funnels.
2M acetic acid (115 cm³ of glacial CH_3CO_2H per litre).
Butan-1-ol or cyclohexane.
Titration apparatus and 10 cm³ safety pipette.
0·5M sodium hydroxide.
Solvent residue bottles.

Prepare mixtures in dry 100 cm³ separating funnels according to Table 11 with exactly 25·0 cm³ of organic phase and exactly 50·0 cm³ of aqueous phase in each mixture. Use burettes to measure out the volumes and label the separating funnels appropriately.

Table 11

Organic solvent (butan-1-ol or cyclohexane)/cm³	*2M acetic acid /cm³*	*Water /cm³*
25·0	10·0	40·0
25·0	25·0	25·0
25·0	35·0	15·0
25·0	50·0	none

Shake each mixture well for five minutes, and leave until the two layers have *completely* separated (this may take some time). Efficient distribution of the acetic acid depends on the rate at which the two layers move *relative to one another*; shaking too vigorously is liable to produce an inefficient emulsion. Separate the layers carefully into different dry boiling tubes.

Take 10·0 cm^3 aliquots of the aqueous layer with a safety pipette and titrate with 0·5M sodium hydroxide using phenolphthalein indicator.

Repeat the titration with aliquots of the organic layer. If cyclohexane was used, it should be titrated with 0·01M sodium hydroxide, prepared by precise dilution of the 0·5M sodium hydroxide in a standard flask.

The only reaction occurring in the titration is:

$$NaOH + CH_3CO_2H \longrightarrow CH_3CO_2Na + H_2O$$

so the titres are directly proportional to the concentration of acetic acid in each phase.

Plot a graph of *concentration of acid in organic layer* against *concentration in the aqueous layer*, and another graph against *square of concentration in the aqueous layer*.

What interpretation can be made of the results? Water and butan-1-ol are polar solvents while cyclohexane is non-polar. Consult a textbook about partition coefficient.

2. Adsorption on surfaces

The adsorption of gases by solids plays a crucial role in many industrial catalytic processes; many organic compounds are purified by adsorption of impurities on charcoal, and the froth-flotation method involves the selective picking up of mineral particles at liquid–gas interfaces.

Surface adsorption is also basic to many chromatographic separations.

Introductory test tube experiments

(*a*) Dilute bromine water with an equal volume of water in a test tube. To half the diluted solution add roughly 0·1 g of activated charcoal and shake vigorously. Filter and compare the colour intensity of the filtrate with the untreated bromine water.

(*b*) Add a few drops of an indicator solution such as screened methyl orange to water. Add activated charcoal, shake and filter. Test the filtrate with acid and alkali.

(*c*) Add about 1 cm^3 of ink to about 5 cm^3 of water; retain a portion and treat the remainder with activated charcoal as above. Compare the filtrate colour with the untreated portion.

(*d*) Repeat (*c*) using a chromatographic adsorbent such as alumina instead of the activated charcoal.

Write a general account of these experiments.

The adsorption of oxalic acid by activated charcoal. This experiment is best carried out by two students working together; one prepares solutions, the other weighs charcoal.

Requirements

Titration apparatus and a second burette.
10 cm^3 and 50 cm^3 safety pipettes.
Six or ten *dry* conical flasks.
Standard 0·2M oxalic acid (25·21 g l^{-1} of $H_2C_2O_4.2H_2O$).
0·1M sodium hydroxide or 0·02M potassium manganate (VII).
Activated charcoal.

Place standard oxalic acid in one burette and pure water in another, and prepare the following dilutions in the dry conical flasks, which should be appropriately labelled.

Table 12

Concentration of acid/M	*Composition/cm³*
0·2	100·0 acid
0·14	70·0 acid + 30·0 water
0·08	40·0 acid + 60·0 water
0·02	10·0 acid + 90·0 water
Blank determination	about 100 water

To each flask add an accurate 2·00 g portion of activated powdered charcoal. Use a carefully dried weighing bottle; not all of the charcoal will be removable but a check should show that the remaining weight is not significant. *Unless you take care, the weighing of charcoal is very messy.*

Cork each flask securely and shake intermittently for at least thirty minutes (the experiment can conveniently be left until the next laboratory period at this point).

Remove the charcoal by filtration through a *dry* filter paper into a *dry* conical flask; do not collect the first few cm³. Why are these precautions necessary? If you are short of dry conical flasks, the empty flasks should be cleaned and dried quickly in an oven.

Determine the concentration of oxalic acid in each filtrate by titration.

Caution: oxalic acid is a poison, so do not pipette it by mouth.

Titrate 10 cm³ aliquots with 0·02M potassium manganate (VII) after adding 2M sulphuric acid, or titrate with 0·1M sodium hydroxide using phenolphthalein indicator; when the titre values are low take larger aliquots or change to a 10 cm³ burette. Full details of the titration procedures are on p. 163 for manganate (VII), and on p. 156 for weak acid-strong base.

Standardize the potassium manganate (VII) or sodium hydroxide by titration against the original 0·2M oxalic acid.

Correct the titration values for the 'blank' determination.

Calculate the concentration of oxalic acid in solution after charcoaling (M moles/litre) and, by difference from the original concentration, the amount of oxalic acid adsorbed on the charcoal (m moles).

Plot graphs of M against m, and $\log_{10} M$ against $\log_{10} m$. What can you deduce from the form of the graphs? Consult a textbook about the Freundlich Isotherm.

What is the purpose of the blank determination?

3. Experiments with a strong cation-exchange resin

A strong cation-exchange resin will consist of an insoluble polymer which has reactive sulphonic acid groups as part of its constitution. The acidic hydrogens of the sulphonic acid groups can be exchanged for other cations in the reversible reaction:

$$\text{resin}^-H^+ + M^+Cl^- \rightleftharpoons \text{resin}^-M^+ + H^+Cl^-$$

These experiments are designed to investigate the potentialities of the reaction and possible applications.

Suitable resins, which can be purchased in the hydrogen form, are 'Zeo-Karb' 225-SRC 13 and 'Dowex' 50W-X8. The pre-treatment and recovery of the resin are described on p. 23.

Requirements

Dry hydrogen form of a strong cation-exchange resin.
Titration apparatus.
Standardized 0·1M sodium hydroxide.
0·05M sodium chloride (2·92 g l⁻¹ of NaCl).
Stop-clock.
Pyrex chromatography column (355 mm × 14 mm).

Introductory test tube experiments. Use only a small quantity of resin (0·1 g) for each experiment and throw none away. Return the resin after use to a 'resin residues' container, as it can be regenerated for repeated use.

(*a*) Heat some resin. Is the polymer organic or inorganic? The residue can be thrown away in this case.
(*b*) Add 5 cm³ of water to some dry resin. Observe closely (listen).
(*c*) Add 3 drops of litmus to 5 cm³ of pure water. Then add some resin and shake. Note the colour change and speed of the change.
Repeat using 5 cm³ of 0·05M sodium chloride instead of the distilled water. What explanation of these two results can you offer?
(*d*) Shake some resin with 5 cm³ of 0·1M copper (II) sulphate and note any colour changes.
Repeat using a few other salts with coloured cations. Does the resin adsorb cations?
(*e*) Repeat (*d*) using a few salts with coloured anions. Does the resin adsorb anions?

Write a general account of the results of these experiments.

Exchange capacity of the resin. Weigh accurately a sample of resin (roughly 0·5 g) into a 250 cm^3 conical flask and add about 50 cm^3 of pure water. Now add 3 g of A.R. sodium chloride (a large excess) so that the hydrogen ions of the resin are brought into solution by exchange for sodium ions.

Add phenolphthalein indicator (3 drops), and titrate *slowly* with standard 0·1M sodium hydroxide. While the last portions of resin exchange slowly, an early end-point, which fades, may appear.

If the experiment is repeated, a fresh quantity of resin must be weighed.

Resin capacity is usually quoted as milli-moles of univalent cation (meq) per g of resin:

capacity $= VM/w$, where V = sodium hydroxide titre (cm^3)

M = sodium hydroxide concentration (mol l^{-1})

w = weight of resin (g)

Rate of exchange of the resin. In a 250 cm^3 conical flask place about 50 cm^3 of 0·05M sodium chloride and phenolphthalein (3 drops). Then add exactly 1·0 cm^3 standard 0·1M sodium hydroxide from a burette. Finally, add an accurately known weight of the resin (roughly 0·50 g) and simultaneously note the time or start a stop-clock.

As soon as the phenolphthalein has been decolorized, record the time (do not stop the clock) and add a further 1·0 cm^3 of standard sodium hydroxide. Continue in this manner for five minutes, when sufficient readings will have been taken.

Repeat the experiment using sodium chloride of lower concentrations (e.g. 0·025M, 0·005M) prepared by approximate dilution.

Plot graphs of *volume of sodium hydroxide added* against *time taken for neutralization*. Comment on their shape and the influence of sodium chloride strength on the rate of cation exchange.

Analysis using the resin. The previous experiments establish the correct procedure for using the resin in a column for quantitative analysis.

Weigh out roughly 5 g of the resin (20 meq) and soak in 20 cm^3 of pure water in a beaker. This swells the resin.

Meanwhile clamp the chromatography column (see Fig. 22) upright and pour in 15 cm^3 of pure water. Pour in the slurry of resin. When the resin has settled use a long rod to push a pad of glass-wool down on top of the resin to prevent its being disturbed.

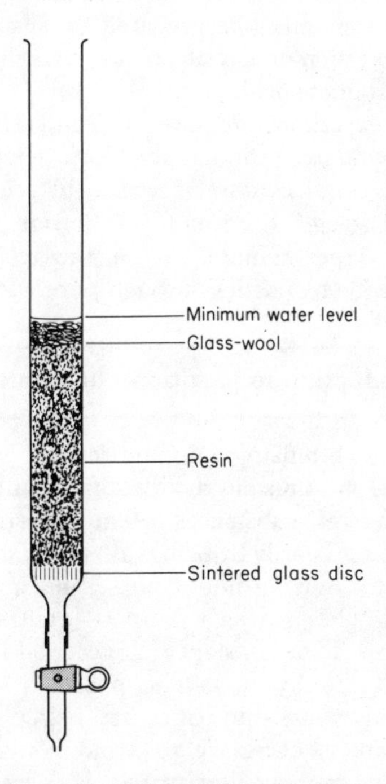

Fig. 22. A Pyrex chromatography column.

Drain off the excess water slowly until the resin is just submerged. Never allow the resin to become uncovered or air pockets will form. Pass through 20 cm^3 portions of pure water until the effluent is neutral; allow the level to drop to the upper glass-wool pad before each addition.

Pipette on to the column a 10·0 cm^3 aliquot of a metal salt solution known to be about 0·1M (1–3 meq). Allow the solution to pass slowly through at a rate of 3 cm^3 per minute, then wash the resin thoroughly with three 20 cm^3 portions of pure water at the same flow rate. Check with litmus paper that the final effluent is neutral. Collect all this effluent in a clean conical flask and titrate with standard 0·1M sodium hydroxide using methyl orange indicator.

Repeat for two further aliquots of metal salt.

Calculate the concentration of the metal salt

using, according to the cation valency, a relationship similar to that quoted in the introduction (p. 108).

Preparation of salts. If the conventional 'acid plus base' method of preparing salts is not possible (because the base is inert or unavailable) then the salt can sometimes be prepared by adsorbing the required cation on a resin column and eluting with the appropriate acid.

This experiment requires M iron (III) chloride (adjusted to pH 1) and M perchloric acid.

Prepare a 5 g column of resin as before, and pass through 20 cm³ of M iron (III) chloride (an excess) at 3 cm³ per minute. Wash the resin free of chloride ion by passing through pure water.

Then pass through about 10 cm³ of M perchloric acid (an insufficiency relative to iron (III) available) followed by two 20 cm³ portions of pure water. Collect this effluent as a solution of iron (III) perchlorate.

Evaporate the solution to dryness on a boiling water-bath, weigh the product and calculate the yield based on the perchloric acid used. The product is very deliquescent.

Other salts can be prepared in the same way. Unusual acids can be obtained by the analysis procedure given above, such as hexacyanoferric (III) acid from potassium hexacyanoferrate (III).

Reference: Salmon, J. E. and Hale, D. K. *Ion Exchange: a Laboratory Manual*. Butterworth, London, 1959.

4. Introduction to partition chromatography on paper

Partition chromatography on paper is used to separate, and thus aid the identification of, micro-quantities of substances. The substances are separated primarily by being differently distributed between a moving liquid phase and a stationary aqueous phase, which is part of the paper's structure. Thus substances are carried forward at different rates by the moving phase.

As only micro-quantities are being separated, extra care is necessary to avoid contamination. Specially graded filter paper is necessary. It should not be touched except on the edges which will not be used and it should be laid down only on a fresh, clean surface.

Solvents can be made up using measuring cylinders. For spraying, very cheap plastic bottles available from chain chemists are adequate, but do not more than quarter fill them. 'Quickfit' make a good spray bottle.

Requirements

Chromatography paper (Whatman Pattern CRL/Paper No. 1).
Two 1-litre beakers, and watch-glasses as covers.
Glass capillary tubes (e.g. organic melting-point tubes).
Solvents and spray reagents (see text).
Samples of inks.
Paper clips.

Separation of ink dyes. Take two chromatography papers and spot them identically with a range of water and ball-point inks. The spots should be no larger than 0·5 cm in diameter and should be placed 1·5 cm from the bottom edge of the paper.

Apply the water inks, after dilution 1 : 3 with water, using a glass capillary tube. Just touch the tube on to the surface of the ink so that only a little is drawn up, then apply the ink to the paper using a quick delicate touch. Each ink is applied with a clean tube to an individual strip (but do not use the outermost strips). Practise the technique on an ordinary filter paper.

Apply the ball-point inks directly from the pen by marking a small spot.

Record the sequence in which the ink spots occur. This can be done in pencil along the top edge of the paper. Allow the spots to dry.

Meanwhile prepare 20 cm³ samples of the two solvents:

(*A*) water (15 cm³), saturated aqueous ammonium sulphate (2 cm³) and ethanol (3 cm³)
(*B*) methanol (16 cm³), concentrated hydrochloric acid (2 cm³) and water (2 cm³)

and pour into their respective covered beakers, which should be labelled.

Now form the papers into cylinders, with the ink

spots at the bottom, by holding the two top corners together with a paper clip.

Stand one paper 'lantern' in each solvent and cover with a watch-glass (see Fig. 23a). Allow the solvents to rise nearly to the top of the slots. Form the paper into a 'lantern' and stand it in the solvent. Cover with a watch-glass. Allow the solvent to rise nearly to the top of the slots. Remove the paper, mark the position of the solvent front, and dry.

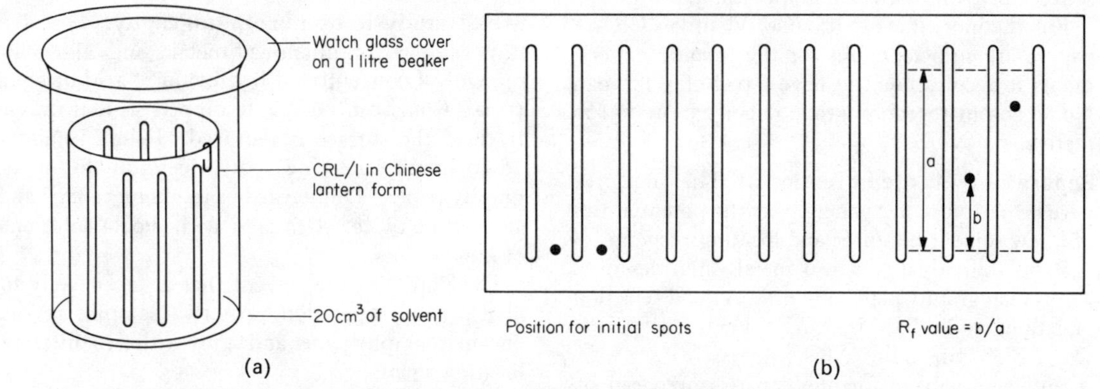

Fig. 23. Apparatus for paper chromatography using Whatman CRL/1 paper.

Remove and dry the papers: not over an open flame, for the solvents have inflammable components.

Mark the positions of the separated components, as some may fade, and write a note comparing the effectiveness of the solvents.

Sensitivity of paper chromatography for analysis. Prepare, by approximate dilution, copper (II) chloride solutions at concentrations of 0·1M, 0·02M, 0·004M, 0·001M.

Draw a horizontal pencil line 1·5 cm from the bottom edge of a chromatography paper. Apply a sequence of copper (II) chloride solution spots on the pencil line: one spot to each vertical paper strip. Repeat the sequence on the remaining strips leaving the central pair unused to test the solvent. If the most dilute solution is spotted first, one capillary tube can be used to apply all the solutions.

Record the sequence of spots and allow them to dry.

Meanwhile prepare 20 cm^3 of the solvent

(*C*) propanone (17·5 cm^3), concentrated hydrochloric acid (1·5 cm^3) and water (1 cm^3),

and pour into a covered, labelled beaker. (N.B. This solvent mixture is unstable and cannot be stored.)

Divide the paper into two parts with one set of spots on each. Spray one part with a 0·1% solution of PAN in ethanol. The paper should be well moistened all over, but not run with excess liquid. Spray the other part with a 0·1% solution of dithio-oxamide in ethanol. Then, *in a fume cupboard*, hold the moist sprayed papers over a basin containing '0·880' ammonia until the change of colour is complete.

Describe your results carefully and calculate the minimum amount of copper ion clearly detected. Is anything detected on the solvent strips?

Determination of R_f values. Every cation has a consistent R_f value for a particular solvent no matter whether the solvent is run for a short or a long distance. If the solvent front has been marked, or is apparent as a 'high tide mark' on any of the above chromatograms, then R_f values may be determined (see Fig. 23b).

Measure the distance from the centre of the initial spot to (i) the centre of the located spot, (ii) the solvent front, and calculate the R_f value using the empirical relationship:

$$R_f = \frac{\text{distance moved by cation}}{\text{distance moved by solvent front}}$$

5. Applications of partition chromatography on paper

Only a limited number of experiments are suggested here; further material is available from the sources listed in the Bibliography, p. 3.

For the inorganic separations Whatman CRL/1 papers are sufficient, but for the organic experiments it is convenient to have a reel of Whatman No. 1 chromatography grade paper 23 cm × 100 metres.

Separation of metal cations. The objective here is to explore in a general way the potentialities of a few solvent mixtures and locating reagents.

Spot, individually, 0·02M metal chlorides onto a chromatography paper and dry. A good selection of cations would be: Mg^{2+}, Al^{3+}, Fe^{3+}, Co^{2+}, Ni^{2+}, Cu^{2+}, Zn^{2+}, Sn^{2+}, Pb^{2+}, Bi^{3+}, Ag^{+}.

Sufficient solvent for one paper is prepared by mixing:

(*C*) propanone (17·5 cm³), concentrated hydrochloric acid (1·5 cm³) and water (1 cm³) or
(*D*) ethanol (18 cm³), concentrated hydrochloric acid (1 cm³) and water (1 cm³) or
(*E*) butan-1-ol (10 cm³), 2M nitric acid (10 cm³) and pentan-2,4-dione (0·2 cm³).

The following locating reagents can be tested initially against cation spots on ordinary filter paper without developing:

(*a*) exposure to hydrogen sulphide gas followed by ammonia fumes,
(*b*) spray of saturated alizarin in ethanol followed by ammonia fumes,
(*c*) spray of 0·1% dithio-oxamide in ethanol followed by ammonia fumes,
(*d*) spray of 0·1% PAN in ethanol followed by ammonia fumes,
(*e*) spray of 1% 8-hydroxyquinoline in ethanol followed by ammonia fumes, the result being viewed in ultraviolet light.

To spot, develop and spray the chromatograms, follow the procedures detailed in chromatography Expt. 4, p. 110.

Mark the position of the cations and the solvent fronts, recording on the paper the solvent and locating reagent used. Calculate R_f values and comment on the relative advantages of the solvents and locating reagents for possible mixtures of cations.

Reference: Pollard, F. H. and McOmie, J. F. W. *Chromatographic Methods of Inorganic Analysis.* Butterworth, London, 1953.

Metal analysis by chromatography. Dissolve tiny samples of unknown metals and alloys in 0·5 cm³ of concentrated hydrochloric acid or aqua regia. Coins can be briefly dipped so that only a trace of the surface is removed. Dilute to about 10 cm³ with water and spot twice onto chromatography paper. Dilute further to about 50 cm³ and again spot twice. Also spot with 0·02M solutions of known salts.

Develop with the solvent that is most likely to give a good separation. Afterwards split up the chromatography paper and spray with two different locating agents.

Mark the positions of located spots and, by comparison with the known cations, determine the composition of the unknown metallic samples.

A foreign 'silver' coin, dipped briefly in aqua regia, gives a suitable solution for analysis. Make comparison spots of Ag^{+}, Ni^{2+}, Co^{2+}, Cu^{2+} and Fe^{3+}. Develop with solvent (*C*) and spray with locating agent (*c*).

Other materials which could be studied include brass, bronze, solder, type metal and foreign coins.

Reference: United States Bureau of the Mint. *Annual Report of the Director of the Mint, 1970.* U.S.G.P.O. (for the composition of coins).

Separation of carotenes and chlorophylls. Grind roughly 1 g of fresh nettle leaves with sand. Add 5 cm³ of propanone and allow to soak for five minutes. Then add 5 cm³ of pure water and decant the extract through a cotton-wool plug in a filter funnel into a separating funnel.

Extract the green colouring into 5 cm³ of petroleum spirit. Run the upper layer into a test tube and dry by shaking with anhydrous sodium sulphate.

Spot onto a 23 cm × 4 cm strip of Whatman No. 1 chromatography grade paper, and develop in a gas jar with the solvent: 40–60°C petroleum spirit (17 cm³) and propanone (3 cm³). If the extract is weak, spot repeatedly at the same place, allowing the paper to dry between spots.

Allow to develop in a cool dark place for as long as possible or until a good separation of coloured spots can be seen. The sequence from the top of the developed chromatogram should be: carotene, phaeophytins, xanthophylls, chlorophyll A, chlorophyll B. Also view in ultraviolet light.

Mark the positions of the spots in pencil for they will soon fade.

Separation of organic acids. Prepare 0·05M aqueous solutions of the sodium salts of aliphatic and aromatic acids. Spot onto 23 cm × 4 cm strips of CRL/1 chromatography paper and develop with 20 cm^3 of the organic (upper) layer from shaking 30 cm^3 of butan-1-ol with 30 cm^3 of 1·5M ammonia.

Detect the acids by spraying *lightly* with bromothymol blue indicator (0·1 g in 1·6 cm^3 of 0·1M sodium hydroxide, diluted to 100 cm^3 with pure water). The acids show as yellow spots on a green background. The position of the sodium ions is revealed by blue spots.

Mark the position of the spots in pencil and comment on the possibility of identifying the acids in natural products.

Separation of amino-acids. The full separation of all the naturally occurring amino-acids requires the use of two-way chromatography: development in one direction with a first solvent is followed by a further development at right angles using a different solvent.

Ninhydrin (indane–trione hydrate) is used to detect the amino-acids and is capable of revealing a microgram (10^{-6} g) of an acid.

Requirements

3 litre unspouted tall form beaker.
23 cm^2 sheet of chromatography paper (Whatman No. 1).
0·1 cm^3 blood pipette (fitted with a teat).
0·02M ninhydrin in propanone (store in a refrigerator).
Amino-acids and solvents (see text).

Prepare 0·01M solutions in propan-2-ol/water of known amino-acids (e.g. B.D.H. Reference Collection).

Prepare amino-acid samples from fruit juices as follows: add 3 cm^3 of ethanol to 1 cm^3 of a fresh fruit juice to precipitate salts and protein, then filter through a small cotton-wool plug in a filter funnel. The filtrate is ready for use.

Prepare amino-acid samples from protein sources (wool, haemoglobin, egg albumen) and nylon as follows: reflux, for one day, roughly 0·5 g of material with 2·5 cm^3 of concentrated hydrochloric acid and 2·5 cm^3 of pure water. After refluxing, add a small portion of charcoal to any discoloured solutions and filter through a small cotton-wool plug in a filter funnel. Evaporate to dryness on a water-bath *in a fume cupboard* and store overnight in a soda-lime desiccator to remove excess hydrochloric acid. Dissolve the residue in 20 cm^3 of pure water. The solution is now ready for use.

Preliminary trials can be carried out on a small piece of chromatography paper. Spot 0·001 cm^3, 0·002 cm^3 and 0·01 cm^3 portions of your samples on the paper (the large portion as a sequence of spots), also a fingerprint and a small drop of saliva. Identify the spots by pencilled reference numbers alongside. Spray with ninhydrin and leave for twenty minutes in an oven at about 110°C. Blue or red spots should appear; for all except the fruit juices 0·001 cm^3 samples should be sufficient. Keep this test piece and note that the spots fade after a few days.

Preliminary runs with the solvents can be carried out with the CRL/1 papers, using 20 cm^3 portions of the solvents listed below. Use the procedure given in the 'Introduction' on p. 110. Spot with appropriate amounts and, after development, spray with ninhydrin and heat as before.

The 23 cm^2 chromatography sheets must be kept absolutely clean if successful results are to be obtained. Cover your bench top with a large, fresh sheet of blotting paper and wear disposable polythene gloves when handling the paper. Note that the solvent will run approximately 10% faster in the machine-direction of the paper (along the reel).

For a one-way chromatogram, mark a line in pencil 2 cm from the bottom edge and spot with samples at 3 cm intervals. For a two-way chromatogram, mark the bottom-right-hand corner in pencil and spot with your sample at a point 3 cm along the diagonal. Identify your samples by reference numbers in pencil along the top of the sheet.

Allow the spots to dry.

Meanwhile prepare the solvent of butan-1-ol (24·0 cm^3), glacial acetic acid (6·0 cm^3) and pure water (10·0 cm^3), and pour it into the 3 litre beaker, which should be sealed with a polythene bag to allow the atmosphere to become saturated with solvent.

Roll the chromatography sheet so that it will stand as a cylinder and clip the two top corners together. The edges should not be allowed to come together except at the top, otherwise solvent will not advance uniformly up the paper. At a convenient time (see below) stand the rolled chromatography sheet in the 3 litre beaker and leave until the solvent has risen nearly the full height of the sheet.

The development time varies with temperature, but is about nine hours. It is best to prepare the experiment one day and carry out the development for the whole of the next day. If a one-way chromatogram is being run in a well sealed beaker, development overnight is possible although solvent will run to the top of the paper.

After development, allow the paper to dry completely *in a fume cupboard*. A warm-air blower can be used to assist the drying, but not a Bunsen flame.

For a two-way chromatogram, re-roll and clip the paper so that the original bottom-right-hand corner becomes the bottom-left-hand corner.

Develop with a second solvent of ethanol (36 cm^3), '0·880' ammonia (2 cm^3) and pure water (2 cm^3). Phenol (30 g) and pure water (7·5 g) is an alternative solvent, but the solution is extremely caustic and its complete removal from the paper presents difficulties. When development is complete, dry the paper *in a fume cupboard*.

Detect the amino-acids by spraying the whole sheet lightly with ninhydrin (an aerosol can is an advantage), and then lay the sheet on clean blotting paper in an oven at about 110°C for twenty minutes.

Mark in pencil the spots which appear, and note the colours. Count the number of components in the various natural products and identify them as far as possible by comparison with reference amino-acids.

The amines from nylon are 1,6-diaminohexane and sometimes 5-aminohexanoic acid from 'type 6' nylon (Clasher, M., Haslam, J. and Mooney, E. F., *Analyst*, 1957, **82**, 101).

To preserve the spots, spray the sheet lightly with a mixture of methanol (19 cm^3), M aqueous copper (II) nitrate (1 cm^3) and 2M nitric acid (a drop), then expose *in a fume cupboard* to fumes from '0·880' ammonia in an evaporating basin. The background should change to blue and the spots to orange.

Keep the finished work in a polythene bag.

6. Column chromatography

Chromatography using columns of adsorbent material is useful for separations on the preparative scale, because gram quantities of material are readily purified.

Many adsorbents are available, but these experiments all use aluminium oxide. After use the aluminium oxide can be recovered (see p. 23) and is still suitable for demonstrations.

Caution: some of the solvents used are highly inflammable, and others are toxic.

Requirements

Pyrex chromatography column (355 mm × 14 mm).
Safety pipette.
Hand bellows.
Aluminium oxide (for column chromatography, Brockmann Grade 1, neutral).

Packing the column

Clamp the glass tube upright and, checking that the tap is closed, half fill with the solvent required for the experiment.

Now slowly pour in roughly 25 g of chromatographic aluminium oxide. Use a filter funnel to guide in the powder and, if a blockage occurs (e.g. just above the solvent level), rock the tube gently. Also tap the tube gently with your fingers to settle the powder uniformly and release any trapped air bubbles.

Push a pad of non-adsorbent cottonwool down the tube to protect the upper surface of the column from disturbance. Drain off the excess solvent until the level falls to the cotton-wool pad; *never let the solvent fall lower*, otherwise the uniformity of the column will be ruined by trapped air bubbles.

The column is now ready for use. Place a sample on the column using a pipette; a precise volume is not required, but the use of a pipette ensures that the sample is placed directly on top of the column and does not drain down the tube walls. Allow the column to drain slowly and wash in the sample by adding small portions of fresh solvent. The sample should now be adsorbed as a narrow band at the top of the column.

Develop the column by running solvent through. Fill up the tube with solvent, then allow solvent to pass through at the rate of about 5 cm³ per minute. Keep the tube topped up, as the liquid pressure will encourage a good flow rate and there will be less danger of letting the column run dry. If the flow rate is too slow, pressure can be applied by attaching a small hand bellows to the top of the tube.

Collect equivolume fractions of solvent draining from the column. Coloured materials are readily seen as they are eluted from the column, but colourless substances must be found by evaporating the fractions to dryness.

Separation of dyes. Prepare a column in 95% aqueous ethanol, and pipette on about 5 cm³ of screened methyl orange. Develop with 95% aqueous ethanol and, if no separation is observed, change to 50% aqueous ethanol.

Methyl orange is screened with xylene cyanol FF or methylene blue. What further experiments are necessary to identify the dye used?

Purification of anthracene. Prepare a column in 60–80°C petroleum spirit and pipette on 0·1 g of technical anthracene dissolved in the minimum of petroleum spirit (about 30 cm³). Develop with 50 cm³ of petroleum spirit, followed by three 25 cm³ portions of petroleum spirit: benzene mixture (4:1), (2:1) and lastly (1:1).

Follow the development by viewing the column in ultraviolet light.

Separation of *o*-nitroaniline and *p*-nitroaniline. Prepare a column in benzene (***caution: toxic***), and pipette on 5 cm³ of a solution ½% in *o*-nitroaniline and ½% in *p*-nitroaniline in benzene.

Develop the column with benzene.

Following the preparation of *p*-nitroaniline on p. 87 the purity of your product can be checked by this procedure.

Purification of benzene-azo-2-naphthol. Benzene-azo-2-naphthol, prepared by the procedure on p. 79, can be purified on a column of roughly 10 g of aluminium oxide made up in ether. The solid product is placed on top of the column and washed through with ether. Impurities remain adsorbed at the top of the column. Collect the effluent with minimum exposure to the air.

The experiment should also be conducted with product which has been 'purified' by recrystallization.

Purification of 2-methoxynaphthalene. 2-Methoxynaphthalene, prepared by the method given on p. 64, can be purified by the same procedure as benzene-azo-2-naphthol above.

Table 13

Mixture	*Adsorbed at top of column*	*Adsorbed at bottom of column*	*Filtrate*	*Activity*
1	*p*-Methoxy-azo-benzene	Azo-benzene		I
1		*p*-Methoxy-azo-benzene	Azo-benzene	II
2	Sudan yellow	*p*-Methoxy-azo-benzene		
2		Sudan yellow	*p*-Methoxy-azo-benzene	III
3	Sudan red	Sudan yellow		
3		Sudan red	Sudan yellow	IV
4	*p*-Amino-azo-benzene	Sudan red		
5	*p*-Hydroxy-azo-benzene	*p*-Amino-azo-benzene		V

Standardization of aluminium oxide. The adsorptive power of aluminium oxide is variable, and a procedure has been developed by H. Brockmann using azo-dyes to grade the adsorptive power from 'highest' (activity I) to 'lowest' (activity V).

Suitable dyestuffs are dissolved in benzene (1 part) and diluted with 60–80°C petroleum spirit (4 parts) to give a 0·1% solution. Two dyestuffs, mixed according to Table 13, are added to the column and activity is rated according to their relative retention by the aluminium oxide.

Prepare columns in a 60–80°C petroleum spirit: benzene mixture (4:1) and pipette on 8 cm³ of a dyestuff mixture in the same solvent. Develop the column with 20 cm³ of the solvent mixture. Deduce the activity of your aluminium oxide from the table.

The influence of pre-treatment of the aluminium oxide can be studied by using material washed with acid or alkali and dried in a hot oven.

Aluminium oxide is progressively deactivated by the addition of water (up to 15%). Mix in a corked flask, shake well and leave for two hours to attain uniformity before use.

7. Thin-layer chromatography

In thin-layer chromatography (T.L.C.) a suitable adsorbent is spread on a glass plate. After activation of the adsorbent by heat, the plate is spotted with a dilute solution of the material under study and then developed with a suitable solvent. When the solvent has risen a convenient distance up the thin film, the plate is dried and treated with a detecting agent.

Silica gel is the preferred adsorbent for T.L.C., although cellulose and alumina thin films are readily prepared.

In all cases the adsorbent must have been specially manufactured for T.L.C. work. Suitable materials are available from B.D.H., e.g. 'Chromedia' C C 41 for cellulose thin films. Alternatively prepared plates can be obtained such as M N polygram SIL N–HR from Camlab or the Eastman 'Chromagram' Kit from Kodak.

T.L.C. is faster than the other techniques in general and sharper separations are possible, but to master the method you will have to work with care and ensure your apparatus is properly cleaned.

Requirements

Adsorbents for T.L.C. (see Introduction above).
Microscope slides.
100 cm³ beakers and covers.
Glass plates (15 cm × 5 cm).
1 litre beakers and covers.
Large sheets of blotting paper.
0·1 cm³ blood pipette (fitted with a teat).

Preparing the plates

(*a*) A slurry of roughly 30% w/v of silica gel in trichloromethane is kept in a well-sealed wide-mouthed bottle, and microscope slides are coated by dipping them into the slurry.

The slurry bottle should be placed on a large sheet of blotting paper *in a fume cupboard.*

Shake the slurry bottle and then dip in two well-cleaned microscope slides, held together at the top by crucible tongs. Dip in and lift out the slides in a continuous movement; do not coat the top 1 cm of the slides. Allow to drain briefly. Handling the edges only, ease the slides apart and lay them, thin film uppermost, on the blotting paper for five minutes to dry.

Activation is not necessary.

If the film is not uniform, the microscope slide was not clean.

(*b*) Prepare a slurry of 1 g of cellulose in 5 cm³ of propanone by mixing well in a small glass mortar.

Hold a 15 cm × 5 cm glass plate over a sheet of blotting paper and pour the slurry on to one end of the plate. By gently rocking the plate, spread the thin film uniformly over the plate, then lay it down for five minutes to dry.

Activation is not necessary.

By use of the same technique it is possible to spread on 15 cm × 5 cm plates slurries of alumina or silica gel (1 g in 2·5 cm³ of 85% aqueous ethanol; if the slurry proves too thick or too thin, vary slightly the volume of solvent).

Allow to dry at room temperature, then activate in an oven at about 120°C for thirty minutes.

Spotting the plates

Thin films must be handled and spotted with extra care because of their fragile nature.

Spot the plates with not more than 0·002 cm^3 of 0·1–0·01M solutions from a blood pipette or fine wire loop. If possible, solutions should be prepared in the same solvent that will be used for development of the chromatogram.

As many as three separate spots can be placed on a microscope slide, if channels are scratched in the thin film with the edge of a spatula and surplus material is cleaned from the edges of the slide.

Development and detection of the chromatograms

Mark the slides 6 cm above the spot, and the plates 10 cm above, and develop until the solvent front reaches these marks.

To develop the slides, place 3 cm^3 of solvent in a 100 cm^3 (unspouted plastic) beaker, lower in the slide gently and cover with a watch-glass.

To develop the plates, place 20 cm^3 of solvent in a 1 litre beaker lined with a filter paper to help saturate the atmosphere. Lower in the plate gently and cover with a polythene bag.

When fully developed, remove and dry the chromatograms.

Detection can be by specific reagents or by the general reagents iodine and 2′,7′-dichlorofluorescein.

For iodine, place the chromatogram in a covered beaker containing a *few* crystals of iodine. Dark spots should appear in a few minutes.

For 2′,7′-dichlorofluorescein, use a fine spray of a 0·1% methanolic solution and view under ultraviolet light. Dark spots should show against a fluorescent background.

Purity of anthracene. Spot a silica gel slide with a solution of anthracene and develop with benzene. Carry out the experiment *in a fume cupboard* because benzene vapour is toxic.

View the chromatogram in ultraviolet light and consult a textbook about the nature of the impurities.

Analysis of butter and margarine. Spot a silica gel slide with solutions of butter and margarine, and develop with benzene *in a fume cupboard*.

Expose the chromatogram to iodine vapour. Can you distinguish between butter and margarine?

If other natural fats such as lard are available, use a silica gel plate to obtain a fuller development.

Separation of phenols. Spot silica gel slides with solutions of closely related phenols (e.g. cresols, nitrophenols, dihydroxybenzenes) and develop with trichloromethane.

Can the phenols be separated and distinguished? Use the technique on the products of your phenol preparations (p. 86).

Separation of ink dyes. Treat 1 cm^3 of ink with 5 cm^3 of ethanol, spot silica gel plates and develop with 95% aqueous ethanol.

Compare with the result from Chromatography Expt. 4, p. 110.

Separation of carotenes and chlorophylls. Obtain an extract of nettle leaf by following the procedure given under Chromatography, Expt. 5, p. 112.

Spot silica gel slides or plates, and develop with benzene (14 cm^3) and propanone (6 cm^3), in a refrigerator if possible.

The sequence of spots, in descending order, should be: carotene, phaeophytins, chlorophyll A, chlorophyll B, xanthophylls. View the slides or plates in ultraviolet light also.

To obtain a better separation of the chlorophylls, use less acetone in the developing solvent.

Separation of fatty acids. Spot silica gel plates with fatty acids (saturated straight chain, 6–18 carbon atoms) dissolved in ether. Develop with butan-1-ol which has been saturated with water. Dry the plates thoroughly, then spray lightly with 0·1% potassium dichromate in concentrated sulphuric acid. Warm in an oven to char the fatty acids.

Determine the R_f values and plot a graph of *R_f value* against *number of carbon atoms*.

Reference: Singh, E. J. and Gershbein, L. L., *J. Chem. Educ.*, 1966, **43**, 29.

Separations on cellulose thin films. Repeat any of the experiments on paper given under Chromatography, Expts. 4 and 5, pp. 110 and 112.

Separations on alumina thin films. Repeat any of the experiments on alumina columns given under Chromatography, Expt. 6, p. 114.

8. Gas chromatography

Gas chromatography is valuable where gaseous or low-boiling mixtures must be analysed, as in the petrochemical industry. The technique can even be used on the preparative scale, and new standards of purity have become possible.

However, for laboratory use, the technique requires either the purchase of apparatus such as the excellent Gallenkamp Junior Gas Chromatograph or the time and patience to construct home-made apparatus.

The Gallenkamp apparatus is supplied with a booklet of experiments, and articles published elsewhere describe the construction and use of home-made apparatus.

Interested readers are advised to consult the following sources if they wish to construct a gas chromatograph:

Cowan, P. J. and Sugihara, J. M., *J. Chem. Educ.*, 1959, **36**, 246.
McLean, J. and Paulson, P. L., ibid., 1963, **40**, 539.
Silberman, R., ibid., 1967, **44**, 590.
Williams, I. W., *S.S.R.*, 1964, **48**, 402.
Hughes, D. E. P., ibid., 1967–68, **49**, 125 and 725
Thompson, J. J. and Heasman, C. J., ibid., 1969, **50**, 536.
Beaumont, T. P. and King, G., ibid., 1970, **52**, 118.

THE COLLOIDAL STATE

This is the name given to systems where the particles in the disperse phase (corresponding to solute) are in the range 10^{-9} to 10^{-7} m in diameter.

The Tyndall effect

One of the easiest ways of recognizing a colloidal system is to use the characteristic scattering of light originally noted by Tyndall.

Set up a strong beam of light (an ordinary 35 mm slide projector is a convenient source of light for this purpose), with a convex lens to converge the beam to a point within the system to be investigated. A glass or Perspex box with parallel sides is the most convenient container to use (see Fig. 24).

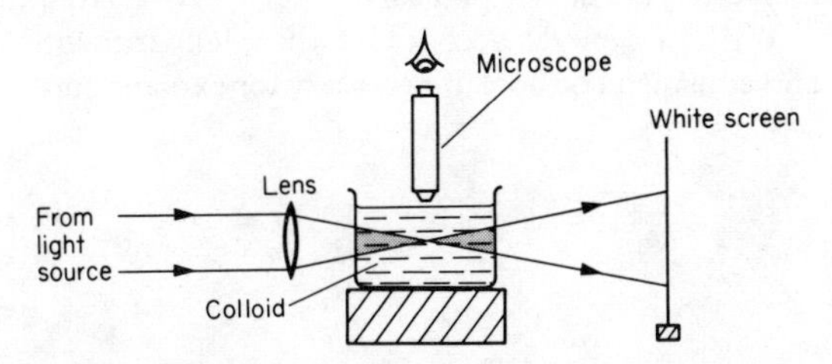

Fig. 24. Apparatus to demonstrate the Tyndall cone. Brownian movement can be observed if the microscope is focused on the brightest point in the cone.

Simple gas-phase colloidal systems which can be investigated this way include smokes and mists (for example the spray from an aerosol can). The sols prepared below can be similarly investigated.

Set up a microscope focused onto the point where the light converges, and note the individual points of light. Observe individual points, and try to explain what you see.

Preparation of liquid-phase colloids (sols)

Colloids in the liquid phase are divided into two categories: lyophilic, which are large molecules which form stable colloid systems; and lyophobic, which are aggregates of a normally insoluble material, and are rather unstable.

Particles of colloidal size may be prepared either by breaking down large particles (dispersion) or by allowing smaller particles to grow to colloidal size but no more (aggregation).

The aggregation methods use dilute solutions, and the results are particularly suitable for the investigation of the Tyndall effect.

Gelatin and starch. These are lyophilic sols prepared by dispersion methods.

Add the solid gelatin to warm water a little at a time, a procedure carried out in making a jelly.

Starch requires boiling to break up the individual grains. Use about 1 g per litre of water.

Cadmium sulphide sol. Peptization is used to disperse lyophobic cadmium sulphide sol. This involves breaking up large particles chemically.

Add '0·880' ammonia dropwise to a 0·25M solution of cadmium sulphate until the hydroxide precipitated has redissolved. Now, *in a fume cupboard*, pass a slow stream of hydrogen sulphide for five minutes

Allow to stand for a few minutes, then filter off the yellow precipitate. The filtrate should be a yellow sol of cadmium sulphide in water.

Red gold sol. Dilute 1 cm³ of 1% chloroauric acid ($HAuCl_4$, most readily made by dissolving gold in aqua regia) to 200 cm³ with water; make alkaline with 2M ammonia, then add dropwise with stirring fresh dilute tannin solution until an intense red colour forms.

Raise to boiling and add another 1 cm³ of 1% chloroauric acid and a little more dilute tannin solution.

Blue gold sol. Dilute 1 cm³ of 1% chloroauric acid to 200 cm³ with water; then add dropwise, with stirring, a 1% hydrazine solution. A blue colour should develop almost at once.

Consult a textbook about the relative size of the particles in the different gold sols.

Silver sol. Use a 0·01M silver nitrate solution in place of the gold compound in the previous two experiments. Note the colours of the silver sols.

Iron (III) hydroxide sol. Heat 400 cm³ of water to boiling and add, dropwise with stirring, 1 cm³ of 0·1M iron (III) chloride solution. A brown sol should develop quite rapidly.

Molybdenum blue. This is a complex compound in which molybdenum has oxidation states of +6 and lower. It is prepared by reducing molybdate ions in acid solution.

Dissolve a small crystal of ammonium molybdate in 200 cm^3 of water with warming. To the hot solution add, with stirring, 0·1M tin (II) chloride. A blue sol should rapidly develop.

A similar experiment can be carried out using a tungstate in place of the molybdate.

Sulphur sol. Sulphur very readily forms colloidal particles, most conveniently from acidified thiosulphate solutions.

Add 20 cm^3 of 0·1M sodium thiosulphate to 100 cm^3 water, then add, dropwise with stirring, 1 cm^3 of 2M hydrochloric acid. The sulphur sol forms slowly (for the kinetics of this reaction see p. 127).

This preparation is a very suitable one for showing the gradual formation of a sol and its effect on a beam of light.

Carry out the preparation in a parallel-sided vessel and put a white screen on the opposite side to the projector.

Explain what you observe.

Precipitation of colloids

Lyophobic sols are usually precipitated by electrolytes.

Investigate the effect of adding equimolar solutions of sodium chloride, magnesium chloride and aluminium chloride dropwise to various sols; for example gold, silver, or cadmium sulphide.

Investigate the effect of adding sodium chloride, sodium sulphate and sodium citrate to iron (III) hydroxide sol.

Look for the turbidity when the sol coagulates.

Also investigate what happens when iron (III) hydroxide sol (which carries a positive charge) is added to gold sol (which carries a negative charge).

Write a general account of these experiments, and consult a textbook if necessary for explanations.

THERMOCHEMISTRY

1. Heat of neutralization

According to Hess's Law the energy change accompanying a reaction is the same no matter what intermediate reactions are used to carry out the overall reaction. This experiment investigates the molar heat of neutralization of acids. It is instructive to study acetic acid and ammonia as one acid-base pair.

The calorimetry technique suggested will not give precise values, so the specific heat capacity of the mixture can be taken as 4·2 J $g^{-1}K^{-1}$, and energy changes due to dilution can also be neglected. The bases are used in excess so that a *definite quantity of acid* will have been neutralized in each case.

Expanded polystyrene cups have negligible thermal capacity as used in this experiment and therefore serve as excellent calorimeters.

Requirements

Two expanded polystyrene cups.
Thermometer, 0–50°C, with 0·1 graduations.
Standard 1M acids.
1·1M bases.
50 cm^3 measuring cylinder.

Measure 50 cm^3 portions of an acid and a base as accurately as possible into separate polystyrene cups (supported in beakers, see Fig. 25). Record the accurate temperatures of the acid and the base.

Tip the base into the acid and stir the mixture well with the thermometer. Record the maximum steady temperature of the mixture.

Repeat the experiment with other acid-base pairs.

From the temperature rise and the quantity of water present, calculate the number of joules liberated (specific heat capacity of water is 4·2 J g^{-1} K^{-1}).

From the quantity of acid present and the joules liberated, calculate the molar heat of neutralization for each acid-base pair.

Write the ionic equation for each neutralization and compare with the molar heats of neutralization. How do you account for the acetic acid–ammonia result?

2. Heat of solution

The heat of hydration of a salt can be deduced from the heat of solution of the salt in its anhydrous and hydrated forms.

Alternatively, the heats of solution of the salts of a Periodic Table group can be studied. The quantities used should be in a definite molar ratio: 1 mole of salt to 200 moles of water.

Expanded polystyrene cups have negligible thermal capacity as used in this experiment and therefore serve as excellent calorimeters.

Requirements

Two expanded polystyrene cups.
Thermometer, 0–50°C with 0·1 graduations.
Group II nitrates, anhydrous and hydrated.
100 cm^3 measuring cylinder.

Weigh accurately a polystyrene cup for use as the calorimeter and measure into it 4 moles of water (about 72 cm^3). Weigh accurately again to find the precise quantity of water in the calorimeter cup.

Weigh roughly 0·02 mole of the chosen salt in

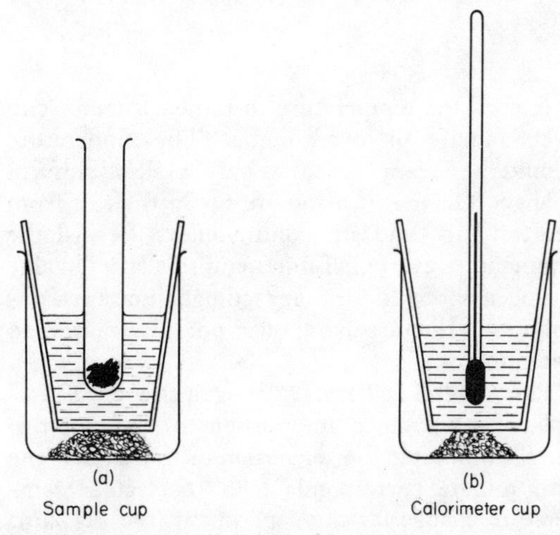

Fig. 25. Apparatus for thermochemistry.

finely ground form, and place in a dry test tube. Weigh accurately the tube and contents. Now suspend the tube and contents in a sample of water in the second polystyrene cup so that it will have the same temperature as the calorimeter cup (see Fig. 25).

The extrapolation procedure compensates for heat gains or losses from the surroundings.

Weigh accurately the almost empty test tube to find the precise quantity of salt added to the calorimeter cup.

From the temperature rise and the quantity of

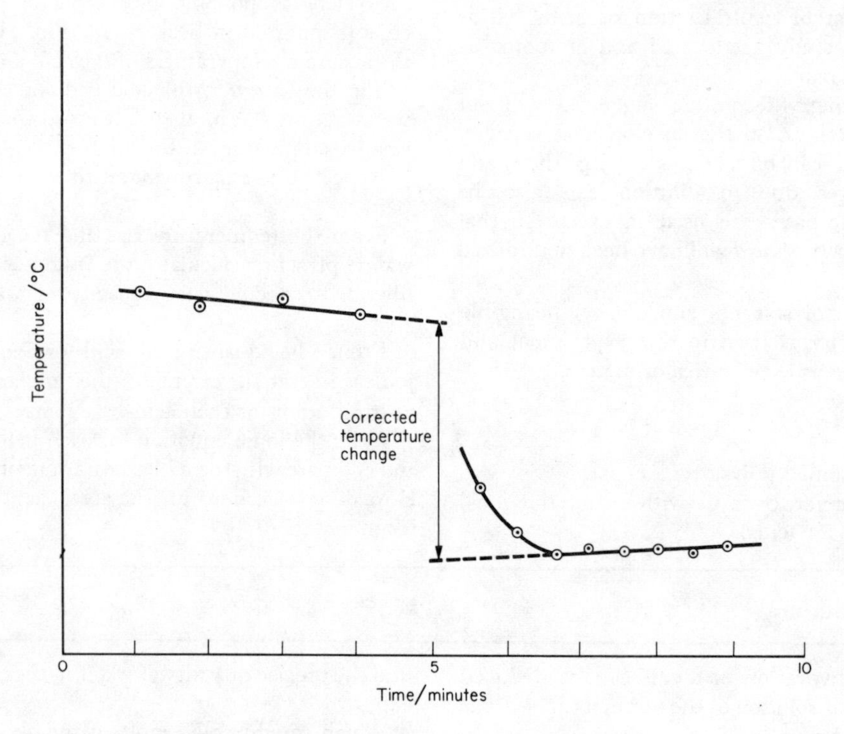

Fig. 26. Graph plotted to obtain a corrected temperature change.

Record the temperature in the calorimeter cup every minute for four minutes. The temperature should be steady or show only a slow uniform change. On the fifth minute tip in the salt from the test tube and stir continuously. Record the temperature every half minute until a steady value has been observed for four readings. For a reasonable result the dissolving must not take more than two minutes.

Plot a graph of *temperature* against *time*. Extrapolate both straight line portions of the graph to the fifth minute; the separation of the lines at the fifth minute corresponds to a 'corrected' temperature change for the experiment (see Fig. 26).

water present, calculate the number of joules liberated (specific heat capacity of water is 4·2 $J\,g^{-1}\,K^{-1}$).

From the quantity of salt present and the joules liberated, calculate the molar heat of solution in joules per mole.

Comment on any progression to be seen in the heats of solution of the periodic group. Look up the heats of solution for other periodic groups, for example the alkali metal chlorides.

To what extent is heat of hydration important in determining the solubility of salts? Consult a textbook about lattice energy.

3. Heat of precipitation of metals

The enthalpy change (ΔH) for the displacement of one metal from solution by another is measured calorimetrically in this experiment, e.g.

$$Zn(s) + Cu^{2+}(aq) \rightarrow Zn^{2+}(aq) + Cu(s)$$

The free energy change (ΔG) of the reaction can be deduced from a measured e.m.f. using the procedure for electrode potentials on p. 134.

Requirements

- An expanded polystyrene cup.
- Thermometer, 0–50°C, with 0·1 graduations.
- Standard 0·25M A.R. copper sulphate (or standard 0·25M silver nitrate).
- Metals in powder form (Fe, Zn, Mg).
- A silver residues bottle.

Weigh roughly 0·02 mole of metal powder (an excess).

Place a 50·0 cm^3 aliquot of the salt solution in the polystyrene cup and record the temperature in the cup every minute for four minutes. The temperature should be steady or show only a slow uniform change. On the fifth minute tip in the metal powder and stir continuously. Record the temperature every half minute until a steady value has been observed for four readings. The replacement reaction should be complete in less than a minute.

Test your final solution for the absence of copper or silver ions. Any deposit of metallic silver should be placed in a silver residues bottle.

Plot a graph of *temperature* against *time*. Extrapolate both straight line portions of the graph to the fifth minute; the separation of the lines at the fifth minute corresponds to a 'corrected' temperature change for the experiment (see Fig. 26). The extrapolation procedure compensates for heat gains or losses from the surroundings.

From the temperature rise and the quantity of solution present, calculate the number of joules liberated. The specific heat capacity of the solutions is approximately 4·2 J cm^{-3} K^{-1} (note the unusual units).

Excess metal was added so, *from the quantity of solution present* and the joules liberated, calculate the molar heat of reaction in joules per mole of copper or silver atoms.

Interpret your results in terms of the electrochemical series of the elements.

CHEMICAL EQUILIBRIUM
1. Introductory study

When a chemical system is in equilibrium, it does not follow that reaction has ceased, only that the products are reverting at the same rate they are being formed from the reactants.

The 'Law of Mass Action' was originally expressed in kinetic terms by Guldberg and Waage in 1864, but it is more properly derived from the laws of thermodynamics, the condition being that the free energy change is zero if the chemicals are mixed in their equilibrium proportions; i.e.

$$\begin{matrix}\text{total free energy} \\ \text{of the reactants}\end{matrix} = \begin{matrix}\text{total free energy} \\ \text{of the products}\end{matrix}$$

The equilibrium law is, however, adequately established on an empirical basis by experimental evidence (e.g. see Wright, P. G., *Educ. Chem.*, 1965, **2**, 14). In dilute solutions the results are adequately expressed in terms of the concentrations of the species contributing to the equilibrium; e.g.

$$Cu^{2+} + 4NH_3 \rightleftharpoons Cu(NH_3)_4^{2+}$$

$$K_{\beta_4} = \frac{[Cu(NH_3)_4^{2+}]}{[Cu^{2+}]\,[NH_3]^4} = 4{\cdot}7 \times 10^{12}\ \text{mole}^{-4}\ \text{litre}^4$$

K_β is the conventional equilibrium constant and the subscript β indicates that the equilibrium of the simple species with the cumulative complex ions is considered.

$$Ag_2CrO_4(s) \rightleftharpoons 2Ag^+(aq) + CrO_4^{2-}(aq)$$

$$K_{s_0} = [Ag^+]^2\,[CrO_4^{2-}] = 8{\cdot}0 \times 10^{-11}\ \text{mole}^3\ \text{litre}^{-3}$$

K_{s_0} is the conventional solubility product, and the subscript zero indicates that the equilibrium of the solid with a saturated solution of the simple ions is considered.

The best source of stability constants is Special Publication No. 17 (1964) of the Chemical Society, compiled by Bjerrum, Schwarzenbach and Sillén (see Bibliography).

Introductory test tube experiments

(*a*) Add potassium chromate solution to solutions of calcium, strontium and barium chlorides. Allow to stand for one minute. Where precipitates occur, shake and divide into two portions. Add 2M hydrochloric acid to one portion and 2M acetic acid to the other portion.

Interpret your results in terms of the solubility of the chromates and the strengths of the acids. What is the cause of the colour change?

(*b*) Add concentrated hydrochloric acid in excess to solutions of silver, lead and copper (II) nitrates. The final solutions are due to the formation of complex ions. Consult a textbook to discover their formulae.

(*c*) Aqueous solutions of the halogens involve the equilibrium:

$$Hal_2 + 3H_2O \rightleftharpoons Hal^- + HalO^- + 2H_3O^+$$

Add excess silver nitrate solution to iodine solution and bromine water.

Interpret your results in terms of displacing the above equilibrium.

(*d*) The relative stability of iron (III) complexes is illustrated in this experiment using the formation of the blood red $Fe(CNS)^{2+}$ complex ion as a reference standard. Add 0·02M iron (III) nitrate solution to a 0·1M solution of each of the following, using the same volumes each time:

sodium chloride
sodium fluoride
sodium oxalate

Now add 0·02M ammonium thiocyanate solution, and note the relative volumes used before the red colour appears. Write the equations of the competing equilibria.

(*e*) The precipitation of cadmium sulphide in Group II of the hydrogen sulphide scheme of analysis often gives difficulty. The competition between the formation of a precipitate and the formation of complex ions is illustrated by this experiment. Add 2 cm³ of cadmium sulphate solution to each of the following:

2 g of sodium chloride in 6 cm³ of water
2 g of sodium iodide in 6 cm³ of water
6 cm³ of 50% aqueous sulphuric acid (diluted in a beaker)
6 cm³ of concentrated hydrochloric acid
6 cm³ of 2M hydrochloric acid
6 cm³ of water (as reference standard)

Now add 2 drops of aqueous hydrogen sulphide solution and compare the results. Add further 2 drop portions comparing the results.

List the solutions in order of their ability to keep cadmium cations in solutions, and interpret your results in terms of the following equilibria:

$$CdS(s) + 2H_3O^+(aq) \rightleftharpoons Cd^{2+}(aq) + H_2S(g) + H_2O(l) \quad K_{S_0} = 1{\cdot}3 \times 10^{-5}$$

$$2Cd^{2+} + 3Cl^- \rightleftharpoons CdCl^+ + CdCl_2 \quad K_{\beta_2} = 4{\cdot}0 \times 10^2$$

$$4Cd^{2+} + 10I^- \rightleftharpoons CdI^+ + CdI_2 + CdI_3^- + CdI_4^- \quad K_{\beta_4} = 1{\cdot}3 \times 10^6$$

(*f*) The stabilities of the trioxalato-complexes of aluminium, chromium, iron and cobalt (prepared by the procedures on pp. 27, 46, 48, 49) are compared in this experiment.

Dissolve, without heating, a small portion of a trioxalato-complex in 2M acetic acid. Add 1 cm³ of potassium manganate (VII) and place the tube in a hot water-bath. Note the relative times taken for the manganate (VII) to be decolorized.

(*g*) Add a few drops of ethyl acetoacetate to 20 cm³ of water in a beaker, followed by a few drops of iron (III) chloride solution. Stir well and note the purple colour of the iron (III)–enol complex. Now add bromine water until the colour fades. Allow to stand, when the colour should return. Add further portions of bromine water.

For an explanation consult Equilibria, Expt. 3, below.

2. Hydrolysis of chlorides

A semi-quantitative study of the hydrolysis of antimony (III) and bismuth (III) chlorides can be made quite simply.

Requirements

Two burettes.
Antimony (III) chloride.
Bismuth (III) chloride.

Add roughly 4·6 g of antimony (III) chloride (0·02 mole) to 2 cm³ of concentrated hydrochloric acid and 2 cm³ of pure water measured from burettes. Warm to dissolve, then cool to room temperature.

Add water from a burette in 0·2 cm³ portions until precipitation occurs—a permanent cloudiness which is best seen against a black background. Without delay add 0·5 cm³ of concentrated hydrochloric acid to redissolve most of the cloudy precipitate. Record the burette readings.

Continue in this way until a total of about 6·0 cm³ of concentrated hydrochloric acid has been added. The experiment can be repeated using 6·3 g of bismuth (III) chloride (0·02 mole).

Plot graphs of volume of *concentrated hydrochloric acid* against *total volume* (i.e. acid plus water) when precipitation occurs.

Interpret your results and write the equations of the reactions.

3. Keto-enol tautomerism of ethyl acetoacetate

Ethyl acetoacetate is an equilibrium mixture of two different molecular structures:

$$\underset{\text{keto form}}{CH_3{\cdot}CO{\cdot}CH_2{\cdot}CO_2C_2H_5} \rightleftharpoons \underset{\text{enol form}}{CH_3{\cdot}C(OH){:}CH{\cdot}CO_2C_2H_5}$$

The rate of interconversion is slow and it is therefore possible to determine quantitatively the enol form in a rapid reaction with bromine. As a direct titration the method is known as the Kurt Meyer titration (*Ann.*, 1911, **380**, 212).

A more accurate indirect method involves the addition of excess bromine to ethyl acetoacetate, the excess being removed immediately by addition of 2-naphthol.

$$CH_3{\cdot}C(OH){:}CH{\cdot}CO_2C_2H_5 + Br_2 \longrightarrow CH_3{\cdot}CO{\cdot}CHBr{\cdot}CO_2C_2H_5 + HBr$$

$$C_{10}H_7OH + Br_2 \longrightarrow C_{10}H_6(Br)OH + HBr$$

The amount of bromoester is determined by reduction with potassium iodide, the liberated iodine being titrated with standard sodium thiosulphate solution:

$$CH_3{\cdot}CO{\cdot}CHBr{\cdot}CO_2C_2H_5 + 2HI \longrightarrow CH_3{\cdot}CO{\cdot}CH_2{\cdot}CO_2C_2H_5 + I_2 + HBr$$

thus overall 1 mole of the enol form yields 1 mole of iodine molecules.

The influence of solvent on the equilibrium may be studied by dissolving the ethyl acetoacetate in solvents of different relative permittivity; e.g. methanol, 32·6; 2-methylpropan-2-ol, 10·9.

This experiment is best attempted on a class basis with different students examining the various conditions.

Requirements

Titration apparatus with safety pipette and a standard flask.
Standard 0·5M ethyl acetoacetate in a suitable solvent (65·08 g l^{-1}).
0·1M bromine in the same solvent (5·15 $cm^3\, l^{-1}$).
1M 2-naphthol in the same solvent (144·2 g l^{-1}).
0·5M aqueous potassium iodide.
Standard 0·1M aqueous sodium thiosulphate.

Prepare 0·4M and 0·2M ethyl acetoacetate solution by accurate dilution.

Place an aliquot of standard ethyl acetoacetate solution in a conical flask, using a safety pipette, and have available in measuring cylinders a volume about equal to the aliquot of 0·1M bromine solution (an excess), and about $\frac{1}{5}$ the volume of 1M 2-naphthol (an excess). Record the temperature of the ethyl acetoacetate solution.

Note the time, and add the bromine solution to the ethyl acetoacetate. Shake well and after exactly thirty seconds add the 2-naphthol solution to remove the excess bromine. Shake well.

Now add about a $\frac{1}{5}$ volume of 0·5M potassium iodide (an excess) and allow to stand for fifteen minutes with occasional shaking. The colour of iodine should appear.

Titrate the entire reaction mixture with standard 0·1M sodium thiosulphate to the colourless end-point. The iodine colour tends to reappear, and the final solution may have a brownish tinge if impure 2-naphthol was used. The blue starch–iodine complex will not form in alcoholic solutions. If necessary, titrate from a 10 cm^3 burette.

Carry out at least two determinations for each ethyl acetoacetate concentration.

Calculate the concentration of enol form from the titration, and determine the concentration of keto form by difference from the original ester concentration; hence determine the equilibrium constant at the working temperature as

$$K = [\text{enol}]/[\text{keto}].$$

Interpret the results.

References: Ward, C. H., *J. Chem. Educ.*, 1962, **39**, 95.
Lockwood, K. L., *J. Chem. Educ.*, 1965, **42**, 481.

REACTION KINETICS

1. Introductory study

Requirements

1M sodium thiosulphate solution.
0·5M nitric acid.
10 cm³ measuring cylinder.
Test tubes (150 mm × 25 mm).
Stop-clock.

Dilute the molar sodium thiosulphate solution so that you have available 5 cm³ portions of $\frac{2}{3}$M, $\frac{1}{3}$M, $\frac{1}{6}$M as well as 1M sodium thiosulphate.

Measure one of the 5 cm³ portions of sodium thiosulphate solution into a large test tube. Add a 5 cm³ portion of 0·5M nitric acid, starting the stop-clock *at the moment of addition*. Shake well to mix the solutions.

Note the time taken for a precipitate of sulphur to appear.

Repeat the experiment with the other sodium thiosulphate solutions in turn. Conclude all the experiments when the same degree of cloudiness is apparent.

Plot a graph of *concentration* of sodium thiosulphate against *time* for precipitation, and a second graph of *concentration* against 1/*time* (1/time is equivalent to the 'speed' of the reaction; explain this).

What form do the graphs take? Can you produce the mathematical expression relating concentration to speed?

2. The Harcourt and Esson experiment

Hydrogen peroxide and iodide ion react in acidic solution releasing iodine

$$H_2O_2 + 2I^- + 2H^+ \longrightarrow 2H_2O + I_2$$

so the rate equation can be expressed:

$$\text{rate} = k[H_2O_2]^m\,[I^-]^n\,[H^+]^o$$

The order of reaction with respect to hydrogen peroxide (m) is found by making up a reaction mixture with a large excess of iodide and acid relative to the hydrogen peroxide used. The rate equation therefore simplifies to:

$$\text{rate} = k'[H_2O_2]^m$$

The time is noted as definite small amounts of iodine are produced. Sodium thiosulphate and starch are added to the reaction mixture; as soon as the iodine being produced has used up all the thiosulphate it will react with the starch. The usual blue-black colour is developed and acts as an excellent indicator. If a second portion of sodium thiosulphate is now added to the reaction mixture the blue-black colour will disappear and the time can be noted when the colour reappears. And so the decrease of hydrogen peroxide with time can be followed by making successive additions of sodium thiosulphate.

Requirements

0·1M hydrogen peroxide (about '1 volume').
1M potassium iodide.
2M sulphuric acid.
0·025M sodium thiosulphate.
Starch indicator.
50 cm³ burette.
5 cm³ pipette.
200 cm³ measuring cylinder.
500 cm³ conical flask.
Stop-clock.

Into a 500 cm³ conical flask put about 150 cm³ of pure water and add about 20 cm³ of M potassium iodide, about 10 cm³ of 2M sulphuric acid, about 1 cm³ of starch indicator and, from the burette, 5·0 cm³ of 0·025M sodium thiosulphate. Then add 5·0 cm³ of 0·1M hydrogen peroxide (a) quickly and start the clock half way through the addition. The reaction does not start until the hydrogen peroxide is added.

Shake the reaction mixture vigorously to ensure complete mixing. Note the time (t) the instant the starch-iodine colour appears. *Without stopping the clock* add a further 5·0 cm³ portion of sodium thiosulphate and note the time when the starch-iodine colour reappears. Repeat the procedure for a total of five or six additions.

Plot a graph of $(a - x)$ against *time*, where a is the initial volume of hydrogen peroxide in the reaction mixture and x is the accumulated volume calculated to have reacted after each 5·0 cm^3 portion of 0·025M sodium thiosulphate has been used up.

$$H_2O_2 + 2H^+ + 2I^- \longrightarrow 2H_2O + I_2$$
$$I_2 + 2S_2O_3^{2-} \longrightarrow 2I^- + S_4O_6^{2-}$$

Does the graph give a straight line? What other graphs might give useful information? The rate equation can be expressed as:

$$dx/dt = k'(a - x)^m$$

Consult Appendix II, p. 203, if necessary.

3. Hydrolysis of esters

Organic esters are slowly hydrolysed by water, forming a weak organic acid and an alcohol. For example, methyl formate:

$$HCO_2CH_3 + H_2O + H^+ \longrightarrow HCO_2H + CH_3OH + H^+$$

If low molecular weight esters are hydrolysed using an acid catalyst, the reaction can be conveniently studied in an hour. As the reaction proceeds there is an increase in free acid concentration which may be followed by withdrawal of aliquots at definite times. The aliquots are titrated with a suitable base.

Since the water will be present as a large excess and the acid catalyst remains nearly constant, they will not affect the measured order of reaction, and the experiment should therefore determine the order with respect to ester.

Requirements

Titration apparatus, including 5 cm^3 pipette.
An ester (methyl formate).
0·25M sulphuric acid.
0·1M sodium hydroxide.
Phenolphthalein indicator.
Ice-cold water. Stop-clock.

Organize the titration apparatus as required for the experiment.

Place 100 cm^3 of 0·25M sulphuric acid in a conical flask. When all is ready, add a 5 cm^3 sample of ester and start the clock. Stopper the mixture and shake well for half a minute.

After a minute withdraw a 5 cm^3 aliquot and run into a conical flask containing about 20 cm^3 of ice-cold water to arrest the reaction. Note the time (t minutes) at some precise moment in this operation, such as the moment of release of the aliquot (only *relative* times are required).

Without delay titrate the chilled aliquot with 0·1M sodium hydroxide, using phenolphthalein indicator (titre V_t). The pink end-point will fade because the hydrolysis continues very slowly.

Withdraw further aliquots at convenient intervals (five to ten minutes) over a period of an hour, noting the precise time and titrating as before.

Finally leave the reaction mixture for two days to reach equilibrium, then titrate a last aliquot (titre V_∞).

The rate law for the hydrolysis is:

$$-d[\text{ester}]/dt = k[\text{ester}]^m[H_2O]^n[H^+]^o$$

In this experiment the water and hydrogen ion concentrations are constant, and the ester concentration is proportional to $(V_\infty - V_t)$ cm^3, so

$$-d(V_\infty - V_t)/(V_\infty - V_t)^m = kdt$$

Plot three graphs of $(V_\infty - V_t)$, $log_{10}(V_\infty - V_t)$ and $1/(V_\infty - V_t)$ against t corresponding to the integrated rate law for $m = 0$, $m = 1$, $m = 2$. The graph corresponding to the correct value for m should be a straight line.

A brief introduction to the mathematics of reaction kinetics is given in Appendix II, p. 203.

4. Halogenation of propanone

The reaction between a halogen and aqueous propanone is acid catalysed:

$$I_2 + CH_3.CO.CH_3 + H^+ \longrightarrow CH_3.CO.CH_2I + 2H^+ + I^-$$

The influence of iodine concentration on the rate of reaction can be studied if the concentrations of propanone and acid catalyst remain constant during the reaction. This is achieved by using a large excess of propanone and acid in the initial reaction mixture.

Requirements

Titration apparatus, including a 10 cm³ pipette.
0·02M iodine (5 g I_2 plus 33 g KI per litre).
1M aqueous propanone (73 cm³ per litre).
1M sulphuric acid.
0·5M sodium hydrogen carbonate solution.
0·01M sodium thiosulphate solution.
Starch indicator.
50 cm³ measuring cylinder.
Stop-clock.

Mix 25 cm³ of M aqueous propanone with 25 cm³ of M sulphuric acid, and note the time when 50 cm³ of 0·02M iodine is added with good shaking.

Withdraw a 10 cm³ aliquot and stop the reaction by running it into 10 cm³ of 0·5M sodium bicarbonate. Note the precise time at which the aliquot is released into the hydrogen carbonate. Titrate the iodine present with 0·01M sodium thiosulphate, using starch indicator (about 1 cm³) added near the end-point.

Withdraw further 10 cm³ aliquots about every seven minutes, and treat them similarly, always noting the precise time at which the aliquot is run into the sodium hydrogen carbonate.

Plot a graph of *titre* against *time*.

Since *titre* is equivalent to iodine concentration, deduce from the form of the graph the order of reaction with respect to iodine (look at Reaction Kinetics Expt. 3, above, if necessary).

What does the order of reaction imply about the influence of iodine on the rate of reaction? What explanation can be made of this result?

Consult a textbook about keto-enol tautomerism.

5. Saponification of esters

The saponification of ethyl acetate can be studied more easily if the initial concentrations of sodium hydroxide and ethyl acetate are identical:

$$CH_3CO_2CH_2CH_3 + OH^- \longrightarrow CH_3CO_2^- + CH_3CH_2OH$$

The hydroxide ion has a high conductivity ($\Lambda(OH^-) = 0{\cdot}174$ ohm⁻¹ m² mol⁻¹ at 18°C) compared to the acetate ion ($\Lambda(CH_3CO_2^-) = 0{\cdot}035$) so there is a decrease in conductivity during the reaction. As the reaction occurs quite quickly it is more convenient to follow the change in conductivity than to titrate aliquots.

Requirements

Conductivity bridge with dip cell, see p. 136.
600 r.p.m. stirrer.
0·05M aqueous ethyl acetate (4·9 cm³ pure ester per litre).
Standardized 0·05M sodium hydroxide.
0·025M sodium acetate (3·40 g of trihydrate per litre).
Two 50 cm³ measuring cylinders.
Stop-clock.

Place 100 cm³ of 0·025M sodium acetate in a 150 cm³ beaker and take readings of conductivity until a steady value can be recorded ($1/r_\infty$).

Prepare 100 cm³ of 0·025M sodium hydroxide by careful dilution of the 0·05M sodium hydroxide. Record its conductivity ($1/r_0$).

Place 50 cm³ of 0·05M sodium hydroxide in a clean 150 cm³ beaker with the stirrer operating and the dip cell clamped in place. When all is ready, quickly add 50 cm³ of 0·05M aqueous ethyl acetate and note the precise time (t_0).

Record the temperature of the reaction mixture near the beginning and end of the reaction; any change should be slight otherwise the kinetic results will be poor.

Take readings of conductivity every minute for

about eight minutes, then every two minutes for the next eight minutes, and finally every four minutes until the experiment has been running for half an hour ($1/r_t$ values). In this way there should be a noticeable drop in conductivity between each reading.

Determine the hydroxide ion concentration, expressed as $(1/r_t - 1/r_\infty)$ ohm^{-1} at each time value (t min.), and plot three graphs of *hydroxide concentration*, *log₁₀ hydroxide concentration* and, finally, *reciprocal of hydroxide concentration* against *time*. The relationship corresponding to the true order of reaction should give a straight line graph. Look at Reaction Kinetics Expt. 3, p. 128, if in doubt.

Deduce the order of reaction and write the rate law in its correct form for this reaction.

Why was it helpful to have the initial concentrations of ester and hydroxide identical?

6. Activation energy

The study of the influence of temperature on reaction rate leads to the concept of activation energy.

It is convenient to study again Reaction Kinetics Expts. 1 and 3, pp. 127 and 128, at temperatures between room temperature and about 45°C. If thermostatically controlled water-baths (see Fig. 27) are available, a great deal of effort is saved and attention can be concentrated on the kinetics.

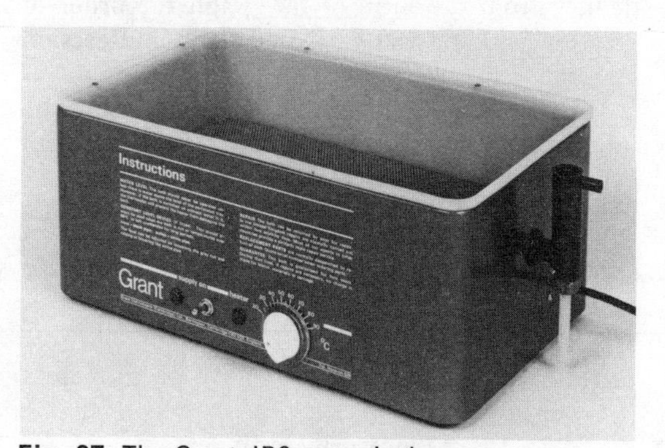

Fig. 27. The Grant JB2 water-bath.

Introductory study (see also p. 127). The reaction between sodium thiosulphate and nitric acid can be studied at different temperatures using a 500 cm³ beaker as a water-bath. Suitable concentrations to study are $\frac{1}{3}$M sodium thiosulphate with 0·5M nitric acid.

Follow the previous instructions, but suspend your 5 cm³ samples in the water-bath for five minutes before mixing so that they will be at the working temperature. Record the temperature of the water-bath, and carry out the experiment at four different temperatures up to 45°C.

Plot a graph of *1/time* (speed of reaction) against *temperature in K*; and a second graph of *log₁₀ (1/time)* against *1/temperature in K*.

What form do the graphs take? Can you produce the mathematical expression relating speed of reaction to absolute temperature?

Ester hydrolysis (see also p. 128). It is advisable to have already attempted this experiment at room temperature. The technique should have been mastered, and a result will already be available of rate of reaction at room temperature.

The hydrolysis should be studied at three temperatures between room temperature and 45°C using thermostatically controlled water-baths.

This experiment is best attempted on a class basis with different pairs of students for each temperature.

Place about 100 cm³ of 0·25M sulphuric acid in a labelled conical flask, and leave overnight in the water-bath to come to constant temperature.

Now follow the previous instructions. The ester sample, being small, need not be pre-heated, but the reaction mixture must be kept in the water-bath except for the initial shaking and should be well stoppered. If you are working at the higher temperatures try to take readings every five minutes. Record the temperature of your water-bath.

To reach equilibrium the reaction mixture should be left at room temperature for two days before the last aliquot is taken and titrated.

For each temperature plot a graph of *log₁₀* $(V_\infty - V_t)$ against *time* in minutes; this should be a straight-line graph whose slope is $(-k/2{\cdot}303)$.

k = first order rate constant at the particular temperature

$2{\cdot}303 = \log_e n/\log_{10} n$

Using the k values at the different temperatures, plot a second graph of $\log_{10} k$ against $1/temperature$ in K: this should be a straight-line graph whose slope is $(-E/2{\cdot}303R)$

E = activation energy of the reaction
R = gas constant ($8{\cdot}314$J mol^{-1} K^{-1})

The Arrhenius equation takes the form

$$k = Ae^{-E/RT}$$

On taking logarithms this becomes, since $\log_{10} A$ is constant:

$$\log_{10} k = \frac{-E}{2{\cdot}303R}\left(\frac{1}{T}\right) + \text{constant}$$

Calculate the value of E in kilojoules for the ester hydrolysis.

What values does E have for other reactions? Why are very large values not found for reactions known to take place? Consult a textbook for a full discussion of the significance of the Arrhenius equation.

7. A study of the manganate (VII)–oxalate reaction

In the titration of oxalates with standard potassium manganate (VII) the titration mixture is heated because the reaction is slow at room temperature. The reaction mechanism is complex but can be studied in part by this kinetics experiment.

The reaction was studied extensively as a school project: see J. Bradley and G. van Praagh, *J. Chem. Soc.* 1938, 1624. A more recent reference is S. J. Adler and R. M. Noyes, *J.A.C.S.*, 1955, **77**, 2036.

Requirements

Titration apparatus and a 10 cm³ pipette.
Measuring cylinders.
0·01M sodium thiosulphate.
0·02M potassium manganate (VII).
0·2M oxalic acid.
0·2M manganese (II) sulphate.
0·1M potassium iodide.
Starch indicator.
Stop-clock.

Prepare a reaction mixture according to Table 14, using measuring cylinders.

Add 50 cm³ of 0·02M potassium manganate (VII) and simultaneously start the stop-clock. Shake the reaction mixture for half a minute to mix well.

Table 14

Solution	*Experiment 1*	*Experiment 2*
0·2M oxalic acid	100 cm³	100 cm³
0.2M manganese (II) sulphate	none	15 cm³
2M sulphuric acid	5 cm³	5 cm³
water	95 cm³	80 cm³

After about a minute withdraw a 10 cm³ aliquot and run it into a conical flask. Note the time and quickly add about 10 cm³ of 0·1M potassium iodide. This stops the reaction and releases iodine equivalent to the residual oxidizing agents. Titrate the liberated iodine with 0·01M sodium thiosulphate, adding 1 cm³ of starch indicator near the end-point.

Remove further aliquots every three to four minutes and determine by the same procedure. Continue until the titre is less than 3 cm³.

Plot graphs of *titre* in cm³ against *time* in minutes and offer any explanation you can of their form.

CATALYSIS

There are many examples of reactants which only give products slowly because the reaction route has a high 'activation energy'. Catalysts are substances which participate in reactions opening new pathways of lower energy; the products are obtained faster and very little of the catalyst is used up in the reaction.

'Negative catalysis' is considered impossible for a reaction will not go by a slow route when a fast route is available. Any apparent examples of this phenomenon need an alternative explanation.

The situation is different in chain reactions because alteration of rate involves full participation in the reaction and the *sensitizer* or *inhibitor* will be quantitatively used up. Furthermore the main reaction pathway, the propagation steps, will remain unchanged.

The significance of catalysis is that, without the use of a catalyst, many major industrial processes would be uneconomic or impossible to realize while, in life-processes, the catalytic action of enzymes is crucial.

Test tube experiments

(*a*) Attempt to decompose 2M hydrogen peroxide by the addition of small portions of metal oxides (MnO_2, Cu_2O, PbO_2, ZnO, MgO). Do the oxides with catalytic power have any common features?

Repeat the experiment with a small piece of liver. Consult a textbook about the biochemical significance of this result.

(*b*) Mix equal volumes of 0·1M sodium oxalate and 2M sulphuric acid. Divide into two portions. Add 1 cm^3 of 0·02M potassium manganate (VII) and 1 cm^3 of 0·1M manganese (II) sulphate to one portion and potassium manganate (VII) alone to the other portion. Allow to stand.

This reaction is studied quantitatively in Reaction Kinetics Expt. 7, p. 131.

(*c*) Heat a small portion of potassium chlorate mixed with 5% by weight of manganese (IV) oxide. Be especially careful to use the correct substances. Note how readily oxygen is evolved.

Now heat a small portion of pure potassium chlorate and, when molten, add *a speck* of manganese (IV) oxide. Stop heating and allow to cool. Can you explain the colour?

(*d*) Mix 10 cm^3 of 0·5M potassium sodium tartrate (Rochelle salt) with 10 cm^3 of 2M hydrogen peroxide *in a 250 cm³ beaker*. Bring to the boil carefully and note any signs of a reaction. Stop heating and add 1 cm^3 of 0·1M cobalt (II) salt. Describe the result fully and identify some of the products.

Confirm that cobalt (II) ions do not change colour with either of the reagents separately.

(*e*) Dissolve a small portion of dimethyl oxalate in cold water and divide into three equal portions. Add 3 drops of 0·02M potassium manganate (VII) to each portion; there should be no change at this stage. Add equal volumes of water, 2M acetic acid and 2M sulphuric acid to separate portions and heat in a hot water-bath.

Note the acid catalysis of the hydrolysis of dimethyl oxalate.

This reaction is used in homogeneous precipitation (see p. 185).

(*f*) Dissolve a portion of sulphamic acid in cold water and divide into three equal portions. Add 0·1M barium chloride to each portion; there should be no change at this stage. Add equal volumes of water, 2M acetic acid and 2M hydrochloric acid to separate portions and heat in a hot water-bath.

Note the acid catalysis of the hydrolysis of sulphamic acid. Sulphamic acid is used as a titrimetric primary standard (see p. 156). Is this a valid use in view of its hydrolysis? State your reasons in quantitative terms.

(*g*) Mix equal volumes of 2M carbamide and 2M sodium hydroxide and divide into two portions. Add a small portion of urease-active meal (an enzyme material) to one of the portions. Wipe the test tube mouths free of sodium hydroxide and place across each mouth a piece of moist red litmus paper. Heat the tubes in a hot water-bath, with the mouths well separated.

Write the equation for this hydrolysis.

This reaction has been studied in acidic media by G. B. Kistiakowsky and R. Lumry (*J.A.C.S.*, 1949, **71**, 2006), who found that the activation energy for the catalysed reaction is

37 kJ mol^{-1}, and for the uncatalysed reaction is 137 kJ mol^{-1}.

(*h*) Styrene is readily polymerized by suitable catalysts. Anhydrous tin (IV) bromide (for preparation, see p. 31), anhydrous aluminium chloride (for preparation, see p. 40; open stock bottles with caution), and lauroyl peroxide are suitable catalysts. Carry out this experiment *in a fume cupboard*.

Mix 5 cm^3 of styrene and 5 cm^3 of 1,2-dibromoethane in a 150 mm × 25 mm test tube and add roughly 0·5 g of a catalyst (weighed quickly). Plug the mouth of the tube with cotton-wool and heat in a hot-water bath for twenty minutes. Allow to cool, dilute with 10 cm^3 of propanone and decant from any solids. Precipitate the polystyrene by adding an excess of ethanol. Compare the physical nature of your products.

This experiment could also be carried out in the apparatus for distillation, with addition, as in Fig. F. After the heating, add 15 cm^3 of 2M hydrochloric acid to dissolve the metal salts, and steam-distil out the solvent and unchanged styrene. At the conclusion of the distillation, empty your flask while still hot and clean with a little propanone.

(Based on an experiment by D. M. Stebbens in the A.S.E. booklet *Tested Experiments for Grammar Schools*.)

ELECTROCHEMISTRY

1. Determination of the Faraday constant

The Faraday constant is the amount of electric charge per mole of electrons, and its value can be estimated by the electrolysis of compounds. A simple example is the electrolysis of copper sulphate solution between copper electrodes. The cathode reaction is:

$$Cu^{2+} + 2e^- \longrightarrow Cu$$

thus 2 moles of electrons are required to deposit 1 mole of copper.

Requirements

Electrical equipment, as in Fig. 28.
Copper foil.

The electrolyte is prepared by dissolving roughly 12·5 g of copper (II) sulphate pentahydrate in 92 cm³ of pure water and adding 3 cm³ concentrated sulphuric acid plus 5 cm³ of ethanol.

The electrodes consist of copper foil (6 cm × 2 cm) with a narrow tongue at one end for making electrical connections.

Clean the electrodes by dipping briefly into 2M nitric acid. Then rinse well with pure water and ethanol before drying in an oven at 100°C. After cooling the cathode, weigh it accurately to the nearest milligram.

Set up the apparatus according to the circuit diagram (Fig. 28), and have it checked before commencing the experiment. The variable resistor should be set to have maximum resistance in the circuit. Place the electrolyte in a 150 cm³ beaker and dip in the electrodes. Hold the electrodes in place by bending their tongues down outside the beaker.

When all is ready, note the precise time and switch on. Quickly adjust the variable resistor so that the current flowing is about 0·2 A. The resistance is adjusted as necessary to maintain constant current throughout the experiment. Record the current value. After about forty minutes switch off the current, noting the precise time.

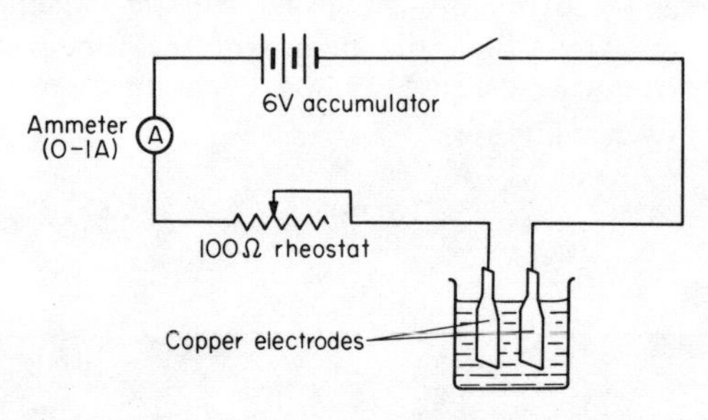

Fig. 28. Circuit diagram for the determination of the Faraday constant.

Carefully remove the cathode, rinse with pure water and ethanol, then dry. After cooling the cathode weigh it accurately to the nearest milligram.

From the relationship

$$\frac{\text{atomic mass of copper}}{2 \times \text{Faraday constant}} = \frac{\text{mass of copper deposited}}{\text{charge passed}}$$

calculate the value of the Faraday constant in *amp-hours mol⁻¹*, and also in *amp-seconds* (*coulombs*) *mol⁻¹*.

From X-ray crystallographic studies the Avogadro constant (L) is given the value $6{\cdot}02 \times 10^{23}$ mol^{-1}; hence calculate the charge on the electron in attocoulombs (10^{-18}C).

2. Electrode potentials

The e.m.f. of a cell is the potential of two combined 'half-cells', but it is not possible to measure the potential of an isolated half-cell. However, a series of different half-cells can be compared by measuring the e.m.f. of cells which have one half-cell in common.

Standard measurements are made using a hydrogen half-cell, $Pt[H_2(g)] \mid 2H^+(aq) \,\vdots$ in specified conditions but for these experiments it will be more convenient to use the saturated calomel electrode $Hg, Hg_2Cl_2 \mid$ sat. $KCl(aq) \,\vdots$ as the common half-cell. After the experiments the saturated potassium chloride of the calomel electrode should be replaced to eliminate any contamination.

The preparation of a hydrogen electrode is described in Appendix I for the benefit of those who wish to use one.

One valve voltmeter and calomel electrode will serve for a group of students.

Requirements

Valve voltmeter (see Fig. 29).
Saturated calomel electrode (or hydrogen electrode).
Metal foil electrodes (Zn, Cu, Mg).
1·0M metal ion solutions ($ZnSO_4$, $CuSO_4$, $MgSO_4$).
25 cm^3 beakers.

Set up the equipment using a piece of metal foil dipping into a 1·0M solution of its sulphate in a 25 cm^3 beaker as one half-cell. Complete the cell by dipping the calomel electrode into the sulphate solution and measure the e.m.f. of the cell by connecting directly to a valve voltmeter after checking the zero adjustment. Record the polarity (+ or −) of the metal foil electrode and the voltage. Repeat for the various metal foil/metal ion solutions available.

Now set up the cell Cu | Cu^{2+}(aq) ¦ Zn^{2+}(aq) | Zn in two 25 cm^3 beakers using as salt bridge between the two beakers a filter paper strip moistened with saturated potassium chloride solution. Record the polarity of the right-hand metal foil electrode and the voltage of the cell. Repeat for all other metal foil pairs that can be arranged.

Do the results suggest, within the limits of these simple approximate experiments that there is a simple relationship between the e.m.f. of cells? For example can the e.m.f. of the Cu | Cu^{2+}(aq) ¦ Zn^{2+}(aq) | Zn cell be deduced from the e.m.f. of the cells calomel ¦ Cu^{2+}(aq) | Cu and calomel ¦ Zn^{2+}(aq) | Zn?

3. Redox potentials

In the section on redox reactions (page 55) it is suggested that redox reactions involve electron transfer. The energy change involved can be measured as an e.m.f. if the reaction can be arranged as a voltaic cell. Thus the electron transfer

$$Fe^{2+} \longrightarrow Fe^{3+} + e^-$$

can be set up as a half-cell by mixing an iron (II) and an iron (III) salt and using an indicator electrode of unreactive platinum

$$¦\ Fe^{3+}(aq),\ Fe^{2+}(aq)\ |\ Pt$$

To complete the voltaic cell a saturated calomel electrode can be used as the other half-cell and the e.m.f. measured using a valve voltmeter.

One valve voltmeter and calomel electrode will serve for a group of students.

Requirements

Valve voltmeter (see Fig. 29).
Saturated calomel electrode.
2 platinum electrodes.
0·1M iron (II) salt.
0·1M iron (III) salt.
0·1M iodine in potassium iodide.
25 cm^3 beakers.

Set up the equipment using as one half-cell a piece of platinum foil dipping into a 25 cm^3 beaker containing a mixture of approximately equal volumes of 0·1M iron (II) and iron (III) salts. Provided the same mixture is used for all measurements its exact composition does not matter in this experiment. As the other half-cell use a saturated calomel electrode and measure the e.m.f. of the complete cell by the procedure of Electrochemistry Expt. 2 above.

Fig. 29. The Heathkit valve voltmeter.

Next set up and measure the e.m.f. of the cell

calomel ¦ I_2(aq), $2I^-$(aq) | Pt

using a solution of iodine in potassium iodide solution.

From the two measurements of e.m.f. what e.m.f. do you predict for the cell?

Pt | Fe^{2+}(aq), Fe^{3+}(aq) ¦ I_2(aq), $2I^-$(aq) | Pt.

What sign (+ or −) do you predict for the voltage? By convention *the polarity of the right-hand electrode of the cell as written down* is taken as the sign of the voltage for the cell.

Test your prediction by connecting together the already prepared half-cells with a filter paper strip moistened with saturated potassium chloride solution, and measuring the e.m.f. of the complete cell.

Consult a textbook about the measurement of standard redox potentials and the Nernst Equation. To study the change of e.m.f. during a reaction carry out Electrochemical titrations Expt. 2, page 180.

4. Conductivity: introductory study

The measurement of solution conductivity is of interest since the method is very sensitive, enabling the nature of solutes at very low concentrations to be investigated.

The actual measurement is of *resistance* using a Wheatstone bridge circuit; alternating current of at least 1000 c.p.s. is necessary to avoid electro-

Fig. 30a. The Grayshaw conductivity bridge.

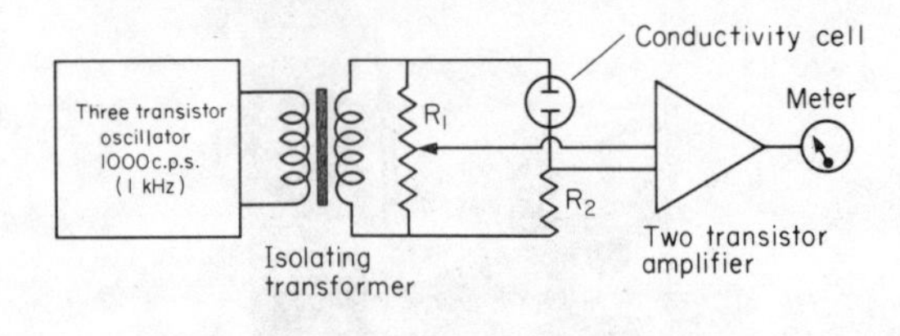

Fig. 30b. Circuit diagram of the Grayshaw conductivity bridge.

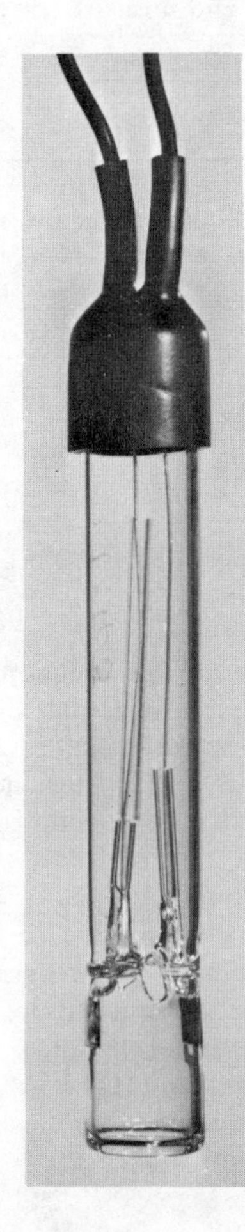

Fig. 31. The M.E.L. conductivity cell.

lysis of the solute, and electrodes coated with 'platinum black' must be used for the same reason. A commercial electronic conductivity bridge in conjunction with a dip-type electrode cell will be found most convenient.

Because of the sensitivity of the method, particular attention should be paid to cleanliness of apparatus and purity of chemicals; very pure water (conductance water) is used to prepare solutions which can be stored in polypropylene bottles. Conductance water can be prepared by running distilled water through a column of mixed anion–cation exchange resins, such as Permutit 'Biodeminrolit' (see Appendix I).

Temperature control is also important because of the rapid variation of the viscosity of water with temperature. The standard temperature for electrochemical measurements is 25°C. A thermostat set to operate at 25°C should be used; if this is not available, work at room temperature and make corrections as suggested in the text.

Requirements

Grayshaw CT50 conductivity bridge (see Fig. 30a).
M.E.L. dip cell type E7591/B (see Fig. 31).
Conductance water.

For these experiments use ordinary laboratory solutions, but rinse the dip cell thoroughly in conductance water between each experiment. Work at room temperature.

(*a*) Measure the conductivity, in $ohms^{-1}$ (1/resistance), of fresh conductance water, distilled water and tap water.
(*b*) Measure the conductivity of hydrochloric acid of the following concentrations prepared by approximate dilution: 0·001M, 0·01M, 0·1M. If the weakest solution is used first, it is not necessary to rinse the dip cell between readings.
(*c*) Repeat (*b*) using acetic acid.
(*d*) Repeat (*b*) using magnesium sulphate.
(*e*) Measure the conductivity of 0·01M hydrochloric acid at approximately 10°C, 20°C, 30°C, 40°C.

Wash out the dip cell *thoroughly* with conductance water before putting away.

Plot your results as graphs of *conductivity* against *concentration*. Interpret the graphs.

5. Determination of cell constant

The electrolytic conductivity of potassium chloride solution has been determined absolutely for a wide range of concentrations using cells of known dimensions. The results can be used to determine the effective dimensions (cell constant) of cells of unknown size.

Requirements

Conductivity bridge and dip cell.
Water-bath at 25°C.
Hard-glass test tubes (150 mm × 25 mm).
A.R. potassium chloride.
Standard flask. Conductance water.

Table 15

Concentration of potassium chloride	*Electrolytic conductivity/ohm^{-1} m^{-1} at 10°C*	*at 18°C*	*at 25°C*
0·1M	0·93	1·12	1·29
0·02M	0·200	0·240	0·277

For all but a new cell, the cell constant will have been determined previously. So for demonstration purposes the experiment can be carried out quickly at room temperature if standard potassium chloride solutions are available.

Prepare standard solutions of potassium chloride of exact concentration by accurate weighing: 0·1M requires 7·456 g l^{-1} of KCl; 0·02M requires 1·491 g l^{-1} of KCl. The solutions must be made up in conductance water.

Clean four hard-glass test tubes by boiling conductance water in them. Dry the test tubes, add two samples of each potassium chloride solution and clamp in a water-bath at 25°C.

Rinse the dip cell in samples of conductance water until a minimum conductivity reading is obtained, showing the dip cell is clean. Then rinse the dip cell in 0·02M potassium chloride and clamp one of the 0·02M samples in the water-bath. Take readings of conductivity every half minute until constant. Repeat the readings in the second sample as a check on any accidental contamination.

Repeat the procedure to determine the conductivity of 0·1M potassium chloride.

Finally, rinse the cell well with conductance water and put away carefully.

Calculate the cell constant for both concentrations of potassium chloride using the relationship:

$$\kappa = 1/r \times l/a$$

where κ = electrolytic conductivity of potassium chloride in $ohm^{-1}\ m^{-1}$

$1/r$ = conductivity in ohm^{-1}

l/a = cell constant, l is length in m, a is area of cross-section in m^2

Remember to subtract the conductivity of conductance water from the readings if significant.

6. Conductivity of strong electrolytes

The variation of conductivity with concentration is studied quantitatively in this experiment using a salt of high conductivity. The results are expressed as *molar conductivities* (Λ $ohm^{-1}\ m^2\ mol^{-1}$) which are equal to the electrolytic conductivity of one mole of the salt at concentrations of M moles per litre (litre = dm^3 = $10^{-3}\ m^3$).

$$\Lambda = l/a \times 1/r \times 1/1000M$$

where l/a is the cell constant in m^{-1}

$1/r$ is the conductivity in ohm^{-1}

$1/1000M$ is the dilution in $m^3\ mol^{-1}$

The experiment may be carried out at room temperature or in a water-bath at 25°C.

Requirements

Conductivity bridge and dip cell.
Water-bath at 25°C (or 600 r.p.m. stirrer).
Hard-glass test tubes 150 mm × 25 mm (or 100 cm³ beaker).
10 cm³ and 50 cm³ burettes.
Standard 0·1M potassium sulphate in conductance water.
Conductance water.

To carry out the experiment at room temperature, set up the apparatus as for a conductometric titration according to Fig. 45 on p. 181. Insert a thin piece of expanded polystyrene sheet between the stirrer and beaker to prevent heat transfer.

Place exactly 40 cm³ of conductance water in the beaker and take a reading of conductivity. Add standard 0·1M conductivity potassium sulphate from the 10 cm³ burette and take conductivity readings when the following total volumes have been added (change to the 50 cm³ burette when appropriate):

Table 16

Total volume of potassium sulphate added/V cm³
1·0
2·5
5·0
10·0
20·0
30·0
40·0

Record the temperature of the solution.

To carry out the experiment at 25°C, clean seven hard-glass test tubes by boiling conductance water in them. Dry the test tubes, add samples according to Table 17, cork and invert to mix, and clamp in a water-bath at 25°C.

Take readings of conductivity, starting with the weakest solution and shaking the dip cell as dry as possible when transferring between solutions. For each sample take readings every half minute until constant. Do not remove the tubes from the

Table 17

0·1 M potassium sulphate /cm³	Conductance water/cm³
none	40
1·5	38·5
3·0	37·0
6·0	34·0
10·0	30·0
20·0	20·0
40·0	none

water-bath, and leave the cell dipping in the solution between readings.

Record the temperature of the water-bath.

Calculate the concentration (M) and molar conductivity (Λ) for each solution; remember to subtract the conductivity of the conductance water from the readings if significant.

Plot a graph of *molar conductivity* against *concentration*, and also plot *molar conductivity* against the *square root of concentration.*

From your graphs determine a value for molar conductivity at infinite dilution (Λ^∞). Does potassium sulphate behave as a strong electrolyte over the concentration range studied? Consult a textbook about the Debye–Hückel theory.

Reference: Clews, C. J. B., *Proc. Phys. Soc.*, 1934, **46**, 764.

7. Conductivity of weak electrolytes

Some acids and bases, especially organic, have very low conductivities and their solutions contain undissociated molecules in equilibrium with the ions. For very weak electrolytes, the dissociation constant (K) can be determined from molar conductivity values (Λ) using the relationship:

$$\Lambda = \Lambda^\infty K^{\frac{1}{2}}/M^{\frac{1}{2}}$$

This experiment can be carried out using acetic acid at room temperature or 25°C:

Table 18

Λ^∞ for acetic acid /ohm⁻¹ m² mol⁻¹	Temperature/°C
0·260	0
0·350	18
0·391	25

Requirements

As for Electrochemistry Expt. 6 opposite.

Standardized 0·1M acetic acid in conductance water.

This experiment is carried out according to the procedure of Expt. 6 opposite, using 0·1M acetic acid instead of 0·1M potassium sulphate.

Calculate the concentration (M) and molar conductivity (Λ) for each solution; remember to subtract the conductivity of the conductance water from the readings if significant.

Plot a graph of *molar conductivity* against 1/*square root of concentration.* According to the relationship in the introduction this should be a straight-line graph:

$$\text{slope} = \Lambda^\infty K^{\frac{1}{2}}$$

Using the appropriate Λ^∞ value, determine the dissociation constant (K) of acetic acid at the temperature of your experiment.

Derive the relationship in the introduction. Consult a textbook about Ostwald's Dilution Law. How is Λ^∞ determined for acetic acid?

Reference: MacInnes, D. A. and Shedlovsky, T., *J.A.C.S.*, 1932, **54**, 1429.
This paper is an excellent example of the care necessary when making reliable electrochemical measurements.

RADIOCHEMISTRY

1. Introduction

In this section a knowledge of radioactivity and the mode of operation of Geiger–Muller tubes and scalers is assumed. A textbook which may be consulted is *Nuclei and Radioactivity* by G. R. Choppin.

This introduction deals with counting methods and corrections using a G.M. tube and scaler (see Fig. 32), and also with the experimental procedures necessary for safety.

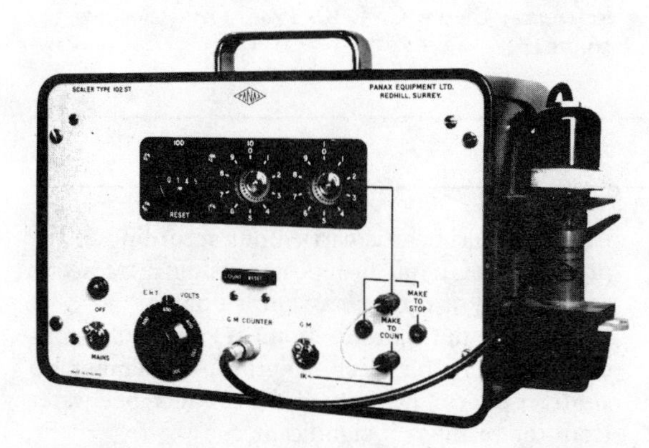

Fig. 32. The Panax 102ST scaler.

Work with radioactive substances in schools must conform to the requirements of the Government department concerned. For example, in England it is the Department of Education and Science's *Administrative Memorandum* 1/65 (or its later editions).

The special apparatus required, and suppliers, are listed in Appendix III.

Counting methods and corrections

The counting method for a particular experiment will be based on two considerations: the half-life of the source, and its activity.

If the count rate is to be determined every minute, a useful procedure is to have the scaler ON for $\frac{2}{3}$ of each minute, then OFF for $\frac{1}{3}$ of a minute to record the number of counts. The cycle is repeated the next minute, the stop-clock of course being allowed to run continuously. The ON–OFF cycle should be not greater than one half-life of the source, and preferably a great deal less.

Because disintegration is a random process, measurements are subject to statistical variations. Of the results recorded for a source of constant activity, 68·3% will be within the 'standard deviation' from the mean value. Thus the possible percentage error in the number of counts recorded will depend on the total recorded (see Table 19).

Therefore counting times are chosen as far as possible to give a total count which will ensure good accuracy. A source of 10 000 c.p.m. would give reasonable accuracy if counted for one minute.

When the count rate has been determined with the best possible accuracy, it must be corrected for 'background' and 'resolving time'.

A background activity exists as a result of natural materials and fall-out, and contamination may add to this. The background activity should be low, and it is determined by a minimum five-minute count. It must be subtracted (in c.p.m.) from the recorded activity.

The resolving time of a G.M. tube and scaler is

Table 19

Number of counts recorded/n	Standard deviation $\sigma \simeq \sqrt{n}$	Possible per cent error in (n) /%
50 000	223	0·45
10 000	100	1
2 000	45	2·25
400	20	5
100	10	10
16	4	25

Table 20

Recorded activity /c.p.m.	*Per cent correction to be made/%*	*Correction /c.p.m.*	*Corrected activity /c.p.m.*
1 000	0·5	5	1 005
2 000	1·0	20	2 020
5 000	2·5	125	5 125

approximately the period of time the tube is inactive after a 'pulse' has been produced due to an ionizing ray. During this time other ionizing rays will not be recorded, so the counts recorded will always be less than they should be.

For a resolving time of 300 microsecond (μs), an approximate correction of 0·5% per 1000 c.p.m. can be made (see Table 20).

Experimental procedures

Experimental work with processed uranium and thorium salts involves a very low radiological health hazard (see Table 21), but no student should receive a dose of more than 50 m rem in any one year as a result of school experiments of all types.

It is suggested that the radiochemical experiments in this book be carried out taking the basic precautions necessary for all open source work:

1. Special laboratory coats should be reserved for radiochemical work and laundering carried out separately from other clothes.

2. Gloves, either disposable polythene or rubber, must be worn to prevent contamination of the skin. Discomfort is avoided by dusting the hands with talc. Use tongs or tweezers to pick up active materials, e.g. planchettes.

3. No mouth operations are permitted. Safety pipettes and squeeze wash-bottles should be available. Labels must be moistened with a wash-bottle.

4. Confine all operations, except counting, to a *fume cupboard* which should be thoroughly cleaned and monitored before being returned to conventional use.

5. Apparatus should be reserved exclusively for radiochemical work. An adhesive tape is available to mark the apparatus.

6. Double containment is used for all active chemicals to avoid the spread of contamination by breakage or spillage. Stock bottles must be stored in metal cans, and all chemical manipulations carried out in a large enamel tray which is lined with bitumen-interleaved paper. Active chemicals and apparatus are kept inside the tray, inactive materials outside (see Fig. 33).

   Spillages should be mopped up with paper tissues.

7. Special containers are necessary for active waste. Disposal of low activity waste of the type resulting from school experiments presents no problem; solid waste can be placed in ordinary dustbins, liquid waste washed down a sink with ample water.

8. Inform your local Fire Brigade and local Health Authority that you are using radioactive materials.

Table 21

Situation	*Source*	*Dose rate*
Bristol, U.K.	Background, limestone	39 m rem/year
St. Ives, U.K.	Background, granite	120 m rem/year
Kerala, India	Background, monazite sand	2 000 m rem/year
10 cm from glass container	500 g thorium nitrate	3·5 m rem/hour
10 cm from glass container	250 g uranyl nitrate	0·8 m rem/hour

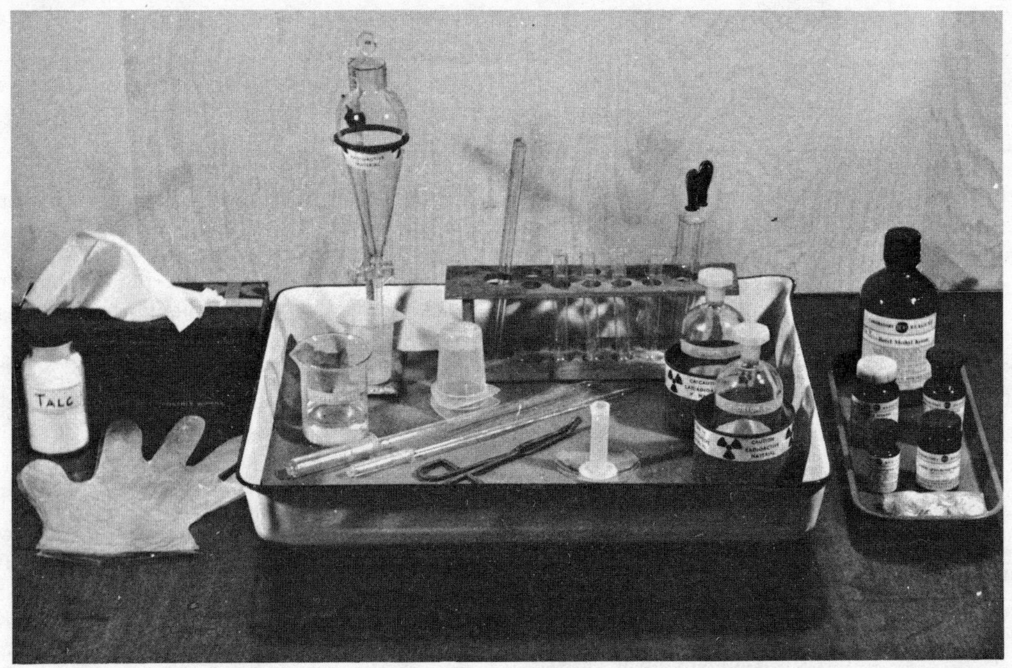

Photograph: K. N. Bawtree

Fig. 33. Layout for radiochemistry.

2. Autoradiography of minerals

The effect of radiation from radioactive materials on photographic plates has a classic place in the discovery and investigation of radioactivity, and the technique remains important in both medical and industrial examinations.

Table 22

Radioactive minerals
Autunite
Davidite
Monazite
Phosphouranylite
Pitchblende
Torbernite

Radioactive minerals are readily purchased (see Appendix III).

Requirements

X-ray film such as Kodirex folder-wrapped 125 mm × 175 mm.
Lead-backed X-ray exposure holder.
Darkroom with safelight filter such as Wratten 6B (brown).
X-ray developer and fixer (see text).
Radioactive minerals of uranium and thorium.

Working in total darkness or suitable safelight, place a sheet of fast direct-type X-ray film in a lead-backed X-ray exposure holder. Take care to fold the holder correctly so that all light will be excluded. The experiment can now be continued in daylight.

Place the exposure holder where it need not be disturbed for at least a day. Wearing polythene gloves (for protection against contamination by solid material), put three mineral specimens and one salt specimen in a glass vial directly on top of the holder. Note the time. A drawing of the shapes and relative positions of the specimens should be made.

After a day remove the specimens, and note the total exposure time. Discard the polythene gloves into the special container for radioactive waste. The film can be developed when convenient.

To make 1 litre of X-ray developer: dissolve in pure water 2·2 g of metol (*p*-methylaminophenol sulphate) and 72 g of anhydrous sodium sulphite. Then add and dissolve 8·8 g of quinol, 48 g of anhydrous sodium carbonate and 4·0 g of potassium bromide. Dilute to 1 litre.

To make 1 litre of X-ray hardening fixer: dissolve in 750 cm³ of pure water at 35°C 12·5 g of 'chrome alum' [chromium (III) potassium sulphate], 12·5 g of anhydrous sodium metabisulphite and 6·25 g of anhydrous sodium sulphite. Then add and dissolve 400 g of sodium thiosulphate. Dilute to 1 litre and use within a few days.

Work in total darkness or suitable safelight while developing and fixing the film.

Totally immerse the film in about 250 cm³ of X-ray developer and gently agitate while held by a suitable clip (as the emulsion is soft when wet). Continue the agitation for five minutes. Then transfer the film, after allowing it to drain, to about 250 cm³ of *fresh* X-ray fixer, and similarly agitate for five minutes. Conclude the experiment in daylight by soaking the film in a large volume of fresh water for at least ten minutes. Leave the film overnight to dry.

Write a full description of the effect of each mineral specimen. Consult a textbook about the relative effect on photographic plates of the different radiations from radioactive materials.

3. Thorium stars in photographic emulsion

Radium-224 may be separated from thorium compounds using a barium carrier. If photographic plates are exposed to such a radium solution, then the successive α-particles emitted in the decay chain (see Table 24) will make a four-pointed track in the photographic emulsion.

The bismuth-212 decays by two methods but the β–α decay leads to the emission of an α-particle whose energy is 20–25% greater than any other α-particle of the decay chain. Thus the star caused by decay by this route will have one exceptionally long track.

Stars with less than four tracks are due to the fact that development of the plate has taken place before those particular atoms had completely decayed.

A few stars may be found with only three concurrent tracks: the start of the first track is a fraction away due to radon diffusing through the emulsion before decaying to continue the star.

The separation of radium-224 is not essential, and thorium salts can be used without pre-treatment.

Nuclear emulsion plates are unstable; they should be used within two months of arrival, meanwhile being stored in a refrigerator.

Requirements

Nuclear emulsion plates (Ilford KO emulsion, thickness 50 μ, 75 × 25 mm plates).
Thorium nitrate.
Developer (as previous experiment).

Dissolve 0·05 g of thorium nitrate and 0·01 g of barium nitrate in about 15 cm³ of water. Add a few drops of '0·880' ammonia to this solution, boil for a few minutes, and filter off the thorium hydroxide precipitate using a porosity 3 sintered-glass crucible.

The filtrate contains the soluble hydroxides of barium and radium. Boil the filtrate gently to expel excess ammonia and make it faintly acid with 2M nitric acid.

Carry out the remainder of the experiment in total darkness or suitable safelight.

Soak pieces of nuclear emulsion plate in the acidified filtrate for fifteen minutes, then drain and allow to stand in the dark for about three days. It is important when unwrapping the plates to identify the emulsion side correctly. The plates are wrapped in groups of four with spacers, so that only the glass makes contact with the wrapping. When held near the safelight the glass will give a sharp reflection, the emulsion a diffuse reflection. When treating the plates ensure that the emulsion side is uppermost.

Develop the plates using the developer (diluted 1 : 1 water) described in the previous experiment. Soak the plates in pure water for ten minutes, transfer to developer for fifteen minutes, then wash in running water for ten minutes.

Soak in fixer of 30% sodium thiosulphate for 50% longer than the *clearing time* (the time taken for the initially opaque plates to become transparent). The plates can be examined in daylight now. Finally, wash the plates in running water for one hour and allow to dry overnight in a warm room.

Mount in the usual manner for a microscope slide, and examine with a magnification of about ×450.

4. Decay products of uranium

Chemical reagent uranium salts have been so processed that virtually the only uranium isotope present is $^{238}_{92}U$. This is a long-lived isotope and will therefore establish a 'secular' equilibrium with its daughter products (see Table 24).

Thus four months are needed, after the preparation of a uranium salt, to establish equilibrium as far as protactinium-234 (five half-lives of the daughters) but, beyond uranium-234, equilibrium can only be expected in unprocessed minerals.

These experiments study the separation of protactinium-234 by solvent extraction and ion-exchange. Carry out the experiments with the full precautions detailed in the introduction to Radiochemistry, p. 141.

Requirements

Scaler and G.M. liquid counter.
0·6M uranyl nitrate.
2-Methylpentan-4-one (isobutyl methyl ketone).
Stop-clock.

Solvent extraction of protactinium-234. Determine the background count by filling the liquid counter with pure water and counting for five minutes. Record the result in counts per minute (c.p.m.).

Place 15 cm³ of 2-methylpentan-4-one, 15 cm³ of 0·6M uranyl nitrate and 35 cm³ of concentrated hydrochloric acid in a separating funnel, and shake well for two minutes. Only the protactinium is extracted into the organic layer.

Without delay run the lower, aqueous, layer into a beaker, and run the organic layer into a liquid counter. Start counting as soon as possible; let the stop-clock run continuously and record the results as in Table 23.

Plot graphs of *count rate* against *mid-time of count*, and *log*$_{10}$ (*count rate*) against *mid-time of count*, after correcting the count rate for background and resolving time. What deductions can be made from the graphs?

The uranyl solution and organic solvent should be placed in residue bottles and recovered for re-use.

Growth of protactinium-234. Place 50 cm³ of 2-methylpentan-4-one, 4·5 cm³ of 0·6M uranyl

Table 23

Counting period	No. of counts	Mid-time of count/sec	Count rate /c.p.m.
0–20 sec		10 sec	
30–50 sec		40 sec	
60 sec–1 min 20 sec		70 sec	
1 min 30 sec–1 min 50 sec		100 sec	
etc.			

nitrate and 11·5 cm^3 of concentrated hydrochloric acid in a separating funnel.

Proceed as before, but run the lower, aqueous, layer into the counter and start counting *immediately*. Take 20-second counts every thirty seconds (see Table 23).

Plot a graph of *count rate* against *mid-time of count*. Interpret your result; only the energetic β particles from protactinium-234 decay can penetrate the glass walls of the Geiger tube and be counted.

Ion-exchange separation of protactinium-234

Additional Requirements

Ion-exchange resin (Zeo-Karb 225–SRC 13).
Chromatography column (355 mm × 14 mm).

Slurry 5 g of a strong cation-exchange resin, in hydrogen form, with water and set up a column as described on p. 108.

Pass through the column, at 5 $cm^3\ min^{-1}$, 15 cm^3 of 0·6M uranyl nitrate diluted with 15 cm^3 of pure water. All the cations will be adsorbed.

Now elute the uranyl cations by passing through the column 150 cm^3 of 0·05M sulphuric acid. Rinse the column with 50 cm^3 of pure water. The column now contains thorium-234 in transient equilibrium with protactinium-234, and can be 'milked' for protactinium-234 at ten-minute intervals over several half-lives of the parent thorium-234.

During one minute, pass a 15 cm^3 portion of 2M hydrochloric acid through the column (low activity material is eluted and possibly some uranium), followed by a sufficient portion to fill the G.M. liquid counter.

Start counting immediately, taking 40-second counts every sixty seconds. Alternatively, record the cumulative count every thirty seconds.

After seven minutes there will be a residual activity due to other isotopes. Count for five minutes to determine a 'background' activity.

Plot a graph of log_{10} (*count rate*) against *mid-time of count*, after correcting the count rate for background and resolving time. Deduce the half-life of protactinium-234.

Table 24

Decay sequence from uranium-238	*Decay sequence from thorium-228*
$^{238}_{92}U$ (half-life 4·5 × 10^9y)	$^{228}_{90}Th$ (half-life 1·9y)
↓ α	↓ α
$^{234}_{90}Th$ (24·5d)	$^{224}_{88}Ra$ (3·64d)
↓ β	↓ α
$^{234}_{91}Pa$ (1·14m)	$^{220}_{86}Rn$ (54·5s)
↓ β	↓ α
$^{234}_{92}U$ (2·69 × 10^5y)	$^{216}_{84}Po$ (0·158s)
↓ α	↓ α
further products	$^{212}_{82}Pb$ (10·6h)
	↓ β
	$^{212}_{83}Bi$ (60·5m)
	↙ α ↘ β
	(3·1m) $^{208}_{81}Tl$ $^{212}_{84}Po$ (3 × 10^{-7}s)
	β ↘ ↙ α
	$^{208}_{82}Pb$

5. Decay products of thorium

The separation of thorium from its minerals gives chemical reagent thorium salts containing $^{232}_{90}Th$ and $^{228}_{90}Th$ isotopes.

The decay series is shown in Table 24; it will take about $5 \times 1{\cdot}9$ years for the transient equilibrium based on thorium-228 to die out and $5 \times 6{\cdot}7$ years to re-establish the secular equilibrium of radium-228. Experiments can be carried out using thorium salts in any state of equilibrium, although 'vintage' salts will give higher activities.

These experiments involve the separation and half-life determination of members of the thorium-228 decay series.

Fig. 34a. The Stephen sensitive ionization chamber dosimeter.

Separation of radon-220

Requirements

A sensitive ionization chamber dosimeter (see Fig. 34).
A 'radon bottle'.
Two 120-volt dry batteries.
Stop-clock.

Connect the dosimeter to a suitable D.C. voltage so that when the charge button is pressed the scale wire is brought to the zero mark (see manufacturer's leaflet). The charge should not leak more than very slowly; study the scale so that you understand the graduations.

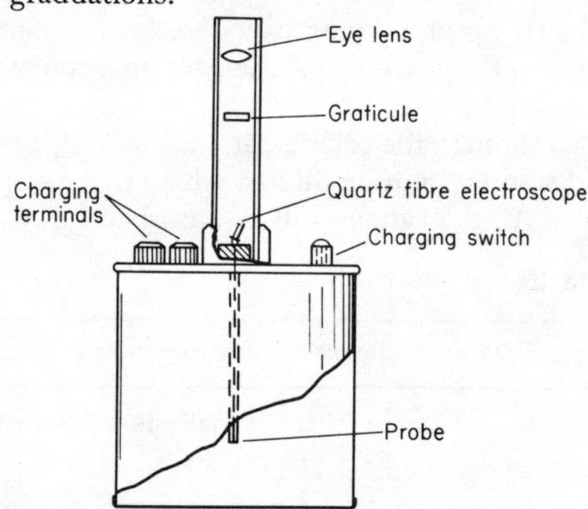

Fig. 34b. Dosimeter (sectional).

Now quickly puff radon gas from the bottle of 'vintage' thorium hydroxide into the ionization chamber and replace the bottom of the container.

Press the charge button and start the stop-clock simultaneously; note approximately how long is required for charge to leak away equivalent to a full-scale deflection.

Repeat the procedure until just longer than twenty seconds is needed. Readings may now be taken.

Take the reading of the scale deflection, using the same timing procedure as for protactinium in the previous experiment, i.e. allow the stop-clock to run continuously, press the charge button every thirty seconds, and take a reading of scale deflection exactly twenty seconds later.

The experiment can be repeated after five minutes, when secular equilibrium will have been re-established in the radon bottle.

Plot a graph of *log*$_{10}$ (*scale deflection*) against *mid-time of reading*, correcting for leakage if necessary, and determine the half-life of radon gas.

Explain how the decay of radon gas allows the dosimeter charge to leak away.

Separation of lead-212 and bismuth-212

Requirements

Scaler and G.M. end window counter.
G.M. castle.
Thorium 'cow'.
120-volt dry battery.
Nickel and aluminium planchettes.

A thorium 'cow' must be established two weeks before use. Slowly add excess 2M ammonia to 150 cm^3 of 0·2M thorium nitrate, collect the precipitate of thorium hydroxide by filtration at the pump, wash with water and dry overnight in an oven at 100°C. Grind the dry product to a fine powder using a pestle and mortar.

Set up a 'cow' according to Fig. 35, and leave for two weeks. For success high humidity must be maintained in the 'cow'. The delay is necessary because radium-224 is not co-precipitated with the thorium hydroxide and it must be allowed to grow back into equilibrium.

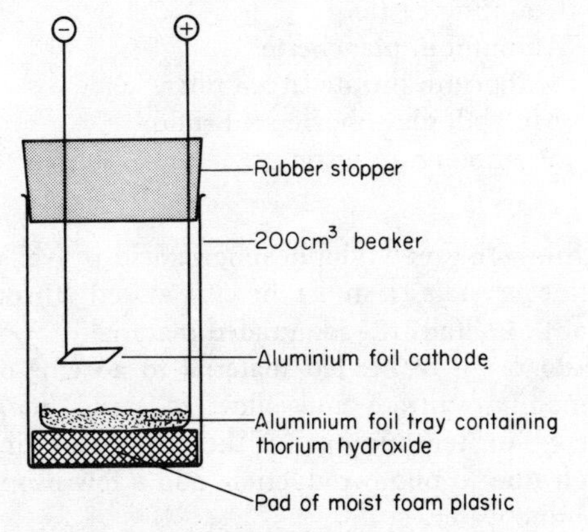

Fig. 35. A thorium 'cow'.

The process of collection is based on 'alpha recoil'. Radon gas escapes from the surface of the finely divided thorium hydroxide and decays in the atmosphere to polonium-216, which in turn rapidly decays by α emission. Now the emission of the α particle results in the simultaneously formed lead-212 nucleus having a recoil momentum. As a result, the lead-212 nucleus loses some of its orbital electrons and is ultimately produced as a positively charged ion. Thus the lead-212 can be collected on the aluminium cathode.

After establishment the 'cow' may be milked every two days until the surface of the thorium hydroxide has aged.

Determine the background count by counting for ten minutes and record the result in counts per minute.

Meanwhile remove the aluminium foil cathode and replace with a fresh piece of foil.

Place the foil cathode in a test tube with 1 cm^3 of water, and add 2 drops of 2M nitric acid.

Boil gently for one minute to dissolve tracer amounts of lead-212 and bismuth-212, then add 1 drop of 0·1M bismuth nitrate to act as carrier.

Transfer the solution, using a dropping tube, to a clean nickel planchette which has been placed on a watch-glass. Warm on a water-bath, or by radiant heat, until a black deposit of bismuth is seen.

Transfer the solution from the planchette to a centrifuge tube, and treat the planchette with further portions of the foil extract, if available. Wash the planchette and dry by warming gently.

Place the planchette on the top shelf of a counter castle, or clamp an end window counter just above the planchette. Determine the activity, taking 10-minute counts every fifteen minutes. Continue for one to two hours.

Plot a graph of *log*$_{10}$ (*count rate*) against *mid-time of count*, after correcting the count rate for background and resolving time. Determine the half-life of bismuth-212 and comment on the efficiency of separation from lead-212.

The solution in the centrifuge tube contains lead-212. Add 1 cm^3 of 0·1M lead nitrate as carrier, and precipitate all the lead by addition of 3 drops of 2M sulphuric acid. Prepare a balance tube and centrifuge for a minute.

Remove as much supernatant liquid as possible with a dropping tube, then stir up the deposit of lead sulphate, and transfer it as a suspension to an aluminium planchette. Dry by radiant heat. Add 2 drops of cellulose acetate solution and dry again. This holds the lead sulphate in position.

Place the planchette on the top shelf of a G.M.

castle and take a 1-minute count. It is not possible to determine the half-life of lead-212 until transient equilibrium, with its daughter bismuth-212, has been established. Take 10-minute counts every possible hour for a period of twenty-four hours, or longer if possible. (If the planchette is covered with an aluminium adsorber, to adsorb weak β-particles from lead-212, the growth curve of bismuth-212 may be obtained.)

Plot a graph of *log*$_{10}$ (*count rate*) against *mid-time of count*, after correcting the count rate for background and resolving time. Determine the half-life of lead-212.

Separation of thallium-208: method one. Bismuth-212 can be adsorbed on an anion-exchange column, and daughter thallium-208 'milked' off when transient equilibrium has been established.

Requirements

Scaler and G.M. liquid counter.
Ion exchange resin (De-Acidite FFIP-SRA70).
10% w/v thorium chloride solution.
2-Mercaptobenzothiazole.

Slurry 1 g of a strong anion-exchange resin (52-100 mesh) in chloride form with M hydrochloric acid and set up a column in a glass tube 50 cm × 0·5 cm as described on p. 108.

Pass through the column 100 cm^3 of a 10% w/v thorium chloride solution at 3 cm^3 min^{-1}. The anionic bismuth chloride complex will be strongly held. Wash the column with 10 cm^3 of M hydrochloric acid to remove thorium and lead.

Wait ten minutes for transient equilibrium to be established between bismuth-212 and thallium-208. Then elute the thallium with 5 cm^3 of M hydrochloric acid directly into a G.M. liquid counter, and count for 1-minute every two minutes. Continue until the count rate is steady to $\pm 2\sigma$; this should take at least twenty minutes.

The results can be improved by developing the resin with 2-mercaptobenzothiazole. In acidic solution this is a specific reagent for bismuth and lead. It holds them firmly in the column. After percolation of the thorium chloride solution, pass in turn through the column:

(*a*) 10 cm^3 of M hydrochloric acid,
(*b*) 10 cm^3 of 0·01M hydrochloric acid,
(*c*) 5 cm^3 of fresh 1% ethanolic 2-mercaptobenzothiazole,
(*d*) 5 cm^3 of M hydrochloric acid.

Wait ten minutes before eluting the thallium and counting as before. The residual activity due to bismuth-212 and lead-212 should be greatly reduced.

Plot a graph of *log*$_{10}$ (*count rate*) against *mid-time of count*, after subtracting the residual count rate and correcting for resolving time.

Reference: Abrão, A., *J. Chem. Educ.*, 1964, **41**, 600.

Separation of thallium-208: method two. Ammonium molybdophosphate (A.M.P.) functions as an ion-exchange resin which selectively and powerfully adsorbs thallium cations, other ions being washed off by nitric acid.

Requirements

Scaler and G.M. end window counter.
Counting castle.
Aluminium planchette.
M thorium nitrate in 2M nitric acid.
Molybdophosphoric acid solid.
M ammonium nitrate.

Unless the molybdophosphoric acid is available as fine crystals, it must be dry sieved (through nylon) to obtain 14–200 graded material.

Add 0·3 g of graded material to 20 cm^3 of M ammonium nitrate and allow to stand, *without stirring*, for ten minutes. If the crystals are tinged green due to photo-reduction, add a few drops of bromine water.

Decant colloidal particles and 'fines' three times, using 2M nitric acid, then collect in a small (2 cm) Buchner funnel by gentle filtration at the pump.

Without allowing the A.M.P. to be sucked dry, wash with 10 cm^3 of 2M nitric acid, followed immediately by 10 cm^3 of M thorium nitrate which should pass through at 5 cm^3 min^{-1}. Finally wash with two 3 cm^3 portions of 2M nitric acid and suck dry. It is important to pass through the thorium solution at the correct rate. Start with gentle suction from the pump and increase gradually to maintain the flow rate.

Transfer the filter paper and A.M.P. to a planchette, holding the filter paper down with glue.

Place the planchette in a G.M. castle, and count for 30 seconds every minute. Continue for fifteen minutes, then make a 5-minute background count.

Plot a graph of *log*$_{10}$ (*count rate*) against *mid-time of count*, after correcting for background and resolving time.

References: Broadbank, R. W. C., Dhabanandana, S. and Harding, R. D., *Analyst*, 1960, **85**, 365.
Smith, J. van R., *J. Inorg. Nucl. Chem.*, 1965, **27**, 227.

VI Titrimetric Analysis

TITRIMETRIC TECHNIQUE
1. Use of the pipette

THE pipette is designed to deliver a definite fixed volume of liquid when correctly filled to its calibration mark. The commonest size is 20·0 cm^3 or 25·0 cm^3.

Before use, a pipette must first be washed out with the solution it is to measure. Pour some solution into a clean, dry beaker and suck sufficient solution into the pipette to fill its lower stem. Now hold the pipette horizontal and rotate it gently so that the solution washes the bulb and upper stem up to the calibration mark. Let the wash solution drain away into a sink.

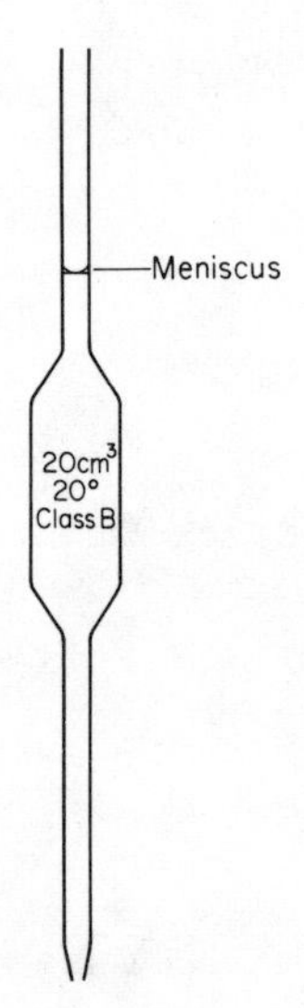

Fig. 36. A pipette.

To fill a pipette, suck solution up to a few centimetres above the calibration mark. Retain the solution by placing a *dry* forefinger on the mouthpiece. Remove the pipette from the beaker of solution and allow surplus solution on the outside to drain away. To bring the solution level down to

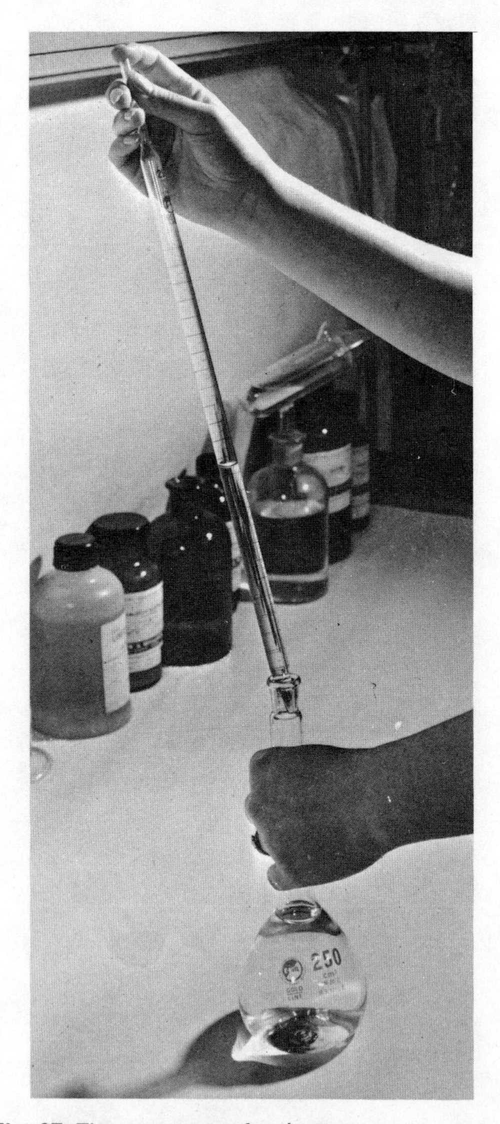

Fig. 37. The correct use of a pipette.

the calibration mark, let your forefinger relax and rotate the pipette slowly. Drops of solution should slowly form on the pipette tip; continue until the meniscus *just touches* the calibration mark (Fig. 36), then tighten your forefinger grip. Remove any partially formed drop from the pipette tip.

For a titration, the contents of the pipette is run out into a conical flask which has been well swilled out with pure water. Allow the pipette to drain for twenty seconds, *keeping the tip in contact with the side of the conical flask for three seconds after movement of the meniscus has ceased.*

Precautions you should observe: do not allow a solution of precise concentration to become diluted by placing in a water-wet container *before* it has been measured; keep the pipette tip well below the liquid surface when sucking (*if you get a mouthful, spit—do not swallow*); to pipette solutions above 0·1M or volatile or unfamiliar liquids, use a safety attachment (see Fig. 38).

In a titration the volume of solution delivered by the pipette is known as 'the aliquot'.

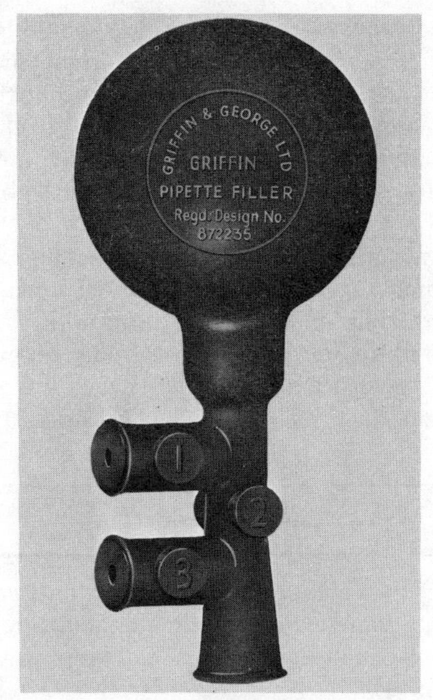

Fig. 38. The Griffin and George pipette filler.

2. Use of the burette

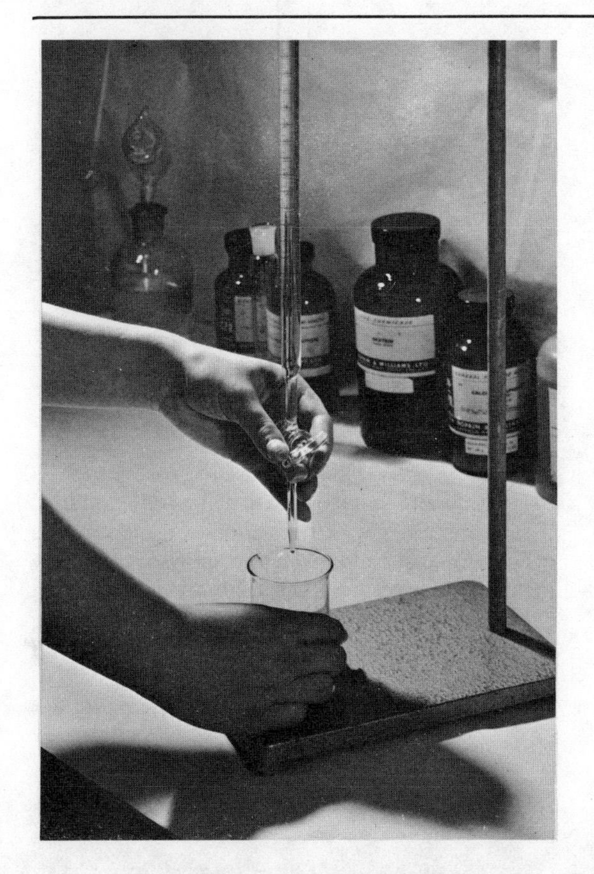

Fig. 39 The correct use of a burette.

The burette (Fig. 39) is designed to deliver definite but variable volumes of liquid. The commonest burette is graduated from 0·0 cm^3 to 50·0 cm^3 in 0·1 cm^3 units.

First, rinse out the burette with the solution it is to contain. Pour in about 10 cm^3 of solution and tilt the burette almost horizontal, rotating it gently; drain the solution into a sink. Sometimes a burette tip is blocked with solid or excess grease from the tap; do not attempt to clear the tip yourself but hand the burette in for professional cleaning. Rinse a second time. Now clamp the burette vertically in a stand: the stands are usually wooden and the jaws of the clamp should be well protected with cork.

Fill the burette carefully, using a beaker; open the tap briefly to fill the burette below the tap (why?). Do not attempt to fill a burette from a large container such as a Winchester, always decant solution into a clean, dry, *labelled* beaker first.

Adjust the meniscus to a definite graduation mark; the bottom of the meniscus should just touch the graduation mark and your eyes should be on a level with the graduation mark. You may find the zero graduation too high for comfort, but there is no objection to adjusting the meniscus to the 10·0 cm^3 mark for example. To observe the meniscus most clearly, hold a white card behind the burette. Record the volume reading in a notebook at once.

When performing a titration, the conical flask containing the aliquot is placed on a white tile under the burette so that the burette tip is inside the mouth of the conical flask. The burette tap should be turned with the left hand, leaving the right hand free to shake the conical flask. Firstly, take an approximate reading; run the solution in from the burette continuously while shaking the flask vigorously. When the end-point is observed, close the tap quickly. The burette reading will give you a rough idea of the volume to be added.

Now repeat the titration with a fresh aliquot. As the rough end-point volume is approached, add solution from the burette one drop at a time. Do not be impatient, otherwise the end-point will be overshot.

The volume run out from the burette to reach the end-point is known as the 'titre'.

3. Use of the standard flask

To prepare a solution of precisely known concentration, a definite amount of material (solute) must be dissolved in solvent to give a definite volume of solution.

The concentration of a solution is defined as the number of moles of the solute dissolved in 1 litre (1 dm^3) of solution.

A mole is the amount of a substance containing the same number of molecules (or atoms, or ions, as the case may be) as there are carbon atoms in exactly 0·012 kilogram of carbon-12.

The definite amount of material is measured by accurate weighing (see p. 6) and the definite volume of solution is prepared in a standard flask (Fig. 40).

A standard flask contains a definite volume when correctly filled to the calibration mark at the temperature stated on the flask. It will not deliver that volume when emptied, and it must never be allowed to get hot.

Tip your solid from the weighing bottle into a large beaker (200 cm^3) and add about 50 cm^3 of pure water. Stir well to dissolve. Heat only if essential because the solution must be cooled afterwards to room temperature. Take care to lose no solution; remember to wash solution off the stirring rod back into the beaker.

Swill out the standard flask twice with pure water: if the flask is clean, the water should drain uniformly and not gather in drops.

Pour your cold solution into the flask through a clean funnel. Wash out the beaker several times and add all washings to the flask. Remove the funnel, washing it well, including the outside of the stem. Every drop of solution should now have been successfully transferred to the standard flask.

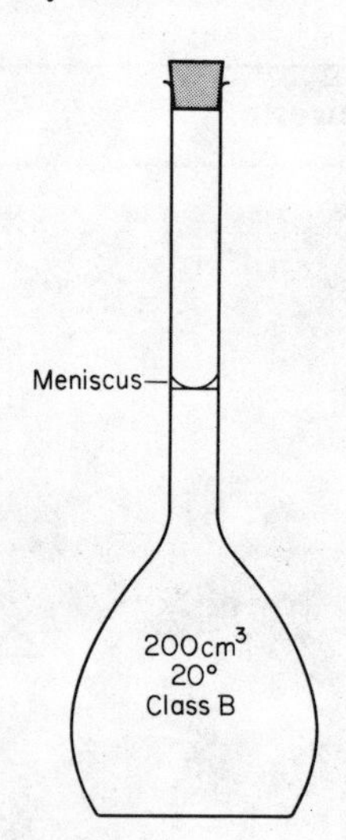

Fig. 40. A standard flask.

Now fill the standard flask with pure water to within about 1 cm of the calibration mark on its neck. Finally, add water dropwise until the meniscus just rests on the calibration mark (Fig. 40); add the last portions of water from a dropping tube. Work patiently at this stage otherwise you will overfill the flask.

Seal the standard flask with a plastic or rubber

bung and invert a dozen times to mix the contents thoroughly. Pour out into an ordinary *dry* flask, seal the flask and label with the following information: name and weight of solute per litre, date of preparation, your name.

Exercise. Transfer a little unweighed potassium manganate (VII) from a weighing bottle to a standard flask following the above procedure. The intense colour of the solution will enable you to appreciate the need for careful washing at every stage.

4. Primary standards

In order to determine the concentration of solutions, it is necessary to prepare standard solutions of known and reliable concentration. To be suitable for use as a primary standard, a compound must meet an exacting specification: it must be available pure, be stable in air, be readily soluble giving a stable solution, have a large molecular weight and titrate reproducibly in a known reaction. When a compound meets this specification, its solutions can be relied on to follow exactly the relationship:

grams of reagent per litre =
concentration (mol l^{-1}) × molecular weight of compound.

For the highest calibre work, silver of 99·9999% purity obtained by vacuum distillation is used; all other standards are calibrated against the silver.

Small professional laboratories will use simpler schemes based on materials such as constant-boiling hydrochloric acid or anhydrous sodium carbonate.

In student work the specification for a primary standard is less exacting since its purity need only match the accuracy of the apparatus used. Also, for convenience, it should be possible to weigh out primary standards without any pre-treatment, and their solutions should be stable over a long period of time. Suitable materials are given in Table 25.

Alternatively, solutions of known concentration can be purchased commercially.

Table 25

Primary standard (A.R. grade)	*Relative Molecular mass*	*Purity available /%*	*Standardization uses*
Sodium carbonate, anhydrous	105·99	99·9	Acids
Potassium hydrogen phthalate	204·23	99·9	Bases
Potassium dichromate	294·22	99·9	Reducing agents
Potassium iodate	214·01	99·9	Sodium thiosulphate
Sodium chloride	58·45	99·9	Silver nitrate
Ethylenediaminetetraacetic acid, disodium salt, dihydrate (EDTA)	372·25	99·0	Metal salts

5. Accuracy of apparatus

Errors in practical chemistry are likely to have three main causes: mistakes in calculation, faults in laboratory technique, and limitations in the apparatus used.

The first two causes are avoidable, but limitations in the apparatus used are more fundamental. Obvious faults such as chipped pipettes can still be avoided, but limitations in accuracy caused by manufacturing process are unavoidable. In class B titration apparatus (when correctly used) the calibration marks are guaranteed as given in Table 26.

Table 26

Apparatus	*Maximum error* ±
1 litre standard flask	0·8 cm^3
250 cm^3 standard flask	0·3 cm^3
25 cm^3 pipette	0·06 cm^3
50 cm^3 burette	0·1 cm^3 between any two marks
10 g weight	0·0005 g

Since it will be impossible to work at the calibration temperature, further limitations are inevitable (100·0 cm^3 water weighs 99·718 g at 20°C, 99·805 g at 15°C).

The experimental technique should be designed to get the best out of the apparatus. The burette error is minimized by having titre values above 30 cm^3 and quantities should be measured by weight or pipette volume whenever possible.

Nevertheless, your best overall accuracy will be about ½%, and it is absurd to quote final results more precisely.

6. Presentation and calculation of results

The objectives to be borne in mind in the presentation of experimental results are completeness and clarity. Both are essential if the results are to be intelligible to others.

While neatness should not be regarded as an end in itself but rather as a means whereby results are expressed clearly, it is worth while taking trouble over the layout of results.

One style of presenting titrimetric results is illustrated in Fig. 41.

Calculations

Titrimetric results can be calculated in different ways. Detailed guidance on calculations is not given in this book; it is more valuable to work out your own method, or be given personal assistance, than to use a formula. If you have difficulty with calculations, there are booklets available with worked examples and questions for practice.

The molarity method is, however, preferred to the use of equivalents and normalities, and all solution concentrations in the book are quoted in mol l^{-1}. You are recommended to read the articles 'Normality—an Unnecessary Concept' by J. Lee (*Educ. Chem.*, 1965, **2**, 229) and 'The Use of the Molarity Concept in Titrimetric Analysis' by J. G. Stark (*Educ. Chem.*, 1966, **2**, 70).

Consider the reaction represented by the equation:

$$H_3PO_4 + 2NaOH \longrightarrow Na_2HPO_4 + 2H_2O$$

If orthophosphoric acid is titrated with sodium hydroxide to an end-point corresponding to the equation, it follows that:

$$\frac{\text{moles of orthophosphoric acid in an aliquot}}{\text{moles of sodium hydroxide in the titre}} = \frac{\text{moles of orthophosphoric acid in the equation}}{\text{moles of sodium hydroxide in the equation}}$$

Hence, if M = moles litre^{-1} of acid or base
V = volume of acid or base, in cm^3

$$\frac{M_A V_A}{M_B V_B} = \frac{1}{2}$$

By writing the appropriate balanced equations, similar expressions can be derived for any titration calculation.

Acid-base Titration Problem: Analysis of Aspirin Tablets

Mass of 5 tablets + beaker = 7.010 g
Mass of beaker : 5.356 g
Mass of 5 tablets : 1.65(4) g

In 200 cm^3 standard flask : 5 tablets + 25.0 cm^3 of 1.02 M sodium hydroxide
pipette : 10.0 cm^3 aliquots of the above solution
burette : 0.098 M hydrochloric acid
indicator : 2 drops of phenol red

Top burette reading	Bottom burette reading	Titre (cm^3)
10.0	27.1	(17)
10.5	26.9	16.4
10.0	26.4	16.4
		16.4 cm^3

Result: 16.4 cm^3 of 0.098 M hydrochloric acid ≡ 10.0 cm^3 of residual sodium hydroxide

Fig. 41. Presentation of titration results.

ACID-BASE TITRATIONS

1. An introductory exercise

Read the previous instructions in Titrimetric Technique, p. 150, for precise details, then carry out the experiment using the following sequence of operations.

Requirements

1 burette and stand
1 white tile
1 pipette
2 conical flasks
1 pure water wash-bottle
} the titration apparatus.
A range of indicators
Standard acid solution.
Standard alkali solution.

Wash out the burette, pipette and conical flasks with a little pure water.

Wash out the burette with standard acid and fill, remembering to run out some solution to fill below the tap. Record the volume reading.

Wash out the pipette with standard alkali and fill to the mark. Run out the measured alkali (the aliquot) into a water-washed conical flask and allow the pipette to drain adequately. Add 3 drops of indicator.

The alkali is now titrated with the acid. For an accurate end-point the colour change of the indicator should be caused by 1 drop of acid. Record the volume reading (the first reading is a 'rough' one to determine the approximate position of the end-point). Wash out the conical flask with pure water and repeat the experiment with a fresh aliquot of alkali. Obtain three consistent results for the volume of acid added (the titre) and take their mean as the final answer.

Now repeat the experiment using a variety of indicators. One rough and one accurate reading will be sufficient, as the objective now is to record carefully the colour changes of each indicator.

2. Primary standard acids

Read the previous discussion (p. 153) on primary standards before carrying out this experiment.

The objective here is to prepare one or more standard solutions of primary standard acids in order to compare their ease in use. The solution will be used to determine the exact concentration of a sodium hydroxide solution.

Requirements

Titration apparatus and a standard flask.
A.R. potassium hydrogen phthalate.
A.R. sulphamic acid.
A.R. constant-boiling hydrochloric acid (at 760 mm).
A.R. hydrated oxalic acid.
Sodium hydroxide solution (about 0·15M).
Indicator such as phenolphthalein.

Weigh accurately a sample of one of the primary acids into a large beaker and use a standard flask to prepare a solution of precise volume according to the procedures on pp. 6 and 152. The concentration should be suitable for titrating the 0·15M Sodium hydroxide solution.

0·1M potassium hydrogen phthalate requires 20·423 g l^{-1} of $C_6H_4(CO_2K)CO_2\mathbf{H}$.

0·1M sulphamic acid requires 9·710 g l^{-1} of $NH_2SO_3\mathbf{H}$.

0·1M hydrochloric acid requires 3·645 g l^{-1} of $\mathbf{H}Cl$ (16·40 cm^3 of 20·24% constant-boiling HCl per litre).

0·05M oxalic acid requires 6·304 g l^{-1} of $\mathbf{H}_2C_2O_4.2H_2O$.

The hydrochloric acid can be measured out by volume from a burette, but determine the exact amount by weighing, which is a more accurate method of measurement.

Titrate aliquots of the sodium hydroxide solution with the standard acid solution, using a few drops of an indicator suitable for weak acid-strong base titrations. Obtain three consistent titres.

Write the equation of the reaction between sodium hydroxide and the acid used. Calculate the exact concentration (mol l^{-1}) of the sodium hydroxide.

3. A primary standard base

As an alternative to the preparation of standard acid solutions, it is possible to prepare a standard base solution. To do this properly requires the observation of elaborate precautions; this experiment can only outline the technique.

Sodium hydrogen carbonate is purified and converted to sodium sesquicarbonate by crystallization from hot water. Sufficient product is converted to sodium carbonate by heating to constant weight so that a standard solution of sodium carbonate can be prepared of concentration about 0·05M.

The experiment can be shortened by starting with sodium sesquicarbonate.

Requirements

Titration apparatus and a standard flask.
Crucible apparatus.
A.R. sodium hydrogen carbonate.
Sulphuric acid solution (about 0·04M).
Indicator such as methyl orange.
Hard-glass test tube (150 mm × 25 mm).

Dissolve roughly 10·25 g of pure sodium hydrogen carbonate in 40 cm^3 of pure water in a hard-glass test tube which is clamped in a water-bath as close to 86°C as possible. Filter the solution while still hot and cool the filtrate rapidly. When crystallization is complete, collect the sodium sesquicarbonate by filtration at the pump and dry in an oven at about 100°C.

The quantity of sodium sesquicarbonate obtained should be sufficient after ignition for 1 litre of 0·05M sodium carbonate. For standard flasks smaller than 1 litre, use only a suitable fraction of the yield for the next step in the experiment.

When dry, transfer a suitable fraction of the sodium sesquicarbonate to a clean porcelain crucible and heat *gently* until constant weight is attained (see p. 182). Ideally, the heating would be conducted at 270°C in an electric oven, for at least two hours. When heating with a Bunsen burner, an initial heating period of ten minutes is sufficient; then cool the crucible and contents to room temperature and weigh accurately. Repeat the procedure with heating periods of two minutes, until constant weight is attained. Record the weight.

Tip out the solid residue of sodium carbonate into a clean beaker and reweigh the crucible accurately. The change in weight corresponds to the weight of anhydrous sodium carbonate that will be used for making the standard solution (see p. 152).

Dissolve the sodium carbonate in the beaker in pure water and completely transfer with washings to the standard flask. Dilute the solution to a precise volume by filling the flask to the calibration mark with pure water. Mix well before using for the titration.

Titrate aliquots of the standard sodium carbonate solution with the approximate 0·04M sulphuric acid using methyl orange or other suitable indicator. Obtain three accurate titres.

Calculate the concentration (mol l^{-1}) of your sodium carbonate solution from the weight used and the volume of the standard flask. Then calculate the exact concentration (mol l^{-1}) of the sulphuric acid from the titration results:

$$Na_2CO_3 + H_2SO_4 \longrightarrow Na_2SO_4 + CO_2 + H_2O$$

Why is there no effervescence of carbon dioxide during the titration?

Write the equations for the conversion of sodium hydrogen carbonate to sodium sesquicarbonate, and of sodium sesquicarbonate to sodium carbonate.

4. Weak acids and bases

If you test aqueous solutions of salts such as ammonium sulphate or sodium acetate with litmus paper, you will observe that the solutions are not neutral. The salts are said to be hydrolysed; i.e. they have reacted with water, forming some ammonia or acetic acid molecules respectively:

$$NH_4^+ + H_2O \rightleftharpoons H_3O^+ + NH_3$$

When acids and bases exist in a solvent as molecules rather than ionized they are said to be 'weak'.

Thus in the titration of ammonia with sulphuric acid it is necessary to choose an indicator which changes colour sharply, not at the neutral point but when the solution has the degree of acidity (pH) of pure ammonium sulphate.

A considerable number of organic compounds are available which change colour at different degrees of acidity (pH <7) or different degrees of alkalinity (pH >7) and are therefore suitable for the titration of weak bases or weak acids.

Requirements

Titration apparatus.
Sulphuric acid (about 0·05M).
Acetic acid (about 0·1M).
Sodium hydroxide (about 0·1M).
Ammonia (about 0·1M).
Sodium carbonate (about 0·05M).
Methyl orange indicator.
Phenolphthalein indicator.

Titrate aliquots of a base with the sulphuric acid solution using methyl orange indicator (3 drops). Only one accurate titre is necessary. *Note particularly the quality of the end-point;* is the colour change sharp or gradual? Repeat the titration with the other bases in turn.

Now repeat the titration using phenolphthalein indicator.

Finally, repeat the whole procedure using acetic acid solution instead of sulphuric acid.

Tabulate your results, stating which indicator must be used in each case to obtain an accurate titre. Which are the weak acids and weak bases? How can you account for the behaviour of sodium carbonate? In which pH ranges do methyl orange and phenolphthalein change colour?

This topic is developed further in the next experiment, *pH range of indicators*, and in the experiment *Potentiometric titrations* on p. 179.

The heats of reaction can be compared by the experiment on p. 121.

5. pH range of indicators

If a titration is carried out in the presence of a suitable buffer solution there will be no sharp change in pH at an end-point, so indicators will change colour gradually.

A Universal Buffer Solution according to the formula of E. B. R. Prideaux and A. T. Ward (*J. Chem. Soc.*, 1924, **125**, 426) is suitable. If a 25·0 cm³ aliquot of the buffer solution is diluted with 10 cm³ of pure water and titrated with standardized 0·1M sodium hydroxide, the appropriate pH of the mixture at any stage of the titration is given by:

$$\text{pH} = 3{\cdot}1 + 0{\cdot}237V$$

where $V =$ cm³ of 0·1M NaOH added, up to a maximum of 35 cm³.

Requirements

Titration apparatus.
Universal Buffer Solution (available from B.D.H.).
Standardized 0·1M sodium hydroxide.
Indicators (methyl red, thymolphthalein, etc.).

Prepare acid colour standards by mixing 50 cm³ of pure water in a conical flask with 2·5 cm³ of 2M hydrochloric acid and adding exactly 5 drops of indicator.

Prepare alkali colour standards in the same way, but use 2M sodium hydroxide instead of the acid.

Dilute a 25·0 cm³ aliquot of the buffer solution with 10 cm³ of pure water and add exactly 5 drops of indicator. Titrate the buffer mixture with standardized 0·1M sodium hydroxide until the indicator first begins to change from its acid colour. Record the burette reading. Continue the titration until the indicator has just finished changing to its alkali colour. Record the burette reading again. In order to make good colour comparisons, the colour standards should be diluted as necessary with pure water to have approximately the same volume as the titration mixture.

Use the relationship given in the introduction to this experiment to calculate the pH range over which the indicator changed colour.

What type of structural change is related to the colour change? What is the connection between the pH range of an indicator and its choice for a weak acid or weak base titration? Consult a textbook about indicators and titration curves.

6. Determination of concentrated acids

Concentrated materials are best diluted before determination; the main problem is to discover the degree of dilution necessary.

Note that a 'concentrated' acid has a large amount of acid per litre, while a 'strong' acid is one whose molecules are mostly ionized.

Requirements

Titration apparatus and a dry standard flask.
10 cm³ burette.
Standardized 0·1M sodium hydroxide.
Methyl orange indicator.
A fresh sample of concentrated acid.

Run exactly 0·05 cm³ of concentrated acid from a 10 cm³ burette directly into a conical flask. Dilute the sample with about 20 cm³ of pure water and titrate to an approximate end-point with standard 0·1M sodium hydroxide using methyl orange indicator (3 drops).

On the basis of this titration, work out the volume of concentrated acid which must be diluted in your standard flask so that an aliquot will give a reasonable titre.

Run the calculated volume of concentrated acid directly into a dry standard flask, and make up to the calibration mark with pure water. To absorb any heat of dilution, add an initial 50 cm³ portion of water all at once.

Titrate aliquots of the diluted acid with standard 0·1M sodium hydroxide using methyl orange indicator (3 drops). Obtain three consistent titres.

Calculate the concentration (mol l^{-1}) of the concentrated acid and compare with the supplier's specification. If the information is available, calculate the cost per mole of the three bench acids; which is the cheapest to use?

7. Reaction of sodium hydroxide with the atmosphere

Solid sodium hydroxide adsorbs carbon dioxide and water when exposed to the atmosphere. This experiment is designed to discover the extent of carbon dioxide adsorption after a week, *so the initial preparations must be carried out a week before the titration.*

By using suitable indicators it is possible to discriminate between the hydroxide and carbonate content of the exposed material; methyl orange changes colour when both have been fully neutralized, and phenolphthalein when the hydroxide has been neutralized but the carbonate only as far as hydrogen carbonate.

The experiment can be varied by using other hydroxides and other periods of exposure.

Requirements

Solid sodium hydroxide.
Titration apparatus.
A measuring cylinder.
0·05M sulphuric acid.
Phenolphthalein indicator.
Methyl orange indicator.

Calculate the weight of sodium hydroxide required to make sufficient decimolar solution for a titration (ten times your pipette volume). Keep a record of your calculation.

Weigh roughly the calculated amount of sodium hydroxide into a small beaker. Label with full details and leave uncovered in a reasonably dust-free place. ***Take care not to spill or touch solid sodium hydroxide.***

Continue the experiment after a week.

Describe the appearance of the exposed sodium hydroxide, then dissolve in an appropriate volume of water to obtain a solution about decimolar. Mix well by shaking.

Titrate aliquots of the solution with 0·05M sulphuric acid using phenolphthalein indicator (3 drops) initially. At the phenolphthalein end-point record the titre (V_p cm³), then add methyl orange indicator (3 drops) and continue to its end-point. Record the total titre (V_m cm³).

Obtain three accurate pairs of titres.

Calculate the percentage of hydroxide converted to carbonate using the relationship:

$$\text{percentage conversion} = \frac{\text{carbonate content}}{\text{total alkalinity}} \times 100$$

$$= \frac{2(V_m - V_p)}{V_m} \times 100$$

Write the equations of the reactions completed at the two end-points and derive the relationship quoted.

What are the disadvantages of attempting to use sodium hydroxide as a standard base?

8. Neutralization of orthophosphoric acid

This experiment is designed to investigate the stoichiometry of the reactions of orthophosphoric acid with sodium hydroxide. Different indicators are used to detect the end-points of the different reactions.

The experiment can be varied by using organic acids, such as citric acid.

Requirements

Titration apparatus.
0·1M sodium hydroxide.
0·05M orthophosphoric acid (3·1 cm³ of 90% acid l⁻¹).
or 0·05M citric acid (10·51 g l⁻¹ of $C_6H_8O_7.H_2O$).
Bromocresol green indicator.
Phenolphthalein indicator.

Titrate aliquots of 0·05M orthophosphoric acid with 0·1M sodium hydroxide using bromocresol green indicator (3 drops). Obtain three consistent titres.

Titrate further aliquots of orthophosphoric acid with sodium hydroxide using phenolphthalein indicator (3 drops). The end-point can be sharpened by half saturating with sodium chloride (roughly 8 g for a 25·0 cm³ aliquot). Obtain three consistent titres.

Write the balanced equations for the reactions complete at the two end-points, naming the salts produced:

$$n\text{NaOH} + \text{H}_3\text{PO}_4 \longrightarrow ?$$

$$\text{where} \quad n = \frac{\text{moles NaOH in titre}}{\text{moles H}_3\text{PO}_4 \text{ in aliquot}}$$

Record your experimental values for n; if the method is valid n values should be whole numbers.

What are the pH ranges of the indicators? What is the function of the sodium chloride? Consult a textbook about buffer action.

9. Analysis of calcium carbonate materials

Calcium carbonate, being insoluble, has to be determined by a 'back titration' in which it is treated with a definite amount of acid, and then the excess acid is determined by titration with a standard base. Any natural form of calcium carbonate which has been powdered is suitable: limestone, marble, mollusc or egg shells.

$$\text{CaCO}_3 + 2\text{HCl} \longrightarrow \text{CaCl}_2 + \text{H}_2\text{O} + \text{CO}_2$$

Requirements

Titration apparatus, and a 200 cm³ or 250 cm³ standard flask.
Standard M hydrochloric acid.
Standardized 0·1M sodium hydroxide.
Methyl orange indicator.
Powdered form of calcium carbonate ($CaCO_3$, M_r 100·1).

Weigh accurately a sample of the calcium carbonate powder (roughly 1 g, 0·01 mole if pure $CaCO_3$) into a 400 cm³ beaker. Cover the powder with about 20 cm³ of pure water, and treat with a 40·0 cm³ aliquot of M hydrochloric acid (0·04 mole, an excess). When the effervescence is complete, transfer the reaction mixture with washings to a standard flask. Dilute to the calibration mark with pure water, then mix well by repeated inversions of the corked flask.

Titrate aliquots of the diluted reaction mixture with standard 0·1M sodium hydroxide, using methyl orange indicator (3 drops).

Calculate the amount of basic material in the powder, expressed as per cent of calcium carbonate. What other basic compounds might be present? What inert materials are likely to be present?

10. Analysis of aspirin tablets

Aspirin is a compound of two acids—acetic acid and salicylic acid. It can be hydrolysed by alkali and the two acids neutralized simultaneously:

$$CH_3 \cdot CO_2C_6H_4 \cdot CO_2H + 2NaOH \longrightarrow CH_3 \cdot CO_2Na + HO \cdot C_6H_4 \cdot CO_2Na + H_2O$$

The presence of the weak acid salts means that the excess of alkali is best titrated using phenol red indicator (pH range 6·8 to 8·4), although phenolphthalein is satisfactory.

This experiment is based on the procedure in British Pharmacopaeia (1953).

Requirements

Titration apparatus and a 250 cm³ standard flask.
M sodium hydroxide.
Standardized 0·05M sulphuric acid.
Phenol red indicator.
Aspirin tablets (containing $CH_3.CO_2C_6H_4.CO_2H$, M_r 180·16).

Weigh accurately a definite number of aspirin tablets (roughly 1·5 g) into a large conical flask. Add a 25·0 cm³ aliquot of M sodium hydroxide and about the same volume of pure water to the tablets. Simmer gently for ten minutes to hydrolyse the aspirin.

Cool the reaction mixture and transfer with washings to a 250 cm³ standard flask. Dilute to the calibration mark with pure water, then mix well by repeated inversions of the corked flask.

Titrate aliquots of the diluted reaction mixture with standard 0·05M sulphuric acid using phenol red indicator (3 drops).

Standardize the M sodium hydroxide, after suitable exact dilution in a standard flask, with the standard 0·05M sulphuric acid by titration using phenol red indicator.

Calculate the weight of acetylsalicylic acid in each tablet and compare your result with the specification.

11. Available nitrogen in fertilizers

Ammonium salts can be treated with an excess of standard sodium hydroxide and the residual sodium hydroxide determined by titration, but fertilizer materials are liable to contain heavy metal salts which would also react so the procedure would no longer be quantitative.

Therefore, fertilizer samples are treated with an excess of sodium hydroxide and the ammonia gas liberated is trapped in an excess of standard acid. The residual acid is determined by titration.

Requirements

Distillation apparatus (see Fig. F).
Titration apparatus and a 250 cm³ standard flask.
Standardized 0·1M sodium hydroxide.
Standard M hydrochloric acid.
Methyl orange indicator.
Fertilizer sample.

Set up the apparatus for distillation with addition, with a receiver adaptor dipping below the surface of a 50·0 cm³ aliquot of standard M hydrochloric acid in a 150 cm³ conical flask (to trap ammonia distilled over).

Weigh accurately a sample of fertilizer (roughly 1·5 g) into a 50 cm³ distillation flask. Add about 25 cm³ of 2M sodium hydroxide and a boiling stone, then *quickly* connect up to the distillation apparatus.

Heat the flask with a small flame so that the contents come to the boil slowly. *Beware of excessive and rapid sucking back* up the receiver adaptor, especially when the reaction mixture first boils. When sucking back occurs, *briefly* open the tap of the funnel to allow air to enter; do not remove the Bunsen flame.

Boil the contents of the distillation flask gently and steadily until about 10 cm³ of water has been distilled over.

When the distillation is complete, disconnect the still-head before removing the Bunsen burner, and wash any residues in the condenser and receiver adaptor into the 150 cm³ conical flask. Be careful to wash the adaptor completely.

Transfer the entire contents of the conical

flask with washings into a 250 cm^3 standard flask and dilute to the calibration mark with pure water. Mix well by repeated inversions of the corked flask.

Titrate aliquots from the standard flask with standard 0·1M sodium hydroxide using methyl orange indicator (3 drops).

Calculate the weight of ammonia distilled over and hence the percentage of available nitrogen (as ammonia) in the fertilizer.

In the conical flask:

$$NH_3 + HCl \longrightarrow NH_4Cl$$

and in the titration:

$$HCl + NaOH \longrightarrow NaCl + H_2O.$$

At room temperature, in the diluted solution, the ammonium chloride will not react with the standard sodium hydroxide.

The available nitrogen (as nitrate) may be determined using a 250 cm^3 distillation flask. In addition to the fertilizer and 2M sodium hydroxide add roughly 3 g of Devarda's alloy and about 100 cm^3 of pure water. Allow the mixture to stand for thirty minutes before distilling about 30 cm^3 of water into a 50 cm^3 aliquot of standard acid. Then continue as before.

Why is the nitrogen content of fertilizers important? What is the composition of Devarda's alloy, and what is the reaction with nitrate groups?

12. Saponification value of oils

Natural oils are commonly esters of glycerol and long chain fatty acids. Their reacting weights on hydrolysis can be expressed as an empirical 'saponification value':

$$\frac{\text{mg of KOH neutralized by the oil}}{\text{g of oil saponified}}$$

Pure tripalmitin has a saponification value of 208·8.

This experiment is based on British Standard 684 : 1958.

Requirements

Titration apparatus.
2 sets reflux apparatus (see Fig. A).
Standard 0·5M hydrochloric acid.
0·5M ethanolic potassium hydroxide (it should be stored in the dark and not used if darker than pale yellow).
Natural oils (coconut, olive, cod-liver, castor, etc.).
Phenolphthalein or thymolphthalein indicator.

Weigh accurately a sample of natural oil (roughly 2 g) into a dry 50 cm^3 flask. Using a *safety pipette* (or a burette) add a 25·0 cm^3 aliquot of 0·5M ethanolic potassium hydroxide. Set up the flask for refluxing. Apply a *little* grease to the ground-glass joint as alkali is liable to cause seizing of the joint. Heat the flask for thirty minutes in a boiling water-bath, shaking occasionally.

On completion of the refluxing, wash down the inside of the condenser with a little pure water to ensure all the alkali is in the flask. Then disconnect the flask *carefully* and transfer the entire contents, together with washings, to a titration flask.

Titrate the entire mixture, while still warm, with standard 0·5M hydrochloric acid using phenolphthalein indicator (about 1 cm^3). If the contents of the flask are dark in colour, use thymolphthalein as indicator.

Standardize the ethanolic potassium hydroxide by refluxing and titrating a 25·0 cm^3 aliquot as above, but without adding an oil sample.

Calculate the saponification value of your oil by using the relationship given above.

Consult a textbook about the constitution of natural oils, and write the equation of the reaction with ethanolic potassium hydroxide. Why is the alkali dissolved in ethanol to carry out this hydrolysis?

REDOX TITRATIONS

1. Potassium dichromate as a primary standard

Potassium dichromate is a good primary standard for redox titrations; it is obtainable pure, can be dried without decomposition, dissolves readily to give stable solutions and reacts quantitatively in acidic solution:

$$Cr_2O_7^{2-} + 14H^+ + 6e^- \longrightarrow 2Cr^{3+} + 7H_2O$$

The end-point of dichromate titrations is detected by using a suitable redox indicator; a colour change from deep green to intense violet-blue occurs with barium diphenylamine sulphonate.

Requirements

Titration apparatus and standard flasks.
A.R. potassium dichromate solid.*
Technical grade iron (II) sulphate solid.*
0·3% aqueous barium diphenylamine sulphonate indicator.
Concentrated 90% orthophosphoric acid.

Prepare a standard solution of potassium dichromate by accurate weighing. For details of the procedure for weighing see p. 6, and for preparing a standard solution see p. 152. The concentration should be about 0·02M (5·88 g l^{-1} of $K_2Cr_2O_7$). From the weight taken, calculate the exact concentration (mol l^{-1}) of your solution.

* Solutions may be provided if practice is not needed in the preparation of standard solutions.

Prepare a solution of iron (II) sulphate of known concentration by accurate weighing. The half equation for the oxidation of iron (II) to iron (III) ions is:

$$Fe^{2+} \longrightarrow Fe^{3+} + e^-$$

So the iron (II) sulphate solution must be six times more concentrated than the potassium dichromate to be equivalent to it, i.e. about 0·1M (27·8 g l^{-1} of $FeSO_4.7H_2O$). Use M sulphuric acid as solvent to prevent hydrolysis. This will also provide the necessary acidic conditions for the titration.

Titrate aliquots of the acidified iron (II) sulphate with the standard potassium dichromate. Add indicator (0·5 cm^3) and about 2 cm^3 of concentrated phosphoric acid which reacts with the iron (III) ions, producing a complex which does not affect the indicator.

When three accurate titres have been obtained, repeat once without the phosphoric acid addition and once without indicator. Comment on the quality of end-point observed in each case.

Calculate the concentration (mol l^{-1}) of your iron (II) sulphate solution and hence the grams per litre of $FeSO_4.7H_2O$ as determined volumetrically. Compare this with the original concentration of technical iron (II) sulphate, expressing the result as percentage purity.

2. Stoichiometry of potassium manganate (VII) reactions

Potassium manganate (VII) is not suitable for use as a primary standard as its solution does not keep satisfactorily; brown deposits of manganese (IV) oxide are often present. Nevertheless, potassium manganate (VII) reacts quantitatively with many reducing agents, and its intense colour results in a sharp end-point without the use of a redox indicator. It is, therefore, a useful titrimetric reagent.

This experiment is designed to discover the stoichiometry of potassium manganate (VII) reactions.

Requirements

Titration apparatus.

0·02M potassium manganate (VII) solution (3·16 g l^{-1} of $KMnO_4$).

0·05M ammonium iron (II) sulphate solution [19·6 g l^{-1} of $(NH_4)_2SO_4.FeSO_4.6H_2O$].

0·05M sodium oxalate solution [6·70 g l^{-1} of $(CO_2Na)_2$].

Thermometer, 0–100°C.

Introductory test tube experiments

(*a*) Add a little potassium manganate (VII) to the iron (II) salt solution. Warm the mixture.

(*b*) Repeat (*a*) but add 2M sulphuric acid before the potassium manganate (VII).

(*c*) Repeat (*a*) and (*b*) but use the oxalate instead of the iron (II) salt.

(*d*) Warm potassium manganate (VII) solution with 2M hydrochloric acid.

In each case describe the final appearance and comment on the relative speeds of reaction. Deduce the best conditions for a titration.

The titration

(*a*) Treat aliquots of 0·05M ammonium iron (II) sulphate with about 10 cm^3 of 2M sulphuric acid and titrate with 0·02M potassium manganate (VII) at room temperature. It is not easy to see the meniscus in the burette because of the intensity of the manganate (VII) colour; on this occasion only it is permissible to take readings of the top of the liquid level.

One drop of potassium manganate (VII) in excess at the end-point will produce a permanent pink coloration.

If a brown precipitate develops, add more 2M sulphuric acid to dissolve it.

Obtain two accurate titres.

(*b*) Treat aliquots of 0·05M sodium oxalate with about 10 cm^3 2M sulphuric acid and warm to 60°C. Titrate the warm mixture with 0·02M potassium manganate (VII) until 1 drop added produces a permanent pink coloration. The solution may need rewarming in the course of the titration.

After the first titration it will be adequate to estimate the temperature.

Obtain two accurate titres.

For each titration calculate the relative numbers of molecules reacting using the relationship:

$$\frac{\text{molecules of } a \text{ in the equation}}{\text{molecules of } b \text{ in the equation}} = \frac{\text{aliquot} \times \text{mol l}^{-1} \text{ of } a}{\text{titre} \times \text{mol l}^{-1} \text{ of } b}$$

Write the balanced equation for each reaction, remembering that sulphuric acid takes part. Also write the half equation for potassium manganate (VII) as an oxidizing agent in acidic solution.

3. Stoichiometry of the iodine–thiosulphate reaction

The iodine–thiosulphate reaction in neutral or weakly acidic solution is quantitative. As the end-point can be accurately detected, it is commonly used in titrimetric analysis; however, the solutions are insufficiently stable for their use as primary standards.

Requirements

Titration apparatus.
0·05M iodine solution (12·7 g l^{-1} of I_2 plus 40 g l^{-1} of KI).
0·05M sodium thiosulphate solution (12·4 g l^{-1} of $Na_2S_2O_3.5H_2O$).
Starch indicator (see Appendix I).

Firstly, carry out qualitative experiments to observe the behaviour of the indicator. Place separate portions of iodine solution (about 5 cm^3) in boiling-tubes. Then titrate with sodium thiosulphate solution after adding:

(*a*) a few drops of starch: note the disappearance of the deep blue colour when excess thiosulphate is added;

(*b*) any other iodine indicator available;

(*c*) nothing: the disappearance of the iodine colour affords quite a sharp end-point.

Now work quantitatively. Titrate aliquots of 0·05M iodine solution with 0·05M sodium thiosulphate until the completely colourless end-point is reached. If starch indicator (about 2 cm^3) is used, add it near the end-point when the iodine colour is very pale, otherwise iodine is strongly adsorbed on the starch and accuracy reduced. Obtain three accurate titres.

The equation of the reaction is

$$n\text{Na}_2\text{S}_2\text{O}_3 + \text{I}_2 \longrightarrow 2\text{NaI} + ?$$

$$\text{so } n = \frac{\text{moles Na}_2\text{S}_2\text{O}_3 \text{ in titre}}{\text{moles I}_2 \text{ in aliquot}}$$

Record your experimental value for *n*, and write the corresponding balanced equation remembering that *n* should be a whole number. Name the products. Why does the iodine solution contain potassium iodide?

4. Potassium iodate as a primary standard

Sodium thiosulphate is not available sufficiently pure, nor are its solutions stable enough for use as a primary standard. It is, however, a very useful volumetric reagent because of its quantitative reaction with iodine:

$$2Na_2S_2O_3 + I_2 \longrightarrow Na_2S_4O_6 + 2NaI$$

The product $Na_2S_4O_6$ is sodium tetrathionate.

Since potassium iodate is a primary standard, its solution may be used to standardize sodium thiosulphate solutions. It is one of many oxidizing agents which will react with potassium iodide in acidic solution, releasing iodine; titration of the released iodine by thiosulphate is therefore a determination of the original oxidizing agent.

$$IO_3^- + 6H^+ + 5I^- \longrightarrow 3I_2 + 3H_2O$$

From the equations it can be seen that one mole of iodate is equivalent to six moles of thiosulphate.

Because of its instability, sodium thiosulphate solution should be 'aged' for three days before standardization.

Requirements

Titration apparatus and standard flasks.
Sodium thiosulphate solid.*
A.R. potassium iodate solid*
Starch indicator (see Appendix I).
0·5M potassium iodide solution.

Prepare a standard solution of potassium iodate by accurate weighing. The concentration should be about 0·02M (4·28 g l^{-1} of KIO_3). For details of the procedure for weighing, see p. 6; for preparing a standard solution, see p. 152.

Prepare a solution of sodium thiosulphate by rough weighing. The concentration should be about 0·1M (24·82 g l^{-1} of $Na_2S_2O_3.5H_2O$).

Treat aliquots of the standard potassium iodate solution with 0·5M potassium iodide (about 10 cm^3, which must be an excess), and 2M sulphuric acid (about 10 cm^3, which must be an excess). Titrate the liberated iodine with sodium thiosulphate until the iodine colour is discharged. If desired, starch may be added near the end-point to obtain an intenser colour change, from blue-black to colourless.

* Solutions may be provided if practice is not needed in the preparation of standard solutions.

5. Analysis of oxalates

Oxalates are readily determined by titration with standard potassium manganate (VII) in warm acidic solution. If coloured or oxidizable cations are present, the procedure will be more complicated.

$$MnO_4^- + 8H^+ + 5e^- \longrightarrow Mn^{2+} + 4H_2O$$
$$C_2O_4^{2-} \longrightarrow 2CO_2 + 2e^-$$

Suitable oxalates for analysis are the complex oxalates (prepared by the procedures on pp. 27, 46, 48, 49) or iron (II) oxalate.

Requirements

Titration apparatus and a small standard flask.
Standardized 0·02M potassium manganate (VII).
A suitable oxalate.

Prepare a solution of an oxalate (roughly 10 g l^{-1}) by accurate weighing.

Dissolve iron (II) oxalate in 2M sulphuric acid before transferring to the standard flask.

Dissolve the complex oxalates of chromium and cobalt in the minimum of hot water in a 250 cm^3 beaker and add 2M sodium hydroxide (about 5 cm^3) until no more precipitate forms. Boil briefly to coagulate the precipitate. Filter at the pump, washing the filter paper with three small portions of hot water. Cool the filtrate before transferring to the standard flask and dilute to the calibration mark with pure water.

Acidify aliquots of the oxalate solution with about 10 cm^3 of 2M sulphuric acid, heat to about 60°C and titrate with standard 0·02M potassium manganate (VII). One drop of potassium manganate (VII) in excess will produce a permanent pink coloration. If a brown precipitate develops, add more 2M sulphuric acid to dissolve it and reheat the solution.

Obtain two accurate titres.

Calculate the percentage of oxalate in your salt and compare with the percentage predicted from the formula. When calculating for iron (II) oxalate, do not forget that manganate (VII) oxidizes iron (II) to iron (III).

6. Analysis of iron ore or iron (III) salts

The procedure given for the analysis of iron ore is suitable for ores which readily dissolve, e.g. spathose [iron (II) carbonate]. For other ores the procedure is more complicated, and an advanced textbook should be consulted.

The procedure for the analysis of iron (III) salts is similar; they should dissolve readily.

Standard potassium dichromate is used as titrant; the presence of hydrochloric acid makes potassium manganate (VII) unsuitable.

$$Cr_2O_7^{2-} + 14H^+ + 6e^- \longrightarrow 2Cr^{3+} + 7H_2O$$

$$Fe^{2+} \longrightarrow Fe^{3+} + e^-$$

Requirements

Titration apparatus and a 250 cm^3 standard flask.
Standard 0·02M potassium dichromate.
0·3% aqueous barium diphenylamine sulphonate indicator.
$\frac{2}{3}$M tin (II) chloride (150 g l^{-1} of $SnCl_2.2H_2O$ in 6M hydrochloric acid).
Concentrated 90% orthophosphoric acid.
Iron ore or an iron (III) salt.

Weigh accurately a dry, powdered sample of iron ore (roughly 5 g) or an iron (III) salt (roughly 12 g) into a 250 cm^3 beaker and dissolve by warming with about 30 cm^3 of concentrated hydrochloric acid. Carry out the warming *in a fume cupboard.* The iron ore will leave a white residue of silica which must be removed by filtration.

Transfer the cooled (and filtered) solution to the standard flask and dilute to the calibration mark with pure water.

Treat an aliquot of the diluted solution with about 5 cm^3 of concentrated hydrochloric acid and heat to about 90°C.

Add $\frac{2}{3}$M tin (II) chloride dropwise from a separating funnel to the hot solution to reduce iron (III) to iron (II). Continue until the yellow colour has almost disappeared. Conclude the reduction to a colourless or faintly green solution, using the tin (II) chloride solution diluted with 5 volumes of 2M hydrochloric acid. Take care not to add too much tin (II) chloride.

Cool rapidly to room temperature, and add 10 cm^3 of 0·1M mercury (II) chloride to remove excess tin (II) chloride. A silky white precipitate should form; insufficient tin (II) chloride was added if no precipitate forms, too much was added if the precipitate is grey.

Dilute to about 100 cm^3 with pure water and add about 10 cm^3 of 2M sulphuric acid and about 5 cm^3 of 90% orthophosphoric acid. Titrate the entire mixture with standard 0·02M potassium dichromate using barium diphenylamine sulphonate indicator (0·5 cm^3). At the end-point the indicator changes to a violet colour.

Calculate the percentage of iron in your sample and write the equations of the reactions that occur during the analysis.

7. Analysis of a suspected double salt

Iron (II) sulphate and copper (II) sulphate are mixed in solution in non-equimolar proportions and allowed to crystallize out together. The crystals obtained have a distinctive shape, and if the product is a true double salt the cations should be present in a whole-number ratio.

Carry out a quantitative analysis for iron (II) and copper (II) ions in order to establish their molar ratio.

Requirements

Titration apparatus and a standard flask.
Standardized 0·02M potassium manganate (VII).
Standardized 0·1M sodium thiosulphate.
0·5M potassium iodide.
Starch indicator.
The suspected double salt.

The salt is prepared as follows: add 50 g of hydrated copper (II) sulphate to 100 cm^3 of 2M sulphuric acid and heat to 75°C. Then add 45 g of hydrated iron (II) sulphate and, when dissolved, decant from any residue and chill the solution to 0°C. Collect the crystals at the pump, and dry at room temperature.

Prepare a standard solution of the salt (roughly 40 g l^{-1}) by accurate weighing.

Determine the iron (II) by titrating aliquots after acidification with 2M sulphuric acid, with the 0·02M potassium manganate (VII).

Determine the copper (II) by titrating aliquots after treatment with excess 0·2M potassium iodide, with the 0·1M sodium thiosulphate.

Use the equations of the reactions to calculate the molar ratio of iron (II) ion to copper (II) ion:

$$5Fe^{2+} + MnO_4^- + 8H^+ \longrightarrow 5Fe^{3+} + Mn^{2+} + 4H_2O$$

$$2Cu^{2+} + 4I^- \longrightarrow 2CuI(s) + I_2$$

$$I_2 + 2S_2O_3^{2-} \longrightarrow S_4O_6^{2-} + 2I^-$$

Also carry out, and interpret, qualitative tests on the solution of the suspected double salt using:

(*a*) 2M ammonia,
(*b*) 2M sodium hydroxide,
(*c*) ammonium thiocyanate solution.

8. Evaluation of commercial bleaches

Commercial bleaches are usually sodium hypochlorite solutions with an added dye. After suitable dilution, the hypochlorite can be determined by conversion to an equivalent amount of iodine and titration with standardized sodium thiosulphate. A number of bleach samples can be purchased to be analysed by a class, and results compared.

$$NaOCl + 2KI + H_2SO_4 \longrightarrow I_2 + H_2O + NaCl + K_2SO_4$$

$$I_2 + 2Na_2S_2O_3 \longrightarrow 2NaI + Na_2S_4O_6$$

Requirements

Titration apparatus and a standard flask.
Standardized 0·1M sodium thiosulphate.
0·5M potassium iodide.
Starch indicator.
Bottles of bleach solution.
10 cm^3 burettes.

The approximate volume of each bleach sample is measured, using a measuring cylinder, and recorded, together with the cost.

Run exactly 0·05 cm^3 of bleach from a 10 cm^3 burette directly into a conical flask. Dilute the sample with about 20 cm^3 of pure water and add 5 cm^3 of 0·5M potassium iodide, followed by 5 cm^3 of 2M sulphuric acid. Titrate the liberated iodine with standard 0·1M sodium thiosulphate, adding starch indicator (1 cm^3) when the iodine colour fades.

On the basis of this titration, calculate the volume of bleach which must be diluted in your standard flask so that aliquots will give a reasonable titre.

Run the calculated volume of bleach into your standard flask and dilute with 20 cm^3 of pure water. Now add a mixture of 0·5M potassium iodide and 2M sulphuric acid (1 : 1 by volume) until the liberated iodine forms a clear brown solution in an excess of potassium iodide. Dilute to the calibration mark with pure water. Titrate aliquots of the liberated iodine with standard 0·1M sodium thiosulphate.

Calculate the concentration (mol l^{-1}) of the original bleach used. Also calculate the cost per mole of sodium hypochlorite and choose a 'best buy' in commercial bleach solutions.

9. Analysis of lead (IV) oxide

Oxides of metals are often difficult to obtain pure. The technique described here can be used to analyse lead (IV) oxide prepared by the method on p. 30.

Requirements

Titration apparatus, and a colourless, glass-stoppered bottle.
Standardized 0·1M sodium thiosulphate.
0·5M potassium iodide.
Tetrachloromethane.
Sample of lead (IV) oxide.

Weigh accurately a sample of lead (IV) oxide (roughly 0·5 g) into a stoppered bottle. Add about 20 cm^3 of 0·5M potassium iodide followed by about 50 cm^3 of concentrated hydrochloric acid. Shake the mixture until no further reaction is observed. Dilute the mixture with about 50 cm^3 of pure water and titrate the entire sample while still in the stoppered bottle with standard 0·1M sodium thiosulphate. Add tetrachloromethane (about 5 cm^3) to detect the end-point and titrate *with thorough shaking* until the violet iodine colour disappears completely from the lower tetrachloromethane layer.

Deduce equations for the reactions occurring and calculate the purity of your sample of lead (IV) oxide. What other oxides could be analysed by the same technique?

10. Stoichiometry of a hydrazine reaction

Hydrazine is a powerful reducing agent, and in concentrated hydrochloric acid solution one product of the reaction with potassium iodate is iodine monochloride.

This experiment studies the stoichiometry of the reaction.

Requirements

Titration apparatus and a colourless, glass-stoppered bottle.
0·02M hydrazine sulphate (2·60 g l^{-1} of $NH_2NH_2.H_2SO_4$).
0·02M potassium iodate (4·28 g l^{-1} of KIO_3).
Tetrachloromethane (or Amaranth indicator).

Place an aliquot of 0·02M hydrazine sulphate in a colourless glass-stoppered bottle, and add about an equal volume of concentrated hydrochloric acid plus 10 cm^3 of tetrachloromethane (to act as iodine solvent).

Titrate the entire mixture with 0·02M potassium iodate. Iodine is liberated and the tetrachloromethane layer becomes dark purple. Shake *vigorously* during the titration; at the end-point the tetrachloromethane layer becomes colourless.

Alternatively, Amaranth indicator can be used instead of tetrachloromethane, and the titration carried out in a conical flask. The Amaranth is added near the end-point, and the first excess of potassium iodate will irreversibly bleach it.

Carry out test tube experiments, if necessary with stronger solutions, to identify the gas evolved during the reaction.

Deduce the stoichiometry of the reaction between iodate and hydrazine, and write the balanced equation of the reaction after completing the half equations:

$$IO_3^- \longrightarrow I^+$$

$$N_2H_4 \longrightarrow ?$$

How is the free iodine produced during the titration? Consult a textbook about the Andrews Titration.

11. Bromination of aromatic compounds

Some aromatic compounds can be quantitatively brominated using bromine solution. If an excess of a standard bromine solution is used, the excess can be determined by conversion to an equivalent amount of iodine and titration with sodium thiosulphate solution. The number of bromine atoms substituting into the aromatic ring can then be calculated.

Requirements

Titration apparatus and a standard flask, plus a colourless, glass-stoppered bottle.
0·1M sodium thiosulphate.
0·5M potassium iodide.
Standard potassium bromate–bromide solution (equivalent to standard 0·1M Br_2: 5·566 g l^{-1} $KBrO_3$ plus roughly 25 g l^{-1} KBr).
Starch indicator.
Aromatic compounds, such as:
phenol, $C_6H_5 \cdot OH$, M_r 94·1.
aniline, $C_6H_5 \cdot NH_2$, M_r 93·1.
o-, *m*- or *p*-nitroaniline, $NO_2 \cdot C_6H_4 \cdot NH_2$, M_r 138·1.
Sulphanilic acid, $NH_2 \cdot C_6H_4 \cdot SO_3H$, M_r 173·2.

Prepare a standard solution of the aromatic compound (roughly 1 g l^{-1}) by accurate weighing. Use M sulphuric acid as solvent.

Dilute an aliquot of the aromatic compound in a glass-stoppered bottle with about 50 cm³ of pure water. Add an equal aliquot of the standard potassium bromate–bromide solution and stopper the bottle at once. Stand the reaction mixture for fifteen minutes with occasional shaking.

Treat the contents of the stoppered bottle with about 20 cm³ of 0·5M potassium iodide and titrate the complete reaction mixture, while still in the bottle, with 0·1M sodium thiosulphate, using starch indicator (1 cm³) added near the end-point.

Standardize the sodium thiosulphate solution against the standard potassium bromate–bromide. Treat an aliquot of potassium bromate–bromide in sequence, with about 50 cm³ of pure water, about 20 cm³ of 0·5M potassium iodide and about 20 cm³ of M sulphuric acid. Titrate the complete reaction mixture with the sodium thiosulphate solution, using starch indicator.

Calculate the number of bromine atoms adsorbed per mole of organic compound, remembering that hydrogen bromide is produced in bromination reactions, and write the equation of the reaction.

Also write the equations of the other reactions used in the titration. Why is a standard bromine solution not used?

COMPLEXOMETRIC TITRATIONS

1. Introductory study

Ethylenediaminetetraacetic acid, H_4Y, in any form is known as EDTA, and its anion is given the symbol Y^{4-}. Solutions are usually made up using the dihydrate of the disodium salt (i.e. $Na_2H_2Y.2H_2O$).

EDTA, in the form of its anion, can combine with many cations, forming six or four co-ordinate complexes (see Fig. 42). These introductory experiments investigate the behaviour of EDTA and special metal indicators.

Introductory test tube experiments on the formation of complexes

(*a*) Add 0·05M EDTA (in solution as its disodium salt) to a nickel salt solution. Note any colour change suggesting the formation of a complex.

(*b*) Add ammonia dropwise to a copper salt until the azure blue colour of ammonia complexes is *just* formed. Then add EDTA when the colour should fade due to the formation of a pale blue, more stable EDTA complex.

(*c*) Add 4 cm³ of EDTA (an excess) to 1 cm³ of a lead salt, followed by potassium iodide or potassium chromate. Repeat this experiment without adding EDTA in order to observe the usual precipitates.

(*d*) Add sodium sulphate to a barium salt followed by an excess of EDTA. Divide into two portions: keep one portion and add pH 10 buffer to the other, and allow both to stand for five minutes.

(*e*) Add pH 10 buffer and an excess of EDTA to a barium salt. Now add sodium sulphate. This is Expt. (*d*) carried out with a different order of mixing. Is the final result the same? Why does the barium sulphate fail to appear?

(*f*) Add EDTA to a chromium (III) salt. Divide into two portions: keep one portion at room

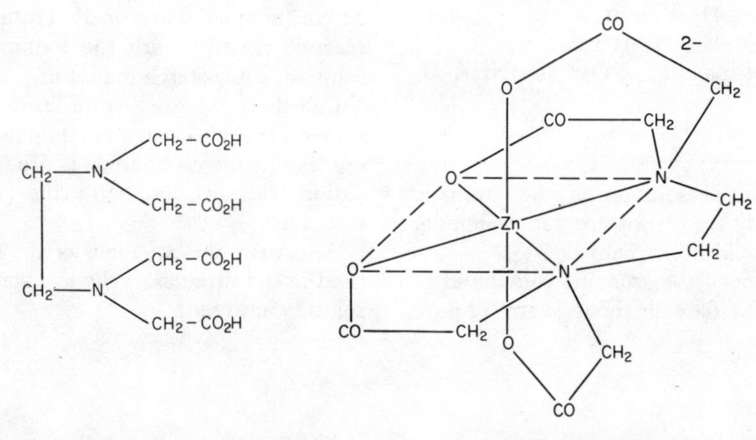

Fig. 42. EthyleneDiamineTetraAcetic acid (EDTA) and the ethylenediaminetetraacetate zinc (II) ion. (After Martell, A. E., *J. Chem. Educ.*, 1952, **29**, 270.)

Requirements

0·05M EDTA (18·61 g l⁻¹ of $Na_2H_2Y.2H_2O$).
Buffers (see Table 27).
Metal indicators (see Table 27).

Table 27

Metal indicators for EDTA, as 1:100 finely ground mixtures	*Buffers for EDTA*
Xylenol Orange in potassium nitrate	Solid glycine gives pH 4 Solid hexamine gives pH 6
Solochrome Black in sodium chloride	1M ammonia and 1M ammonium chloride in equal volumes give pH 8
Murexide in sodium chloride	380 cm³ '0.880' ammonia plus 70 g ammonium chloride per litre gives pH 10

temperature and heat the other until the colour changes. What is the cause of the colour change?

(*g*) Add a drop of methyl orange indicator to separate portions of EDTA and a zinc salt. If a red acid coloration is produced, add 0·1M ammonia *dropwise* to give the yellow colour. Mix the two solutions. Note the change in pH when the zinc complex forms.

These experiments show that cations will form stable complexes with EDTA at a suitable pH although the reaction may be slow. Furthermore, when a complex is formed hydrogen ions are released.

$$Zn^{2+} + H_2Y^{2-} \rightleftharpoons ZnY^{2-} + 2H^+$$

Introductory test tube experiments with metal indicators

(*a*) Investigate the influence of pH on the colour of an indicator by adding a pinch of indicator to each of the following solutions: M hydrochloric acid (pH 0), pH 4 buffer, pH 6 buffer, pH 10 buffer and M sodium hydroxide (pH 14). Compare results with a student who has tested a different indicator.

(*b*) Add a pinch of Solochrome Black to pH 10 buffer followed by a magnesium salt. Then add EDTA until a colour change occurs. Finally add more magnesium salt. Record the colour changes.

(*c*) Repeat Expt. (*b*) using Xylenol Orange indicator, pH 6 buffer and a lead salt.

(*d*) Add only a trace of indicator to pH 10 buffer, and scratch the inside of the test tube with a soft-glass rod. On standing, the indicator should slowly change colour.

These experiments show that suitable compounds are available for use as indicators. They form complexes with metal cations but can be displaced by EDTA with a colour change:

$$\underset{red}{\text{Mg (Solochrome Black)}^-} + H_2Y^{2-} \rightleftharpoons MgY^{2-} + \underset{blue}{\text{H(Solochrome Black)}^{2-}} + H^+$$

However, the indicators are pH sensitive, so titrations will have to be carried out in buffered solution to counteract the hydrogen ion released from EDTA when it forms a complex.

2. Methods for determining twelve cations

Most cations form complexes with EDTA, but only the univalent cation complexes are so weak that a titration procedure is not possible.

All complexes have the same 1 : 1 metal cation to EDTA anion ratio and the reaction between bivalent cations (M^{2+}) and EDTA (disodium salt Na_2H_2Y) can be expressed:

$$M^{2+} + H_2Y^{2-} \longrightarrow MY^{2-} + 2H^+$$

or for trivalent cations:

$$M^{3+} + H_2Y^{2-} \longrightarrow MY^- + 2H^+$$

A quantitative titration with a good end-point depends on the choice of the correct pH and indicator for the particular cation. The stability constant of the cation–EDTA complex is of fundamental importance; the higher the value of the stability constant, the more stable the complex and the lower the pH at which it is dissociated.

The stability constant, K_{MY}, is defined (unlike an equilibrium constant) as:

$$K_{MY} = \frac{[MY^{n-4}]}{[M^{n+}]\,[Y^{4-}]}$$

Some typical values are given in Table 28.

Table 28

Complex	*Log₁₀ K_MY*
NaY^{3-}	1·66
MgY^{2-}	8·69
CaY^{2-}	10·70
PbY^{2-}	18·04
FeY^-	25·10

Requirements

Titration apparatus.
Standard 0·05M EDTA (18·61 g l^{-1} of $Na_2H_2Y.2H_2O$).

Standard 0·05M magnesium sulphate (12·33 g l^{-1} of $MgSO_4.7H_2O$).
Metal indicators (see Table 29).
Buffer solutions (see Table 29).
Hydroxyammonium chloride for Mn.
Salicylic acid for Fe.

Determination of lead (II). Dilute a 10·0 cm^3 aliquot of an approximately 0·1M lead (II) salt solution with about 50 cm^3 of de-ionized water and add roughly 4 g of pH 6 buffer (solid hexamine).

Finally, add a small portion of Xylenol Orange indicator, and titrate with standard 0·05M EDTA until the colour changes from red-violet to lemon yellow.

Determination of cobalt (II). The procedure is the same as for lead, except that the complex forms slowly so the solution should be warmed to 40°C.

Determination of mercury (II). The procedure is the same as for lead. A grey precipitate appears but does not obscure the end-point. The method is limited to mercury (II) in the absence of halides.

Determination of bismuth. The procedure is the same as for lead, except that 2M nitric acid is used to adjust the pH to 1–2.

Determination of zinc. Dilute a 10·0 cm^3 aliquot of an approximately 0·1M zinc salt solution with about 50 cm^3 of de-ionized water, and add 2 cm^3 of pH 10 buffer.

Finally, add a small portion of Solochrome Black indicator, and titrate with standard 0·05M EDTA until the colour changes from mauve to blue.

Determination of magnesium. The procedure is the same as for zinc, except that the magnesium–EDTA complex forms slowly, so the solution should be warmed to 40°C.

Determination of manganese (II). The procedure is the same as for zinc, except that the first addition to the aliquot of manganese (II) solution is solid hydroxyammonium chloride to prevent atmospheric oxidation of the manganese (II) salt. Add sufficient to avoid any brown precipitate.

Determination of copper (II). The procedure is the same as for zinc, except that the pH 8 buffer and Murexide indicator are used. The colour is initially green, fading to a grey tone; at the end-point the colour changes to mauve.

Determination of calcium. The procedure is the same as for zinc, except that pH 12 buffer (5 cm^3 of 2M sodium hydroxide) and Murexide indicator are used. The colour is initially red, changing to purple; at the end-point the colour changes to violet.

Determination of iron (III). The procedure is the same as for zinc, except that pH 4 buffer (solid glycine) and salicylic acid indicator are used. The complex forms slowly, so the solution must be heated to 40°C. The colour is initially dark brown, changing to yellow at the end-point. The method is not easy.

Determination of nickel by back titration. Add a 20·0 cm^3 aliquot of standard 0·05M EDTA. (an excess) to a 5·0 cm^3 aliquot of an approximately 0·1M nickel salt solution. Dilute with about 100 cm^3 of de-ionized water and add 4 cm^3 of pH 10 buffer.

Finally, add a small portion of Solochrome Black indicator, and titrate with standard 0·05M magnesium sulphate until the colour changes from blue to purple.

Determination of aluminium by back titration. The procedure is the same as for nickel, except that the complex forms slowly. Therefore pH 8 buffer is used and the solution boiled for two minutes.

Before adding indicator and titrating, cool to room temperature and add more pH 8 buffer.

3. Hardness of water

Hardness is usually expressed in the arbitrary units of 'parts per million as calcium carbonate'. In this titration the end-point will correspond to the complexing of all cations forming a stable EDTA complex at pH 10. However, if the cations of copper, cobalt or nickel are present, the indicator action is blocked and the method described will fail; small amounts of iron will not interfere.

The presence of magnesium is essential for Solochrome Black does not give a sharp end-point with calcium alone.

The amount of tap water to use will depend on your local water supply. If the hardness is between 10 and 250 p.p.m. use a 100 cm^3 aliquot, if above 250 p.p.m. use a 50 cm^3 aliquot.

Requirements

Titration apparatus.
Standard 0·02M EDTA solution.
Patton and Reeder's indicator, or Solochrome Black.
Murexide indicator.
pH 10 buffer.

Total hardness. Treat a suitable aliquot of tap water with 2 cm^3 of pH 10 buffer and sufficient Patton and Reeder's indicator to develop a good colour.

Titrate with standard 0·02M EDTA until the indicator shows no trace of red but is pure blue (or, occasionally, neutral grey). If the EDTA titre is rather small, use a larger aliquot of tap water for subsequent titrations.

Permanent hardness. Gently boil a suitable aliquot of tap water for two minutes. Filter the aliquot, cool to room temperature, and then add 2 cm^3 of pH 10 buffer and a small portion of Patton and Reeder's indicator. Titrate all the treated filtrate with standard 0·02M EDTA until the indicator changes colour from red to blue.

The boiling precipitates hydrogen carbonate ion as calcium carbonate, which takes no further part in the titration. The loss on the filter paper is not significant.

Calcium hardness. Bring a suitable aliquot of tap water to pH 12 by adding 5 cm^3 of 2M sodium hydroxide. Add a small portion of Murexide, and titrate with standard 0·02M EDTA until the indicator changes colour from red to violet.

The end-point with Murexide is difficult to observe, as a purple tone develops just before the change to violet; at the true end-point an additional 0·1 cm^3 of EDTA should not change the violet colour.

The adjustment to pH 12 precipitates magnesium ion as magnesium hydroxide, which takes no further part in the reaction.

Since the reaction between calcium or magnesium ions and EDTA has a 1 : 1 mole ratio, the hardness is given by the expression:

$$\text{p.p.m. as calcium carbonate} = \frac{\text{titre volume}}{\text{aliquot volume}} \times \text{mol l}^{-1} \text{ of EDTA} \times 10^5$$

Derive this expression yourself.

Calculate the different degrees of hardness by substitution of the appropriate values in this expression. Other types of hardness are obtained by differences, thus:

temporary hardness = (total − permanent) hardness
magnesium hardness = (total − calcium) hardness

In all cases the units are 'p.p.m. as calcium carbonate'.

4. Magnesium salts in pharmaceutical preparations

Many indigestion mixtures are based on magnesium hydroxide, while 'health salts' are based on magnesium sulphate. Since magnesium cations react readily and quantitatively with EDTA, it is possible to determine their concentration in pharmaceutical preparations by titration:

$$Mg^{2+} + H_2Y^{2-} \longrightarrow MgY^{2-} + 2H^+$$

To avoid complications, choose a preparation containing only alkali metals apart from magnesium.

Requirements

Titration apparatus and a standard flask.
Standard 0·02M EDTA.
0·02M hydrochloric acid *or* 0·02M sodium hydroxide.
'Health salts' or indigestion mixture.
Solochrome Black indicator.
Methyl orange indicator.
pH 10 buffer.

The first problem to resolve is the amount of preparation to take.

Weigh roughly 0·1 g of the selected preparation into a conical flask and treat with about 20 cm^3 of pure water. Test the diluted material with litmus paper and then neutralize approximately by titration with 0·02M acid or alkali, as appropriate, using methyl orange indicator. Carry out the titration slowly if solids have to be dissolved.

Now titrate the neutral solution with standard 0·02M EDTA, after adding about 5 cm^3 of pH 10 buffer and a small portion of Solochrome Black indicator. At the end-point the colour changes from red to blue.

Calculate the appropriate amount of preparation to dissolve in your standard flask to give a suitable concentration for titration. Calculate also the volume of acid or alkali required to neutralize the preparation.

Weigh accurately the calculated amount into a 500 cm^3 beaker and treat with about 50 cm^3 of pure water: carefully if effervescent health salts are used. Add the calculated amount of 0·02M acid or alkali to produce a neutral solution, and then completely transfer to the standard flask and dilute to the calibration mark with pure water.

Treat aliquots of the dilute neutral solution with 5 cm^3 of pH 10 buffer, and titrate with standard 0·02M EDTA, using a small portion of Solochrome Black indicator.

Calculate the percentage by weight of magnesium in the original pharmaceutical preparation, and compare with the manufacturer's specification.

5. Analysis of brass

The analysis of alloys is difficult as the metals must either be quantitatively separated or specific analytical methods must be available.

The principal metals in brass are copper and zinc. The copper is determined by iodine titration:

$$2Cu^{2+} + 4I^- \longrightarrow 2CuI(s) + I_2$$
$$I_2 + 2Na_2S_2O_3 \longrightarrow Na_2S_4O_6 + 2NaI$$

and the zinc can then be determined by EDTA titration:

$$Zn^{2+} + H_2Y^{2-} \longrightarrow ZnY^{2-} + 2H^+$$

Requirements

Titration apparatus with two burettes, and a standard flask.
Standardized 0·05M sodium thiosulphate.
0·5M potassium iodide.
M potassium thiocyanate.
Starch indicator.
Standard 0·05M EDTA.
pH 6 buffer (solid hexamine).
Xylenol Orange indicator.
Solid brass.

A sample of brass is required that will give a concentration of 10 g l^{-1} after dissolving and dilution in a convenient standard flask.

Weigh accurately a suitable sample of brass into a 400 cm^3 beaker, and dissolve in the *minimum* quantity of concentrated nitric acid. Transfer, with washings to a standard flask. Make up to the calibration mark with pure water.

Titrate aliquots of the brass solution with standardized 0·05M sodium thiosulphate after treating the aliquot with about 10 cm^3 of 0·5M potassium iodide. When the iodine colour fades, add starch and continue the titration. When the blue-black starch colour fades, add about 10 cm^3 of M potassium thiocyanate, which should revive the starch colour. Titrate rapidly to the true end-point (titre V_{Cu}).

Now add to the titration mixture roughly 4 g of hexamine buffer and titrate with EDTA, using a pinch of Xylenol Orange indicator. At the end-point the colour changes from red to yellow (titre V_{Zn}).

Using the equations given above, calculate the concentrations (mol l^{-1}) of copper and zinc in the diluted brass solution. Hence calculate the percentage composition of the brass. Check the accuracy of your work by calculating the actual weights of copper and zinc in the original brass; any difference from the accurate weight used will be due to other constituents not determined (or to error).

What are the properties of brass that make it a useful alloy?

6. Analysis of solder

The principal metals in solder are lead and tin, and they can be determined in a single sample by titration with EDTA.

When lead and tin (IV) ions in solution are treated with an excess of standard EDTA at pH 6 they will both form complexes:

$$Pb^{2+} + Sn^{4+} + 2H_2Y^{2-} \longrightarrow PbY^{2-} + SnY + 4H^+$$

The excess of EDTA is determined by titration with a lead solution.

If sodium fluoride is now added, it releases EDTA from the tin complex because the more stable hexafluorostannate ion forms:

$$SnY + 6F^- + 2H^+ \longrightarrow SnF_6^{2-} + H_2Y^{2-}$$

The amount of released EDTA, which is equivalent to the tin, is determined by continuing the titration with the lead solution.

The remainder of the EDTA is still complexed to the lead and is determined by difference.

Requirements

Titration apparatus and a 250 cm³ standard flask.
Standard 0·05M EDTA solution.
0·01M lead nitrate solution [3·312 g l⁻¹ of $Pb(NO_3)_2$].
pH 6 buffer (hexamine solid).
Sodium fluoride (solid).
Xylenol Orange indicator.
A lead–tin solder.

Weigh accurately a sample of powdered solder (roughly 0·2 g) into a 250 cm³ beaker. Add 5 cm³ of concentrated hydrochloric acid and 1 cm³ of concentrated nitric acid. Cover with a watch-glass, and simmer gently *in a fume cupboard* for five minutes to dissolve the solder and expel gaseous reaction products. Crystalline lead chloride will separate.

Add a 50·0 cm³ aliquot of standard 0·05M EDTA, and boil gently for one minute to complex the tin and lead ions. If it appears that not all your solder has dissolved, the residue must be recovered, dried and weighed. Cool the solution and transfer, with washings, to a 250 cm³ standard flask, retaining any residue of solder. Dilute to the calibration mark with pure water. This solution should be clear but may be very slightly cloudy due to hydrolysis of tin (IV) ions.

The diluted solution is not stable, so the titration must be carried out without delay.

Treat 25·0 cm³ aliquots of the diluted reaction mixture with roughly 4 g of hexamine buffer and about 100 cm³ of pure water. Add a small portion of Xylenol Orange indicator, and titrate with lead nitrate solution. At the end-point the colour changes from yellow to mauve (titre V_{excess}).

Now add roughly 2 g of sodium fluoride (***caution: toxic***) to the titration flask, and shake to dissolve. The EDTA complexed to the tin (IV) ions is released and the indicator should change back to yellow. Continue the titration to the second end-point (titre V_2). This end-point is subject to some uncertainty, as the mauve colour will appear and then fade as more EDTA is released.

Finally, treat a 5·0 cm³ aliquot of standard EDTA with 100 cm³ of pure water, 4 g of hexamine buffer and Xylenol Orange indicator, and titrate with lead nitrate solution (titre V_3).

Calculate the concentrations (mol l⁻¹) of the cations in the standard flask solution using the relationships (provided the volumes were exactly as stated):

$$\frac{\text{mol l}^{-1}\text{ of solder tin}}{\text{mol l}^{-1}\text{ of lead titrant}} = \frac{(V_2 - V_{excess})}{\text{aliquot volume}}$$

$$\frac{\text{mol l}^{-1}\text{ of solder lead}}{\text{mol l}^{-1}\text{ of lead titrant}} = \frac{(V_3 - V_2)}{\text{aliquot volume}}$$

Hence calculate the weight of lead and tin in the solder and the percentage composition. Any difference from the original weight will be due to other constituents not determined (or to error).

Derive for yourself the relationships quoted above.

How do the properties of solder differ with changing composition? Consult a textbook about the lead–tin phase diagram.

PRECIPITATION TITRATIONS

1. Solubility of potassium halides using Fajans' method

The precipitation of halides from solution by silver nitrate is a simple quantitative procedure when adsorption indicators are used to pinpoint the end-point:

$$Ag^+ (aq) + Hal^- (aq) \longrightarrow AgHal(s)$$

In neutral conditions the silver halides are very insoluble, so the completion of precipitation corresponds accurately to the equivalence point. During the titration the silver halide precipitate has a surface layer of adsorbed anions, but at the equivalence point there is a change to adsorbed cations.

To make the end-point visible, a dyestuff is added which is adsorbed with the cations, changing colour at the same time. The introduction of suitable dyestuffs is due to Fajans.

Table 29

Silver halide	*Dyestuff*	*Colour change*
All	Dichloro-fluorescein	Green fluorescence to deep pink
Chloride	Fluorescein	Green fluorescence to pink
Bromide and iodide	Tetrabromo-fluorescein (eosin)	Orange to deep red

Silver nitrate is available very pure and may be used as a primary standard, but it is expensive so silver salt residues should be retained for recovery of the silver (see p. 21).

This experiment is to determine the solubility of a potassium halide.

Requirements

Titration apparatus and a standard flask.
Standard 0·05M silver nitrate (8·495 g l^{-1} of $AgNO_3$).
Adsorption indicators (0·1% in 70% aqueous ethanol).
Saturated solutions of potassium halides.

Calculate the volume of saturated potassium halide solution (5–10M) to dilute, in your standard flask, in order to obtain a solution suitable for titration by the 0·05M silver nitrate. As this is a solubility determination it is necessary to weigh the halide solution.

Weigh accurately the sample of saturated potassium halide solution and transfer completely, with washings, to a standard flask. Make up to the calibration mark with pure water.

Titrate aliquots of the diluted halide solution with standard 0·05M silver nitrate, using a suitable adsorption indicator (10 drops). Just before the end-point the precipitate of silver halide will coagulate while, at the end-point, the dyestuff will be adsorbed with a change of colour.

Calculate the concentration (mol l^{-1}) of the diluted halide solution, and hence the weight of potassium halide in the original sample. The weight of water in the sample is determined by subtraction from the original accurate weight.

Finally, calculate the solubility of the potassium halide in g of salt per 100 g of water at room temperature.

Describe the mode of behaviour of the indicator in more detail. Consult a textbook about colloids.

2. Concentration of constant-boiling halide acids using Mohr's method

In Mohr's method for titration with silver nitrate, potassium chromate is used as indicator in neutral or slightly alkaline solution, and a blood-red precipitate of silver chromate appears at the end-point. The method is suitable for the determination of chloride and bromide, but other silver salt precipitates adsorb ions giving a poor end-point. By calculation from solubility products it can be shown that virtually all halide ions are precipitated before silver chromate appears, provided that the correct small amount of potassium chromate indicator is used.

K_{S_0} for AgCl $= 5{\cdot}6 \times 10^{-9}$ mol^2 l^{-2}
K_{S_0} for AgBr $= 2{\cdot}0 \times 10^{-12}$ mol^2 l^{-2}
K_{S_0} for $Ag_2CrO_4 = 3{\cdot}4 \times 10^{-11}$ mol^3 l^{-3}

The experiment is to determine the concentration of a constant-boiling halide acid.

Requirements

Titration apparatus, plus a 10 cm^3 burette and a standard flask.
Standard 0·05M silver nitrate (8·495 g l^{-1} of $AgNO_3$).
Constant-boiling halide acid.*
A.R. calcium carbonate (or A.R. sodium hydrogen carbonate).
Neutral chromate indicator (42 g l^{-1} of K_2CrO_4 plus 7 g l^{-1} of $K_2Cr_2O_7$).

Calculate the volume of constant-boiling acid (about 5M) to dilute, in your standard flask, in order to obtain a solution suitable for titration by the 0·05M silver nitrate.

Measure the sample of constant-boiling acid from a 10 cm^3 burette directly into a standard flask. Dilute to the calibration mark with pure water.

Treat aliquots of the diluted acid with small spatula loads of A.R. calcium carbonate until an undissolved residue remains. Titrate the neutralized solution with standard 0·05M silver nitrate, using potassium chromate indicator (1 cm^3). Near the end-point a red precipitate of silver chromate will appear but redissolve; at the true end-point the previously white silver chloride will become permanently pink.

Calculate the concentration (mol l^{-1}) of the diluted acid solution:

$$2HHal(aq) + CaCO_3(s) \longrightarrow CaHal_2(aq) + H_2O(l) + CO_2(g)$$

$$2AgNO_3(aq) + CaHal_2(aq) \longrightarrow 2AgHal(s) + Ca(NO_3)_2(aq)$$

and hence the concentration (mol l^{-1}) of constant-boiling acid at the atmospheric pressure recorded.

Does the constant-boiling mixture ever vary in composition? What are the properties of an azeotropic mixture? Consult a textbook if necessary.

* Constant-boiling hydrobromic acid can be prepared by the procedure on p. 42. Alternatively, constant-boiling hydrochloric acid can be purchased.

3. Chloride in cheese using Volhard's method

For the determination of halides in acidic solution an excess of silver nitrate is used. The excess of silver nitrate is determined by titration with a thiocyanate solution:

$$Ag^+(aq) + SCN^-(aq) \longrightarrow AgSCN(s) \quad K_{S_0} = 1 \times 10^{-12}$$

An iron (III) salt is used as indicator to detect the end-point. The first excess of thiocyanate will react with iron (III) ions to produce a deep red colour:

$$Fe^{3+}(aq) + SCN^-(aq) \longrightarrow Fe(SCN)^{2+}(aq) \text{ etc.} \quad K_1 = 1 \times 10^3$$

This experiment to determine the chloride in cheese is based on British Standard 770:1963.

Requirements

Titration apparatus.
Standard 0·05M silver nitrate solution (8·495 g l^{-1} of $AgNO_3$).
0·04M potassium thiocyanate solution (3·9 g l^{-1} of KSCN).
Indicator of potassium iron (III) sulphate solution (50 g in 90 cm^3 water plus 10 cm^3 2M nitric acid).
Nitrobenzene.
Samples of cheese.

First standardize the potassium thiocyanate using the silver nitrate. Measure an aliquot of standard 0·05M silver nitrate into a conical flask, dilute with about 50 cm^3 of pure water and acidify with about 10 cm^3 of concentrated nitric acid. Titrate the entire mixture with the potassium thiocyanate solution, using iron (III) alum indicator and nitrobenzene (1 cm^3), until there is a permanent orange tint. Calculate the exact concentration (mol l^{-1}) of the potassium thiocyanate.

Weigh accurately a sample of cheese (roughly 1·5 g) into a conical flask.

Heat the flask to about 75°C in a water-bath and add, in sequence, 10 cm^3 of pure water, a 25·0 cm^3 aliquot of standard 0·05M silver nitrate (an excess) and about 10 cm^3 of concentrated nitric acid. Digest the cheese curd by boiling gently for ten minutes; carry out this part of the experiment *in a fume cupboard* to avoid strong cheese odours. The digestion is complete when the liquid is clear lemon and the fat layer free from solid, except for granular silver chloride.

Cool, and dilute the reaction mixture by adding about 50 cm^3 of pure water. Determine the excess of silver nitrate in the reaction mixture by titration with the standardized potassium thiocyanate solution, using iron (III) alum indicator (2 cm^3) and nitrobenzene (1 cm^3). The end-point is an orange tint which should persist for fifteen seconds.

The experiment can be repeated with different cheese for comparison; for each calculate the percentage of chloride (expressed as NaCl) in the cheese. Why does cheese contain salt?

Why is nitrobenzene added to the titration mixture?

ELECTROCHEMICAL TITRATIONS

1. Acid-base titrations using a pH meter

The change in pH during an acid-base titration may be followed with a pH meter, which is essentially a potentiometer for measuring the e.m.f. of the cell:

Calomel electrode ¦ Solution | Glass electrode

The calomel electrode retains constant potential, while the glass electrode varies its potential with the hydrogen-ion concentration of the solution. Hence this specialized potentiometer may be calibrated directly in pH units.

The operating instructions for your particular pH meter should be studied and closely followed.

Requirements

- pH meter and electrodes (see Fig. 43).
- 0·1M sodium hydroxide.
- 0·1M acids (HCl, CH_3CO_2H, H_3PO_4).
- 600 r.p.m. stirrer.

Titrate an aliquot of 0·1M acid with 0·1M sodium hydroxide, using a suitable indicator to obtain an approximate end-point.

Check the pH meter after rinsing the electrode assembly in pure water by dipping the electrodes into a standard buffer solution. Adjust the pH meter so that its pH reading agrees with the stated pH of the buffer solution.

To obtain pH values during a titration, place an aliquot of 0·1M acid in a 200 cm^3 beaker and dip in the electrodes, adding distilled water until they are properly immersed. Take a pH reading.

Run in 4·0 cm^3 of 0·1M sodium hydroxide from a burette and, after stirring well, take another pH reading.

Take pH readings after each 4·0 cm^3 addition of sodium hydroxide until about 5 cm^3 from the approximate end-point, when 1·0 cm^3 additions should be made, and finally 0·2 cm^3 additions when close to the end-point. Make further additions until about 10 cm^3 beyond the end-point.

Plot graphs of *pH* against *volume of sodium hydroxide added*, note the pH at each end-point, and relate it to the pH range of the indicator recommended for that titration. Comment on the shape of the pH curves.

When the experiment is complete, the electrode assemblies must be left immersed in distilled water, and a pH meter which is mains-operated should be left switched on.

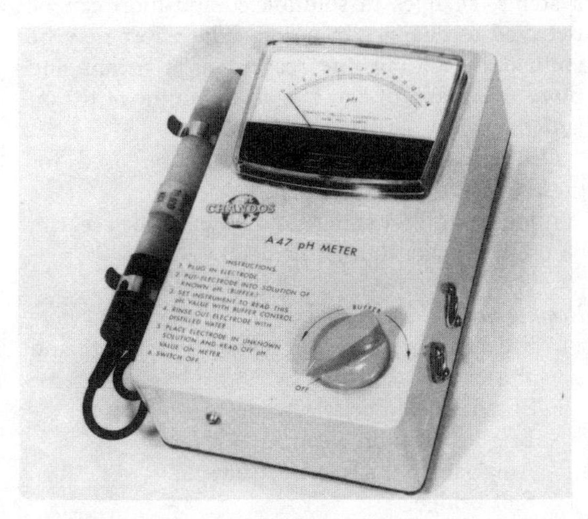

Fig. 43a. The Chandos pH meter.

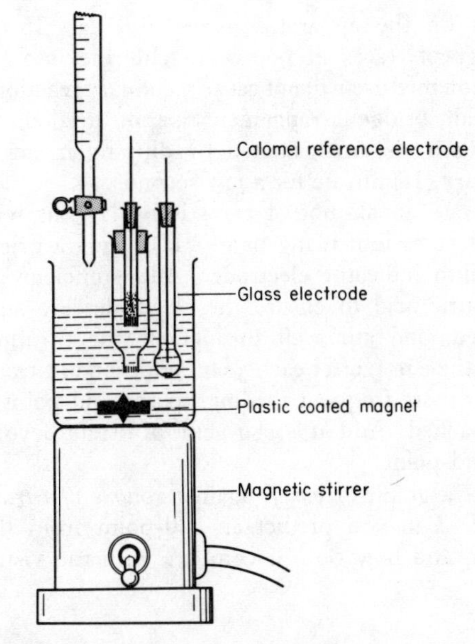

Fig. 43b. Titration using a pH meter.

2. Potentiometric titrations

Experiments on electrode potentials (see p. 135) show that when metals are introduced into solutions of their ions an e.m.f. will be developed which is dependent on the nature of the metal and the concentration of the ions. Thus, in reacting systems, changes in solution composition can be detected as changes in e.m.f. The effect may be studied by adding one reagent as a titrant and plotting a graph of e.m.f. against volume of titrant added.

Ordinary reagent solutions may be used for these exploratory experiments.

A length of filter paper from a 1 cm wide reel can be used to make the salt bridge.

Requirements

Valve voltmeter or pH meter (see pp. 135 and 179).
Standard calomel electrode.
Bright platinum electrode.
Amalgamated silver electrode.
Salt bridge.
Titration apparatus.
600 r.p.m. stirrer.
Solutions (see text).

Set up the apparatus as in Fig. 44. If the leakage of traces of potassium chloride into the titration mixture will not cause secondary reactions, the salt bridge arrangement is not needed. A silver wire is amalgamated by dipping in acidic mercury (II) nitrate for a few seconds.

Titrate an aliquot of 0·1M iron (II) salt with 0·02M potassium manganate (VII), using a bright platinum indicator electrode. Add sufficient 2M sulphuric acid to ensure the electrodes are submerged, and stir well throughout the titration. Note the e.m.f. after each 3 cm^3 addition of titrant, taking more frequent readings as the end-point is approached. Add at least 10 cm^3 of titrant beyond the end-point.

Plot a graph of *e.m.f.* against *volume of titrant added.* Can you predict an end-point from the graph, and how does it compare with the visual end-point? The experiment may be repeated with 0·02M potassium dichromate. What advantages can you see for the potentiometric method?

Titrate an aliquot of 0·05M silver nitrate with a 0·1M halide solution, using an amalgamated silver wire as the indicator electrode and the salt bridge arrangement.

Dilute the silver nitrate aliquot with sufficient water to submerge the electrodes, and carry out the titration as above.

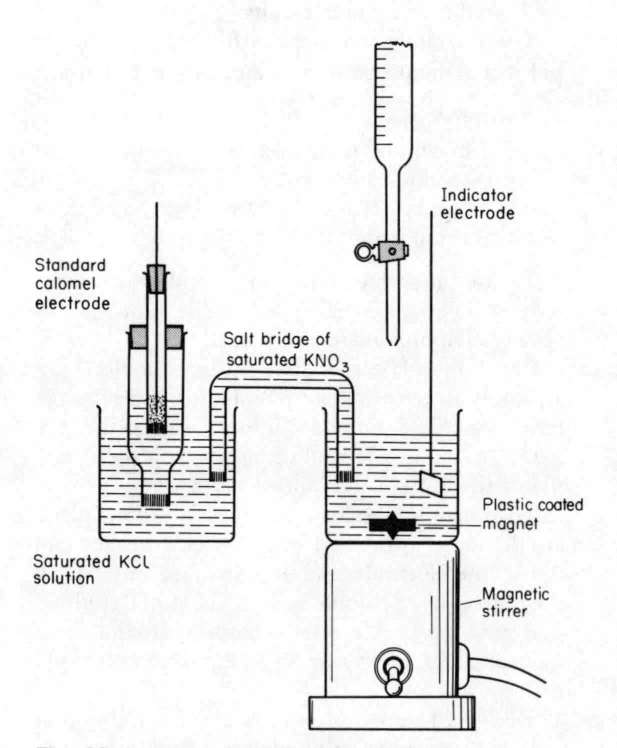

Fig. 44. Apparatus for potentiometric titrations.

Plot a graph of *e.m.f.* against *volume of titrant added.* The experiment may be repeated with the other halides and the results compared. Can you account for any differences in the e.m.f. at the end-points?

3. Conductometric titrations

In an acid-base titration, changes in conductivity will be mainly due to changes in hydrogen or hydroxide ion concentration. This is because the conductivity of hydrogen or hydroxide ions is appreciably greater than the conductivity of other ions.

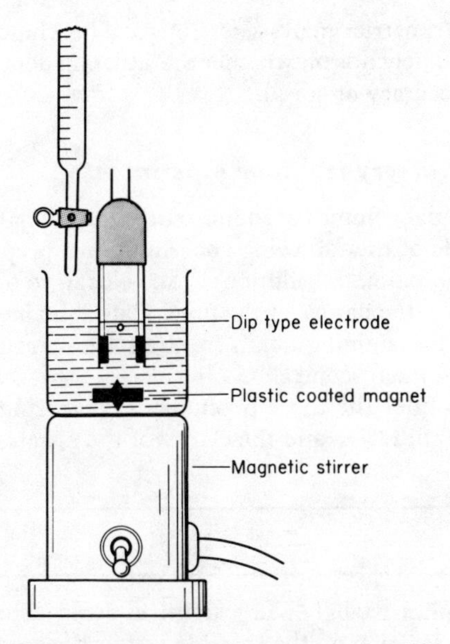

Fig. 45. Apparatus for conductometric titrations.

Dilution during the experiment is minimized by using a titrant twenty times more concentrated than the solution being titrated.

Solutions need not be made up using conductance water, as only relative values of conductivity are required.

Requirements

Conductivity bridge with dip cell.
Titration apparatus using 10 cm³ burette and 5 cm³ safety pipette.
2M acids and bases.
2M ammonium acetate solution.
600 r.p.m. stirrer.

Fig. 45 shows a suitable arrangement of the apparatus; acid is added to the diluted base with continuous stirring, and conductivity readings taken after each addition.

Place a 5·0 cm³ aliquot of a 2M base in a beaker and dilute with 95 cm³ of pure water. Stir the mixture well, and take a conductivity reading.

Add exactly 1·0 cm³ portions of 2M acid from the burette to a total of 9·0 cm³; take a conductivity reading after each addition.

Suitable acids and bases are given in Table 30.

Table 30

Base in the pipette	*Acid in the burette*
(*a*) 2M sodium hydroxide	2M hydrochloric acid
(*b*) 2M ammonia	2M hydrochloric acid
(*c*) 2M ammonium acetate	2M hydrochloric acid
(*d*) 2M ammonia	2M acetic acid

The example (*a*) should also be titrated in the conventional manner using methyl orange indicator.

When the experiment is complete, rinse the conductivity cell well with conductance water and put away carefully.

Plot a graph of *conductivity* in ohms^{-1} against *titre* in cm³ of acid; the graph can be taken as two straight lines which intersect at an 'end-point'. What explanation can you offer of the relative slopes of the two lines? Determine the titre value at the 'end-point'; does it correspond to the conventional indicator end-point?

VII Gravimetric Analysis

1. Introduction

GRAVIMETRIC analysis involves weighing a pure compound of known composition, thereby determining the total quantity present of one of its components. The component being determined is usually removed from solution as a precipitate, collected by filtration, and accurately weighed after drying or ignition.

For the technique to be successful precipitation must be complete, so the substance must be very insoluble; filtration must be easy, so the substance must be crystalline rather than colloidal or gelatinous; and the final product must be pure, so the reaction should be specific.

Gravimetric analysis is capable of great accuracy, but titrimetric analysis is faster. The choice of method depends on whether the situation demands high accuracy or speed.

Introductory test tube experiment

Mix equal volumes of sodium sulphate and barium chloride of the following concentrations, prepared by approximate dilution: 1M, 0·02M, 0·001M. Describe the nature of the precipitates obtained, if possible examining a thin smear of the suspension under a microscope.

Also filter the three precipitates comparing the rates of filtration and the clarity of the filtrates.

2. Loss in weight on heating

If a pure substance changes to another pure substance on heating, the nature of the change can be elucidated in part by studying the weight changes quantitatively.

A simple example occurs when a salt loses water of crystallization on heating:

$$\underset{\text{original weight}}{BaCl_2.nH_2O(s)} \longrightarrow \underset{\text{residual weight}}{BaCl_2(s)} + \underset{\text{loss in weight}}{nH_2O(g)}$$

hence n can be determined.

Requirements

- Porcelain crucible.
- Silica triangle.
- Clean tongs.
- Compounds, such as the hydrates of barium chloride, magnesium sulphate, sodium tetraborate; also sodium hydrogen carbonate, sodium sesquicarbonate.

Select a clean crucible, free from cracks, and support it above a small blue Bunsen flame in a clean silica triangle. Be careful to avoid deposits of soot forming on the crucible. After the crucible is warm, raise briefly to red heat using a larger flame (but not roaring). The crucible is now clean and dry, so it must be handled with clean tongs and placed only on clean surfaces.

Allow the crucible to cool *in situ* for two minutes, then transfer, using clean tongs, to a clean 100 cm³ beaker and allow to cool completely to room temperature. A desiccator is not necessary (see p. 14).

Finally, stand the crucible in the beaker near the balance for two minutes; for accurate weighing the balance and crucible must be at the same temperature.

Weigh accurately the clean empty crucible (W_1). Remove from the balance and half fill with the chosen compound. Reweigh accurately (W_2).

Replace the crucible with contents in the silica triangle and heat for ten minutes. Heat with a small blue flame for the first two minutes so that the *temperature rises gradually*; material is less likely to spit out and the crucible itself is less likely to crack. For the remaining eight minutes heat at just below dull red heat; at higher temperatures reaction with the porcelain glaze is possible.

As before, allow the crucible to cool completely and then weigh accurately (W_3).

To test that the loss in weight is complete, reheat for three minutes, cool and reweigh accurately (W_4). If (W_3) and (W_4) are not identical to ± 5 mg, further periods of heating are necessary until constant weight is obtained.

The number of molecules of water of crystallization can be calculated from the relationship:

$$\frac{n \times M_r \text{ of water}}{M_r \text{ of anhydrous salt}} = \frac{\text{loss in weight}}{\text{residual weight}} = \frac{W_2 - W_3}{W_3 - W_1}$$

When the experiment is complete observe the effect of adding water dropwise to the cold residue.

3. Determination of chlorides

Chlorides can be analysed by conversion to an equivalent amount of silver chloride which is very insoluble and readily collected by filtration:

$$Cl^-(aq) + AgNO_3(aq) \longrightarrow AgCl(s) + NO_3^-(aq)$$

Silver chloride is photosensitive so this experiment should not be conducted in a strong light.

Requirements

Dry sintered-glass crucible (porosity 3).
0·5M silver nitrate solution.
Samples of pure metal chlorides (Group IA or IIA of the periodic table).
Buchner filtration apparatus.
Glass rod with rubber 'policeman'.
50 cm³ beaker.
Silver residues bottle.

Weigh accurately a sample of pure dry metal chloride (roughly 0·5 g for light elements, 1 g for heavy elements, W_1) into a 500 cm³ beaker. Also weigh accurately a dry sintered-glass crucible (W_2).

Dissolve the metal chloride in about 20 cm³ of pure water, warm if necessary, then dilute to about 100 cm³ with pure water. Do not lose any of this solution (e.g. on a glass rod used for stirring) or the final result will be inaccurate.

Acidify the solution with about 1 cm³ of concentrated nitric acid, then add about 25 cm³ of 0·5M silver nitrate *slowly with stirring*. This should be an excess and will cause the silver chloride precipitate to coagulate. When the precipitate has coagulated, test the supernatant liquid by the addition of 1 drop of silver nitrate solution. If an excess was present, no further precipitate should form.

Now transfer the precipitate to the pre-weighed sintered-glass crucible by filtration at the pump (for full details of procedure, see p. 14). Slow and careful work is essential at this stage. Wash the precipitate in the crucible with several portions of 0·01M nitric acid to remove all traces of soluble salts; continue until the filtrate remains clear when 2M hydrochloric acid is added. Finally, wash with two portions of methanol to remove water and acid.

Place the crucible and contents in a clean labelled beaker for protection and identification, and dry to constant weight in an oven at about 120°C. The drying is conveniently carried out overnight, but the beaker must be covered with a watch-glass. Allow the crucible and contents to cool to room temperature, then weigh accurately (W_3).

$$\text{The ratio: } \frac{M_r \text{ of metal chloride}}{M_r \text{ of silver chloride}} = \frac{W_1}{W_3 - W_2}$$

can be used to calculate the formula of the chloride.

4. Determination of nickel

Nickel cations can be quantitatively removed from solution by butanedione dioxime. For an accurate result the final solution must be cold and low in ethanol content, and must contain free ammonia. Several metals can interfere: bismuth, palladium and gold must be absent; iron (III), aluminium and chromium (III) must be complexed first by citrate; cobalt, manganese and zinc require the use of ammonium acetate rather than ammonia to precipitate the complex.

$$Ni^{2+} + 2\left(\begin{array}{l} CH_3-C{=}N-OH \\ \quad\;\; | \\ CH_3-C{=}N-OH \end{array}\right) \longrightarrow$$

$$\begin{array}{ccccc} & O & -H\cdots & O & \\ & | & & \uparrow & \\ CH_3-C{=}N & \searrow & & \swarrow & N{=}C-CH_3 \\ | & & Ni & & | \\ CH_3-C{=}N & \nearrow & & \nwarrow & N{=}C-CH_3 \\ & \downarrow & & | & \\ & O & \cdots H- & O & \end{array} + 2H^+$$

Requirements

- Dry sintered-glass crucible (porosity 3).
- 0·1 M butanedione dioxime in ethanol.
- A nickel salt.
- 500 cm^3 beaker.

Weigh accurately a sample of a nickel salt (sufficient to contain 0·1 g Ni) into a 500 cm^3 beaker. Also weigh accurately the dry sintered-glass crucible (W_2).

Dissolve the nickel salt in about 20 cm^3 of pure water and 20 cm^3 of 2 M hydrochloric acid, then dilute with pure water to about 200 cm^3.

Heat the solution to about 75°C, then add 50 cm^3 of 0·1 M ethanolic butanedione dioxime (an excess, 1 cm^3 for every 2 mg Ni). Now add 2 M ammonia dropwise, while stirring well, until there is a permanent red precipitate; also add a further 5 cm^3 to give a slight excess. The total ammonia volume will be about 30 cm^3.

Test a clear sample of solution with butanedione dioxime and ammonia to ensure that precipitation was complete.

Allow the solution to cool slowly to room temperature (an hour). Now transfer the precipitate to the pre-weighed sintered-glass crucible by filtration at the pump (for full details of procedure, see p. 14). Wash the precipitate in the crucible with five separate portions of pure water to remove soluble salts.

Dry the crucible and contents in an oven at about 120°C for at least an hour (overnight is satisfactory), cool to room temperature and weigh accurately (W_3).

Calculate the weight of nickel (W) in the original sample from the relationship:

$$\frac{A_r \text{ of nickel}}{M_r \text{ of bis(butanedione dioximato) nickel (II)}} = \frac{W}{W_3 - W_2}$$

hence calculate the purity of the original nickel compound.

5. Determinations using homogeneous precipitation

Precipitation that occurs from concentrated solution is very liable to result in a product contaminated by co-precipitation of other substances. Even the slow mixing of dilute solutions with efficient stirring may not give satisfactory quantitative results because of local concentrations at the point of mixing.

Homogeneous precipitation is a method in which no precipitation occurs on the initial mixing. Subsequent reaction, or change of conditions, will gradually and uniformly result in a precipitate forming. An excess of reagent cannot appear in solution until precipitation is complete.

The slow precipitation also results in a more crystalline product which filters well.

These experiments can be studied qualitatively if desired, using about 20 cm^3 of 0·1 M metal salts.

Requirements

- Dry sintered-glass crucibles (porosity 3; porosity 4 for Ca).
- 500 cm^3 beakers.
- Glass rod with rubber 'policeman'.
- 0·1 M butanedione (8·75 cm^3 per litre) for Ni.
- 0·2 M hydroxyammonium chloride (13·9 g l^{-1} adjusted to pH 7·5 with ammonia) for Ni.
- Sulphamic acid for Ba.
- Dimethyl oxalate for Ca.
- 5 M acetic acid (287 cm^3 glacial acetic acid diluted to 1 litre) for Ca and Pb.
- 5 M ammonium acetate (385 g l^{-1}) for Ca.
- 0·5 M sodium acetate (41·0 g anhydrous l^{-1}) for Pb.
- 0·1 M potassium bromate (16·7 g l^{-1}) for Pb.
- 0·1 M chromium nitrate (40·0 g l^{-1}) for Pb.

Nickel. Weigh accurately a sample of nickel salt (sufficient to contain 0·1 g Ni) and dissolve in about 100 cm³ pure water. Add 2M ammonia to adjust the solution to pH 7·5, as judged by narrow-range indicator paper or a pH meter.

Conduct the remainder of the experiment *in a fume cupboard* because of butanedione odour.

Add 70 cm³ of 0·1M butanedione solution (a two-fold excess), followed by 80 cm³ of 0·2M hydroxy-ammonium chloride (at pH 7·5, a slight excess).

Check the pH of the reaction mixture and stand until precipitation starts, then heat on a boiling water-bath for one hour. Cool, filter, using a pre-weighed sintered-glass crucible, wash, and dry at about 140°C.

Weigh the crucible and product of anhydrous bis(butanedione dioximato) nickel (II), and calculate the purity of the original nickel salt.

Write the equations of the reactions occurring.

Reference: Salesin, E. D. and Gordon, L., *Talanta*, 1960, **5**, 181.

Barium. Weigh accurately a sample of a barium salt (sufficient to contain 0·2 g Ba) and dissolve in about 100 cm³ pure water.

Add roughly 1 g of sulphamic acid and heat on a boiling water-bath for thirty minutes. Filter, using a pre-weighed sintered-glass crucible; use a glass rod protected with a rubber 'policeman' to remove any solid clinging to the beaker. Wash with hot water and ethanol, and dry at about 120°C.

Weigh the crucible and product of anhydrous barium sulphate, and calculate the purity of the original barium salt. Drying at 120°C, rather than igniting at 900°C, leads to slightly high results.

Write the equations of the reactions occurring.

Reference: Wagner, W. F. and Wuellner, J. A., *Analyt. Chem.*, 1952, **24**, 1031.

Calcium. Weigh accurately a sample of a calcium salt (sufficient to contain 0·2 g Ca) and dissolve in about 100 cm³ pure water. As pH 4 buffer, add 50 cm³ of 5M acetic acid and 50 cm³ of 5M ammonium acetate.

Add roughly 5 g of dimethyl oxalate (if impure, recrystallize from methanol), after dissolving in about 40 cm³ pure water, and heat on a boiling water-bath for two hours. Cool, filter, using a pre-weighed sintered-glass crucible of porosity 4 (filtration may be slow), wash with water and propanone, and leave overnight in a desiccator to dry.

Weigh the crucible and product of calcium oxalate monohydrate, and calculate the purity of the original calcium salt.

Write the equations of the reactions occurring.

Reference: Gordon, L. and Wroczynski, A. F., ibid., 1952, **24**, 896.

Lead. Weigh accurately a sample of a soluble lead salt (sufficient to contain 0·2 g Pb) and dissolve in about 20 cm³ pure water. Then add 10 cm³ of 5M acetic acid and 10 cm³ of 0·5M sodium acetate to act as buffer.

Add 20 cm³ of 0·1M chromium (III) nitrate and 40 cm³ of 0·1M potassium bromate.

Heat on a boiling water-bath until the supernatant solution is clear yellow (thirty to forty-five minutes); conduct this part of the experiment *in a fume cupboard* as bromine will be evolved.

Cool, filter, using a pre-weighed sintered-glass crucible, wash with 0·1M nitric acid and dry at about 120°C.

Weigh the crucible and product of anhydrous lead chromate, and calculate the purity of the original lead salt.

Write the equations of the reactions occurring.

Reference: Hoffman, W. A. and Brandt, W. W., ibid., 1956, **28**, 1487.

VIII Qualitative Inorganic Analysis

1. General procedure

In this section, the final purpose will be to identify the ions present in a mixture containing not more than two anions and two cations. It is assumed that the cations will be in different groups of the conventional analysis table, and that it will not be necessary to perform the 'phosphate separation'. For the theory behind the methods used, and for a more advanced scheme, you are referred to Vogel: *A Textbook of Macro and Semi-micro Qualitative Analysis* or Moeller: *Qualitative Analysis* (see Bibliography, p. 3).

Two cation analysis schemes are described; the first uses the traditional hydrogen sulphide method, while the second introduces a method developed by the B.D.H. Laboratory Chemicals Division research department based on the use of Emdite (ethylammonium ethyldithiocarbamate).

Normal test tube scale analysis can be carried out quite satisfactorily so long as no filtration is required, hence it is adequate for the identification of single salts. When a mixture is being analysed and it is necessary to separate a precipitate, it is much quicker to work on a semi-micro scale, and to use a centrifuge instead of filtering.

It is better to use *small quantities* for all tests, and to repeat a test if an ambiguous result is obtained the first time. In testing for gases, it is important to remember that all the gas may be evolved very quickly. Remember that a negative result to a test is important and should not be ignored.

It is assumed that the following gases can be recognized: hydrogen, oxygen, carbon monoxide, carbon dioxide, ammonia, nitrogen dioxide, hydrogen sulphide, sulphur dioxide, chlorine, hydrogen chloride, and the vapours of water, bromine and sulphur trioxide.

2. Identification of anions

The anions are divided into three categories:

(*a*) Those that evolve gases or volatile liquids with 2M hydrochloric acid: carbonate, hydrogen carbonate, sulphite, thiosulphate, sulphide, nitrite, hypochlorite, acetate.

(*b*) Those that do not react with 2M hydrochloric acid, but do evolve gases or volatile liquids with concentrated sulphuric acid: nitrate, halides, oxalate, formate.

(*c*) Those that react with neither: sulphate, phosphate, chromate.

Anions which react with 2M hydrochloric acid

Add a little 2M hydrochloric acid to the solid, and warm if necessary.

Carbonate or hydrogen carbonate: carbon dioxide evolved with effervescence. To distinguish between the two, see if the solid dissolves in water. If it does not, it is a carbonate. If it does, heat it dry, and test for the evolution of carbon dioxide. If carbon dioxide is evolved, with water vapour, it is likely to be a hydrogen carbonate of an alkali metal.

Sulphite: sulphur dioxide evolved on warming.

Thiosulphate: sulphur dioxide evolved on warming, and sulphur precipitated.

Sulphide: hydrogen sulphide evolved (though not from some very insoluble sulphides; see below).

Nitrite: nitrogen oxides evolved with effervescence.

Hypochlorite: chlorine evolved with effervescence. To make sure that the chlorine comes from the anion, repeat the test using 2M nitric acid.

Acetate: acetic acid evolved on warming. Not always easy to recognize. (See below.)

Anions which react with concentrated sulphuric acid

Warm a little of the solid with about 2 cm^3 of the acid. *Do not heat strongly*. Remember that anions in group (*a*) will also react here.

Nitrate: nitric acid vapour (pale yellow) evolved. Add a small piece of copper: brown fumes evolved. Confirm nitrate by one of the following tests:

(i) to a little of the solid add some Devarda's alloy and 2M sodium hydroxide and warm: ammonia will be evolved;

(ii) to a solution of the suspected nitrate add iron (II) sulphate solution, and then, carefully, add concentrated sulphuric acid to form a lower layer: a brown ring will be formed between the layers.

Chloride: hydrogen chloride evolved.

Bromide: hydrogen bromide and bromine evolved.

Iodide: hydrogen iodide, iodine, and hydrogen sulphide evolved.

To confirm halides, use the following tests:

(i) to a solution add 2M nitric acid and 0·05M silver nitrate:

white ppt, soluble in a little 2M ammonia
= chloride

cream ppt, soluble in an excess of 2M ammonia
= bromide

yellowish ppt, insoluble in 2M ammonia
= iodide;

(ii) to a solution add about 3 cm^3 of tetrachloromethane to form a lower layer, then add a little fresh chlorine water and shake. Chlorides will give no colour in the lower layer, bromides will turn it brownish-yellow, and iodides will turn it violet.

Fluoride: hydrogen fluoride evolved, and the glass looks greasy. Add a little powdered silica and hold a glass rod with a drop of water on it in the gases evolved: the water will go turbid as silicon tetrafluoride is hydrolysed to silica.

Oxalate: carbon monoxide and carbon dioxide evolved. Add a little of the solid to potassium manganate (VII) solution acidified with 2M sulphuric acid, and warm: the manganate (VII) will be reduced to manganese (II).

Formate: carbon monoxide evolved.

Acetate: acetic acid vapour evolved. Add to the reaction mixture 1 cm^3 of ethanol, and warm. Cool and add about 10 cm^3 of water: characteristic smell of ethyl acetate.

Anions which do not react with acids to give volatile products

Sulphate: dissolve in water, add 2M hydrochloric acid and 0·1M barium chloride: white ppt of barium sulphate. (If the solution gives a ppt when the hydrochloric acid is added, the metal has an insoluble chloride, hence barium nitrate must be used for the test.)

Phosphate: dissolve in water, or nitric acid, and add ammonium molybdate solution, and a little concentrated nitric acid if this was not used for dissolving the solid. Warm the solution to about 50°C: a yellow ppt will form slowly. (Note that arsenic compounds give a similar ppt on boiling.)

Chromate: dissolve in water: yellow solution. Add 2M sulphuric acid: colour changes to orange. Add 1 drop of 2M hydrogen peroxide: transient blue colour, turning green with evolution of oxygen.

If none of the above tests gives a positive result, the solid is possibly an insoluble *sulphide*. Warm the solid with zinc and 2M hydrochloric acid: hydrogen sulphide will be evolved; test with lead acetate paper. If this test, too, is negative, it is possible that an *oxide* may be present.

Summary

Carry out the tests in the following order:

(*a*) Warm with 2M hydrochloric acid.
(*b*) Warm with concentrated sulphuric acid.
(*c*) Add 2M hydrochloric acid and 0·1M barium chloride.
(*d*) Add concentrated nitric acid and ammonium molybdate, warm.
(*e*) Add 2M sulphuric acid and 1 drop of hydrogen peroxide.
(*f*) Confirm any ions suspected as a result of tests (*a*) and (*b*).
(*g*) If no result from (*a*) to (*f*), warm with zinc and 2M hydrochloric acid.

Tests (*c*), (*d*), (*e*) and (*f*) are carried out on aqueous solutions.

3. Identification of cations in solution

The cations included in this scheme are those normally included in GCE A level courses: silver, lead, mercury, copper, bismuth, tin, cadmium, antimony, aluminium, iron, chromium, zinc, cobalt, nickel, manganese, calcium, strontium, barium, magnesium, sodium, potassium and ammonium. It is assumed that no more than one of the cations in any one group is present.

Table 31. Cation[1] analysis

<table>
<tr><td colspan="5">Warm the solution with 2M sodium hydroxide. Ammonia evolved: NH_4^+</td></tr>
<tr><td colspan="5">To a solution of the substance add 2M hydrochloric acid. Cool thoroughly.</td></tr>
<tr><td rowspan="5">Precipitate
(Group I)
$PbCl_2$ white
$AgCl$ white
Hg_2Cl_2 white</td><td colspan="4">Filtrate Make sure the solution is just acid with 2M hydrochloric acid; then pass hydrogen sulphide, using a capillary tube. Add water to the solution.</td></tr>
<tr><td rowspan="4">Precipitate
(Group II)[2]
HgS black
CuS black
Bi_2S_3 brown
SnS brown
SnS_2 yellow
CdS yellow
Sb_2S_3 orange
PbS, black, from Group I, may also be formed
Sulphur[3] may be formed as a yellowish ppt from H_2S</td><td colspan="3">Filtrate Boil off H_2S. Add 1 cm^3 conc. nitric acid and boil to oxidize Fe (II) to Fe (III). Add 2M ammonium chloride and 2M ammonia until alkaline.</td></tr>
<tr><td rowspan="3">Precipitate
(Group III)
$Al(OH)_3$ white
$Cr(OH)_3$ green
$Fe(OH)_3$ brown</td><td colspan="2">Filtrate Add 2M ammonia, 2M ammonium chloride, and pass H_2S</td></tr>
<tr><td rowspan="2">Precipitate
(Group IV)
CoS black
NiS black
MnS pink
ZnS white</td><td>Filtrate Boil off H_2S, add 2M ammonia, 2M ammonium chloride and 2M ammonium carbonate</td></tr>
<tr><td>Precipitate
(Group V)
$CaCO_3$ white
$SrCO_3$ white
$BaCO_3$ white</td></tr>
</table>

The **filtrate** from Group V will contain the cations in *Group VI*: Mg^{2+}, Na^+ and K^+.

1. If it is known that only one cation is present, the original solution can be used for each group test until a precipitate is obtained. If more than one is present, it is essential that the first cation is all removed from the solution by precipitation, followed by filtering or centrifuging, before successive group tests are carried out for the second cation.
2. If the pH of the solution is too low, cadmium and tin may not be precipitated. The addition of water to the solution after hydrogen sulphide has been passed often causes the precipitates to form.
3. The commonest causes of sulphur precipitation are iron (III) ions, chromate ions or sulphite ions.

Confirmatory tests

Many of the cations can be recognized readily from the colour of the group precipitates, so long as it is known that only one member of the group is present. However, the presence of a particular cation should always be confirmed by a specific test. If more than one cation is present, this confirmatory test must be carried out on the precipitate, which should be dissolved in a suitable acid. If only one cation is present, the original solution can be used for the test.

Group I

Add 2M ammonia to the precipitate.

$PbCl_2$ not affected, remains white.
Hg_2Cl_2 turns black.
$AgCl$ dissolves to a colourless solution.

Group II

Precipitates can be dissolved in nitric acid, except HgS which requires aqua regia (3 parts concentrated HCl: 1 part concentrated HNO_3).

Hg(II) add tin (II) chloride solution: white ppt, turning grey.
Cu(II) add 2M ammonia in excess: deep blue solution.
Bi(III) add water in excess: white ppt.
Sn(II) add mercury (II) chloride solution: white ppt, turning grey.
Sn(IV) add a small piece of zinc and concentrated hydrochloric acid. When the zinc has dissolved, add mercury (II) chloride: white ppt, turning grey.
Cd(II) add 2M sodium hydroxide: white ppt, insoluble in excess.
Sb(III) add water in excess: white ppt.
Pb(II) add potassium chromate solution: yellow ppt.

Group III

Precipitates can be dissolved in 2M hydrochloric acid.

Al(III) add 1 drop of alizarin solution, and 2M ammonia: pink lake.
Cr(III) add hydrogen peroxide and 2M sodium hydroxide: yellow colour.
Fe(III) add ammonium thiocyanate solution: deep red colour.
Test for the presence of iron in the original solution:
Fe(II) add fresh potassium hexacyanoferrate (III) (ferricyanide) solution: dark blue ppt (Turnbull's Blue).
Fe(III) add potassium hexacyanoferrate (II) (ferrocyanide) solution: dark blue ppt (Prussian Blue).

Group IV

Precipitates can be dissolved in hydrochloric or nitric acid.

Co(II) add concentrated hydrochloric acid: blue colour, pink on addition of water.
Ni(II) add 2M ammonia and butanedione dioxime solution: pink ppt.
Zn(II) add 2M sodium hydroxide: white ppt, soluble in excess.
Mn(II) add concentrated nitric acid and sodium bismuthate: purple solution.

Group V

Precipitates can be dissolved in 2M hydrochloric acid. Make neutral with 2M ammonia.

Ba(II) add 0·1M potassium chromate: yellow ppt, insoluble in 2M acetic acid.
Sr(II) add 0·1M potassium chromate: yellow ppt, soluble in 2M acetic acid.
Ca(II) add 0·1M potassium chromate: no ppt; add 0·1M ammonium oxalate: white ppt.

Group VI

If the filtrate from previous groups is used, destroy ammonium salts by evaporating carefully to dryness, driving off all fumes, and dissolving the cold residue in 2M acetic acid.

Mg(II) add 1 drop of magneson I (*p*-nitrobenzene-azo-resorcinol) and 2M sodium hydroxide: sky-blue lake.
K(I) add fresh sodium hexanitrocobaltate (III) solution: yellow ppt.
Na(I) add uranyl zinc acetate solution: yellow crystalline ppt; *or* add to potassium antimonate (V) solution: faint white ppt.

Neither of the tests for sodium is very satisfactory, and a flame test should be carried out on the original substance in confirmation: persistent yellow flame.

4. Analysis of a single salt

The analysis of a salt can conveniently be divided into four parts:

(*a*) preliminary tests,
(*b*) identification of the anion,
(*c*) making a solution of the salt,
(*d*) identification of the cation.

Parts (*b*) and (*d*) have already been dealt with above. The purpose of the preliminary tests is to give general pointers to the identity of the salt. They should not be omitted, but not too long should be spent on them. In particular, the action of heat on the solid should not be allowed to occupy too much time.

Preliminary tests

Appearance: observe the colour, the state of the solid (large or small crystals, or powder), and whether it appears to be deliquescent.

Action of heat: heat a small portion in a test tube, at first gently, then strongly. Observe colour changes, and try to identify the gases or vapours evolved.

Flame test: put a little of the solid into a small watch-glass and moisten it with concentrated hydrochloric acid. Dip into it either a clean test tube containing cold water, or a platinum or nichrome wire loop cleaned by heating in a hot flame until it gives no colour. Put the test tube or wire into the edge of a non-luminous flame. Colours which may be seen are:

persistent yellow:	sodium
brick-red, or orange:	calcium
crimson:	strontium
lilac:	potassium
green:	barium or copper
bluish:	several heavy metals

2M *sodium hydroxide:* warm the mixture and test for ammonia gas. Its presence indicates ammonium. Also observe colour changes or precipitates.

Making a solution of the salt

This can be the most difficult part of the analysis if it is not approached systematically. The only solvent used is water. Many salts do not dissolve appreciably in water, so they must be converted into salts which do. This is usually done by treatment with an acid. It is often sufficient to add a little acid to the suspension obtained with water.

First, attempt to dissolve a little of the salt in water. Do not yet attempt to dissolve all the sample to be used in the tests. If it does not dissolve on warming, try the following acids in turn, cold, then warm: 2M hydrochloric acid, 2M nitric acid, concentrated hydrochloric acid, concentrated nitric acid. If it will not dissolve in any of these, try aqua regia (3 parts concentrated hydrochloric acid to 1 part concentrated nitric acid). Concentrated acids are unsatisfactory solvents and should be avoided wherever possible. If nitric acid has to be used, excess should be avoided, otherwise there may be trouble with precipitated sulphur when hydrogen sulphide is used.

If the salt will still not dissolve with any of these acids, it will be necessary to boil it with a concentrated solution of sodium carbonate. After filtration, the filtrate can be used for anion tests, and the residue dissolved in 2M acid and used for cation tests.

When a suitable solvent has been found, make up a solution for the cation identification. If heating is required, cool the solution before starting the group tests.

Conclusions

When the anion and cation have been identified, it is useful to confirm your findings on the original salt.

5. Analysis of a mixture

The procedure described for the analysis of a single salt can be followed for a mixture up to the cation identification. Some anions interfere with the tests for others, for example a nitrite must be destroyed before a nitrate is tested for, but on the whole these are avoided in examination exercises. You are referred to a specialized textbook of analysis for further details.

The same cation analysis scheme as for single salts is satisfactory so long as there is only one cation in any one of the analytical groups. The confirmatory tests must be carried out on the precipitates, not on the original mixture (except for the tests to distinguish between iron (II) and (III)).

If it is not known whether the cations are in different analytical groups, or if a phosphate is present, you must refer to a more advanced textbook, such as those mentioned at the beginning of this chapter.

6. Identification of cations using Emdite

Emdite is an analytical reagent developed by the B.D.H. Laboratory Chemicals Division research department. It is available as a 50% solution, and is used in 5% solution. It is considered possible, but is not yet certain, that it furnishes thiocarbonate (CS_3^{2-}) ions. It has the advantages that it lacks the smell and toxic nature of hydrogen sulphide and, being in solution, can be used in controlled amounts.

It is necessary to control the pH of solutions fairly carefully; methyl violet papers (which turn yellow at pH 0·5), as well as litmus papers, are used. These are prepared by soaking filter paper in an ethanolic solution of methyl violet, and drying in the air at room temperature.

It is assumed that centrifuging rather than filtering will be used.

The cations are divided into five groups, lettered A to E. Some of these groups are the same as in the hydrogen sulphide scheme and, once separated, they can be treated in the same way. These are *Group A* (= *Group I:* lead, mercury (I) and silver), *Group D* (= *Group V:* calcium, strontium and barium), and *Group E* (= *Group VI:* magnesium, potassium and sodium).

The rest of the cations come down in *Group B*,

Table 32 [1, 2] *Separation into Groups.*

<table>
<tr><td colspan="5">Unless it was used as the solvent, add 2 drops concentrated hydrochloric acid. Cool thoroughly. Centrifuge and separate.</td></tr>
<tr><td rowspan="4">Residue
(Group A)
$PbCl_2$ white
Hg_2Cl_2 white
AgCl white</td><td colspan="4">Supernatant liquid Add 0·2 g of ammonium chloride, dissolve and cool. Add 2 drops of concentrated nitric acid, then '0·880' ammonia dropwise until just alkaline to litmus. Mix well, then centrifuge and separate.</td></tr>
<tr><td rowspan="3">Residue
(Group B)
$Al(OH)_3$ white
$Sb(OH)_3$ white
$Bi(OH)_3$ white
$Cd(OH)_2$[3] white
$Cr(OH)_3$ green
$Fe(OH)_3$ brown
$SnO_2.H_2O$ white
$HgNH_2Cl$ white
maybe:
$Pb(OH)_2$[4] white
$Mn(OH)_2$[5] white, turning brown</td><td colspan="3">Supernatant liquid Add glacial acetic acid until just acid to litmus. Add 1 cm³ 5% Emdite and mix. Centrifuge. Make the supernatant liquid alkaline to litmus with ammonia, add 1 cm³ 5% Emdite and mix. Centrifuge and separate.</td></tr>
<tr><td rowspan="2">Residue
(Group C)
Ppts from
Cu brown-greenish
Co brown-green
Ni brown-green
Zn white
Mn white-brown</td><td colspan="2">Supernatant liquid If turbid, boil and centrifuge. Evaporate supernatant liquid to destroy Emdite, and reduce to 2 cm³. Add 1 drop '0·880' ammonia, and 2M ammonium carbonate until precipitation is complete. Warm, then allow to settle. Add 1 drop of 2M ammonium carbonate, centrifuge and separate.</td></tr>
<tr><td>Residue
(Group D)
$BaCO_3$ white
$SrCO_3$ white
$CaCO_3$ white</td><td>Supernatant liquid May contain Group E ions. Evaporate to dryness and heat to volatilize ammonium salts. Dissolve in 2M acetic acid, and test for Mg, Na, K.</td></tr>
</table>

1. Ammonium ions must be tested for before the group tests, by warming with 2M sodium hydroxide and testing for ammonia gas by smell or with litmus paper.
2. No phosphate ions must be present.
3. If the concentration of ammonium chloride is not high enough, cadmium hydroxide may not be precipitated in Group B.
4. Lead ions may be present in Group B because lead chloride is slightly soluble in cold water and may be incompletely precipitated in Group A.
5. Manganese may be partially precipitated in Group B, but it will come down in Group C as well.

Table 33. *Group B:* treatment of the precipitate to separate into B1, B2 and B3.

<table>
<tr><td colspan="3">Wash the residue with water. Dissolve in 2M hydrochloric acid and make up to 2 cm^3 with water. Add concentrated hydrochloric acid dropwise until acid to methyl violet (goes yellow), and 10 drops in excess. Cool. Add 1 cm^3 5% Emdite and centrifuge. Repeat with two more lots of Emdite to ensure complete precipitation. Centrifuge and separate.</td></tr>
<tr><td rowspan="2">Residue
(B1 and B2)
Ppts from:
Sb off-white
Bi yellow
Cd off-white
Fe grey-black
Sn yellowish
Hg(II) white
maybe:
Pb white</td><td colspan="2">Supernatant liquid Ignore any turbidity. Add 2 drops of concentrated nitric acid and boil. Reduce to 2–3 cm^3. Cool, and add an excess of '0·880' ammonia (go by the smell). Centrifuge and separate.</td></tr>
<tr><td>Residue
(B3)
$Al(OH)_3$ white
$Cr(OH)_3$ green
maybe:
$Fe(OH)_3$ brown
$Mn(OH)_2$ white-brown</td><td>Supernatant liquid
Discard.</td></tr>
</table>

Table 34. *Groups B1 and B2*

<table>
<tr><td colspan="5">Wash the residue (B1 and B2) with water, add 2–3 cm^3 of a solution containing 6·8 g of potassium hydroxide and 10 g of potassium nitrate in 100 cm^3 of water. Warm the mixture, then cool, centrifuge and separate.</td></tr>
<tr><td colspan="4">Residue (B1) Wash twice with water. Discard washings. Add 50% nitric acid and warm; when the reaction ceases, centrifuge and separate.</td><td rowspan="4">Supernatant liquid (B2). Add glacial acetic acid carefully until acid to litmus. Add 1 drop in excess. Centrifuge and separate. Discard supernatant liquid. If there is a residue, dissolve it in hot 2M hydrochloric acid, divide solution into two parts and test as follows.
(a) Add a small piece of aluminium foil and warm in a water-bath for a few minutes. Cool, then centrifuge off the residue if present, and add 2 drops of mercury (II) chloride. White ppt turning grey: Sn. Test the original solution to see if it is tin (II) or tin (IV).
(b) Add 3 drops of 5% sodium nitrite solution and 5 drops of rhodamine-B solution. Lavender-blue colour: Sb.</td></tr>
<tr><td rowspan="3">Residue
If black, Hg (II) is present.
Dissolve in 2 cm^3 of concentrated hydrochloric acid and 3 drops of bromine water. Boil, then add tin (II) chloride solution. White ppt: Hg (II)</td><td colspan="3">Supernatant liquid Cool thoroughly, add an excess of '0·880' ammonia. Mix well, centrifuge and separate.</td></tr>
<tr><td colspan="2">Residue Wash with water. Discard washings. Add 7 cm^3 of 2M sodium hydroxide, warm for 3 minutes. Centrifuge and separate.</td><td rowspan="2">Supernatant liquid Add concentrated hydrochloric acid dropwise until acid to litmus. Add 1 cm^3 of 5% Emdite. Off-white ppt: Cd</td></tr>
<tr><td>Residue Wash with water. Discard washings. Dissolve in 2M hydrochloric acid. Divide into two.
(a) Add 1 cm^3 of sodium stannate (II). White ppt going black: Bi
(b) Add potassium manganate (VII) until just pink, then ammonium thiocyanate. Red colour: Fe</td><td>Supernatant liquid Add glacial acetic acid until acid to litmus, then potassium chromate. Yellow ppt: Pb</td></tr>
</table>

Table 35. *Group B3:* use the B3 residue from the Group B separation.

<table>
<tr><td colspan="3">Residue (B3) Wash with water. Discard washings. Add 4 cm^3 of 2M sodium hydroxide and boil. Centrifuge and separate.</td></tr>
<tr><td colspan="2">Residue Add a pellet of sodium hydroxide and 4 drops of sodium hypochlorite solution. Warm for two minutes, centrifuge and separate.</td><td rowspan="2">Supernatant liquid Make just acid to litmus with concentrated hydrochloric acid, mix, then add 1 drop of alizarin solution, then 2M ammonia until alkaline. Pink lake formed: Al</td></tr>
<tr><td>Residue Dissolve in 50% nitric acid. Cool and divide into two parts.
(a) Add solid sodium bismuthate and mix. Purple colour: Mn
(b) Add 3 drops of potassium hexacyanoferrate (II) solution. Dark blue ppt or colour: Fe</td><td>Supernatant liquid If yellow, chromate is present. Acidify with acetic acid, add 3 drops of lead acetate solution. Yellow ppt: Cr</td></tr>
</table>

precipitated by ammonia in the presence of ammonium chloride, and *Group C*, precipitated by Emdite, possibly as thiocarbonates.

Group B is subdivided after precipitation into *Groups B1* (mercury (II), bismuth, cadmium, iron, and residual lead from Group A), *B2* (tin and antimony) and *B3* (aluminium, chromium, residual iron from B1, and possibly manganese). Since Group B does not correspond to any of the hydrogen sulphide groups, it has to be treated quite differently.

Group C is the same as *Group IV* of the hydrogen sulphide scheme with the addition of copper (i.e. copper, cobalt, nickel, zinc and manganese).

The solution is made up in the same way as for the hydrogen sulphide scheme. A solution of about 0·05 g in 1–1·5 cm^3 of water is required.

If a precipitate is formed in Group A, Group C or Group D, it should be dissolved if necessary and treated as in the hydrogen sulphide scheme. The only additional confirmatory test required is for *copper*:

Cu(II): add concentrated hydrochloric acid: yellow-green solution, turns to deep blue on the addition of an excess of '0·880' ammonia.

The residual solution from Group D contains, possibly, magnesium, sodium and potassium. These should be tested for as before.

Reference: Hart, K. K., Hill, A. G. and Savage, B., *J. Roy. Inst. Chem.*, December 1964, 418.

IX Qualitative Organic Analysis

1. Detection of elements

All organic compounds contain carbon, and nearly all contain hydrogen, so if it is known that a compound is organic, it is not necessary to test for these. Many organic compounds contain oxygen, but there is no simple test for the presence of it in a compound. It is, however, essential to test for the presence of nitrogen, sulphur and the halogens, any or all of which might be present in a common organic compound. Metals may also be present; these are tested for in the usual way if, on strong heating, the compound leaves an involatile residue.

(a) Carbon and hydrogen. The presence of these elements can be discovered as a result of one test. Heat a little of the compound with an excess of pure copper (II) oxide dried beforehand in an oven. Test for carbon dioxide and water vapour evolved on heating.

(b) Nitrogen, sulphur and halogens. Two alternative tests can be used: Middleton's, and the Lassaigne sodium fusion test. Both depend on heating the compound with an excess of a metallic reducing agent (zinc in the first test, and sodium in the second) so that nitrogen forms cyanide ions, sulphur forms sulphide ions, and halogens form halide ions.

The second part of the test, which is almost the same for both tests, is the identification of the ions formed.

Middleton's test

Mix about 0·1 g of the compound with about 1 g of Middleton's mixture (2 parts of zinc to 1 part of anhydrous sodium carbonate) in a small test tube, and heat strongly for two minutes in a hot Bunsen flame. Then plunge the red-hot tube into 20 cm³ of water in a beaker. Boil the mixture to dissolve the sodium salts formed. Filter the solution and divide the filtrate into three portions. Test *the residue* for precipitated zinc sulphide by adding to it 2M hydrochloric acid, and testing at once for hydrogen sulphide with lead acetate paper. A dark brown colour on the paper indicates that the original compound contained sulphur.

Test the solution for cyanide ions

Add to the solution a little solid iron (II) sulphate and a few drops of 2M sodium hydroxide, and boil for one minute to form hexacyanoferrate (II) ions. Add a few cm³ of iron (III) chloride solution and sufficient concentrated hydrochloric acid to dissolve any hydroxides present, then filter the solution. The presence of specks of precipitated Prussian Blue on the filter paper indicates that the original compound contained nitrogen.

Test the solution for sulphide ions

Add to the solution a little fresh sodium nitroprusside solution. A purple colour indicates that the original compound contained sulphur. It is unnecessary to do this test if sulphur is already found.

Test the solution for halide ions

If cyanide or sulphide ions are present, they must be removed by boiling the solution with 2M nitric acid, *in a fume cupboard*, to expel hydrogen cyanide and/or hydrogen sulphide.

To half the solution, freed from cyanide and sulphide ions, add 2M nitric acid and 0·05M silver nitrate. A precipitate indicates the presence of halide ions:

white ppt, soluble in a little 2M ammonia: chloride
cream ppt, soluble in an excess of 2M ammonia: bromide
yellowish ppt, insoluble in 2M ammonia: iodide

If the first test is positive carry out a second test on the other half of the solution. Add about 3 cm³ of tetrachloromethane to form a lower layer, then add a little fresh chlorine water, and shake:

a yellow-brown colour in the lower layer indicates bromide; a violet colour, iodide; and no colour, chloride.

Lassaigne's test

This test is carried out in a similar way to Middleton's, but it is more liable to mishaps if the tube is not heated long enough for all the sodium to react. For this reason it is wise to carry out the plunging of the red-hot tube into water *behind a safety screen.* ***The test is dangerous if carried out on highly chlorinated compounds, such as tetrachloromethane.***

Put a small piece of sodium (about 0·2 g) into a hard-glass ignition tube, and warm gently so that it just melts. Now add about 0·1 g of the compound and heat the tube, gently at first and then strongly for about two minutes. Plunge the red-hot tube into 20 cm^3 of water in a mortar, and crush up the remains with a pestle. Filter the solution and test the filtrate as in Middleton's test. It is unnecessary to test the residue.

(c) Phosphorus. To about 2 cm^3 of filtrate from Middleton's or Lassaigne's test add about 4 cm^3 of concentrated nitric acid and boil carefully to oxidize any phosphorus compounds present to phosphate ions. Cool the solution, then add ammonium molybdate solution and warm to no higher than 50°C. A slowly forming yellow precipitate will appear if any phosphate ions are present. Note that arsenate ions will give a very similar precipitate, but only if the mixture is boiled.

(d) Halogens: Beilstein's test. Wrap a piece of copper gauze round the end of a clean piece of copper wire. Heat it in a non-luminous Bunsen flame until there is no colour in the flame, then dip it in the organic compound and return it to the flame. If the compound contains a halogen, the flame will be coloured green by volatile copper compounds. This is not an entirely reliable test, but if there is no colour, it can be taken that halogens are absent. Certain nitrogen compounds also cause a colour to appear.

2. Assignation of a compound to its class

This procedure differs from simple inorganic analysis in two main respects: the vast number of organic compounds and the fact that there are not, generally, positive confirmatory tests available.

For these two reasons physical properties are important, in particular melting- and boiling-points. It is usual to prepare a solid derivative from a compound and to determine its melting-point. Tables are available listing the melting-points of derivatives from a large number of compounds.

The scheme of analysis which follows is designed only to find out which functional groups are present in a compound, and so to assign it to a particular class.

It is essential to make yourself familiar with the following:

(*a*) The simple properties associated with the common groups.
(*b*) The physical state expected in the various classes. For example amides: solid; esters: usually liquid; etc.
(*c*) The smells of many common compounds. The smell is the greatest single guide (to the experienced nose) to the class of a compound.

The tests below should be carried out on the compound under investigation, all observations being recorded at once. They will furnish clues to the nature of the compound, and further tests can be carried out at the end to confirm the class of the compound. Remember that a negative result is a valuable piece of evidence.

(1) Elements present. The following are the most likely compounds associated with particular combinations of elements.

C H (O)	Alcohols. Phenols. Aldehydes. Ketones. Acids. Esters. Acid anhydrides. Carbohydrates. Ethers. Hydrocarbons.

C H (O) Metal	Salts.
C H (O) N	Ammonium salts. Amides. Nitriles. Amines. Amino-acids. Nitro-compounds.
C H (O) Hal	Halogenated hydrocarbons. Acid chlorides. Chloro-acids. Chloral hydrate.
C H (O) S	Sulphonic acids. Bisulphite compounds. Thio-alcohols. Thio-ethers.
C H (O) N Hal	Salts of amines and amino-acids with halogen acids. Halogenated aromatic amines.
C H (O) N S	Sulphates of amine. Amino-sulphonic acids. Sulphonamides. Thiocarbamide.
C H (O) S Hal	Sulphonyl halides.
C H (O) N S Hal	Sulphates of halogenated aromatic amines. Halides of aminosulphonic acids.
C H (O) S Metal	Salts of sulphonic acids.

(2) **Physical state.** Note carefully the physical state of the compound, including crystal shape. Below is a rough guide.

Solid: Phenols. Aromatic acids. Amides. Amino-acids. Carbohydrates. Salts. Thiocarbamide. Methyl oxalate. Aliphatic acids with two or more carboxyl groups. Sulphonic acids.

Liquids: Alcohols. Esters. Aldehydes. Ketones. Aliphatic acids. Nitriles. Ethers. Chlorinated hydrocarbons.

(3) **Smell.** Note carefully the smell of the compound. If necessary warm it gently. With practice, a large number of compounds can be recognized by their smells.

(4) **Colour.** The majority of simple organic compounds are colourless, but the following coloured ones may be met.

Yellow: Nitro-compounds. Quinones. Tri-iodomethane.

Orange: Nitro-amines.

Red: Azo-compounds.

Blue or green: Copper salts.

Some phenols and aromatic amines are readily oxidized, and may contain pink, brown, grey, or black impurities.

(5) **Solubility in water.** The majority of organic compounds are insoluble in water. Add a small quantity of the compound to a few drops of water and shake.

Solid and soluble: Carbamide. Amino-acids. Carbohydrates. Salts. Low aliphatic amides. Chloral hydrate.

Liquid and soluble: Low alcohols, aldehydes, ketones, acids, nitriles, acid chlorides, amines.

If the compound is a solid and it does not dissolve in cold water, heat the mixture to boiling, then cool it. If the solid dissolves on heating and reappears on cooling it is possibly either an aromatic acid or a substituted amide (e.g. acetanilide).

(6) **Action with litmus.** If the compound dissolves in water, test the solution with litmus paper.

Acidic: Acids. Acid anhydrides. Acid chlorides. Easily hydrolysed esters. Salts of amines.

Alkaline: Aliphatic amines. Sodium or potassium salts of acids.

(7) **Presence of ions.** If the compound is soluble in water, and sulphur or halogen has been detected, test the solution for the presence of sulphate or halide ions.

(8) **Action of heat.** Heat a little of the compound on the end of a metal spatula or in a crucible lid.

Burns with a non-luminous flame: Aliphatic. Low % of carbon.

Burns with a smoky flame: Aromatic. High % of carbon.

Does not burn: Rich in halogen. Salt.

Chars rapidly: Carbohydrate. Hydroxyacid.

Residual ash (not carbon): Metal present; identify it.

Violet vapour: Tri-iodomethane.

Further information may be obtained by heating a little of the compound in a small tube. Note the smell and the general behaviour.

(9) **Action with an alkali.** Treat a little of the compound with cold 2M sodium hydroxide.

Soluble, though insoluble in water: Acids. Phenols.

Soluble, giving bright yellow colour: Nitrophenols.

Ammonia evolved: Ammonium salts.
Fishy ammoniacal smell: Aliphatic amine salts.
Oil liberated: Aromatic amine salts.
Trichloromethane liberated: Chloral hydrate.

If the mixture gives no action in the cold, boil it carefully.

Ammonia evolved slowly: Amides. Nitriles.
Amine-like smell: Secondary amides.
Brown resin formed: Aliphatic aldehydes.

Any other changes should be recorded, since many compounds are hydrolysed in alkaline solution.

(10) Action with sodium carbonate. Treat the compound with 1·5M sodium carbonate, then warm if necessary.

Carbon dioxide evolved: Acids. Acid anhydrides. Acid chlorides. Amine salts. Methyl oxalate. Nitro-phenols.

(11) Action with Fehling's solution. Heat the compound to boiling with Fehling's solution (see Appendix I). If the colour is not deep blue on addition, add a little 2M sodium hydroxide.

Orange, red or brown ppt: Reducing agent present. Probably sugar, aldehyde, tartrate.

(12) Action with iron (III) chloride. Treat a very small portion of the compound with neutral iron (III) chloride solution. This is prepared by dissolving solid iron (III) chloride in a little water and adding drops of 2M ammonia until a faint precipitate is present.

Red colour: Formate. Acetate.
Yellow colour: Hydroxy acid.
Violet colour: Phenol group.
Green colour: Benzene-1,2-diol.
Violet-red colour: *p*-Nitrophenol.
Buff ppt: Benzoate. Succinate.

(13) Action with sulphuric acid. If an acid or its salt is suspected, heat *carefully* a small quantity of the compound with a little concentrated sulphuric acid.

No blackening, CO evolved: Formate.
No blackening, CO and CO_2 evolved: Oxalate.
No blackening, pungent smell: Acetate. Benzoate.
Yellow colour, CO and CO_2 evolved: Citrate.
Extensive charring, effervescence: Lactate. Tartrate. Carbohydrate.

(14) Action with soda-lime. If an amino-acid is suspected, heating with soda-lime will decarboxylate to the corresponding amine. Since heating with soda-lime gives many side reactions, it is not a very reliable test to use except for the confirmation of amino-acids.

(15) Hydrolysis of an ester, or a secondary amide. If the presence of either of these classes is suspected, heat about 0·5 g of the compound under reflux with about 10–15 cm³ of 2M sodium hydroxide. This treatment yields the alcohol and the sodium salt from the ester; neutralization of the solution with hydrochloric acid will liberate the acid, while the amine will be liberated from the amide.

Tests (1) to (12) can be carried out quite rapidly. From the evidence, decide whether it is necessary to do tests (13) to (15). A few other useful tests for particular compounds can be carried out. For example:

Carbonyl compounds will give precipitates with Brady's Reagent (2,4-dinitrophenylhydrazinium sulphate in methanol. See p. 71.)
Primary aromatic amines can be diazotized and coupled with 2-naphthol in alkaline solution to give azo-dyes.
o-Hydroxybenzoates warmed with methanol and a little concentrated sulphuric acid give the distinctive smell of methyl *o*-hydroxybenzoate (wintergreen).
Acetates warmed with 3-methylbutan-1-ol and a little concentrated sulphuric acid give the distinctive smell of 3-methylbutan-1-yl acetate (pear drops).

Conclusions

The evidence from all the tests is weighed up, and tests with doubtful results may be repeated. It should be possible to say which groups are present, even if the compound itself cannot be identified. Organic analysis becomes much easier with experience.

X Observational Problems

THE following nine observational problems are drawn chiefly from the A level Practical Chemistry papers of the Oxford and Cambridge Schools Examination Board.

They call for a knowledge of inorganic and organic chemistry.

OP 1

Carry out the following experiments with substance **A**, and draw what conclusions you can as to its nature and the course of reactions you observe:

(*a*) Heat it.
(*b*) Allow the residue from (*a*) to cool, and then add dilute hydrochloric acid.
(*c*) Treat **A** with dilute hydrochloric acid.
(*d*) To a concentrated solution of **A** add silver nitrate solution, followed by dilute nitric acid.
(*e*) To a solution of **A** add barium chloride solution.
(*f*) To a solution of **A** add iron (III) chloride, then boil gently.
(*g*) To a solution of **A** add sodium carbonate solution.

OP 2

Carry out the following experiments with substance **HA**, and describe all observations. State your conclusions as to the nature of **HA** and the course of the reactions you observe:

(*a*) Ascertain whether it is acidic, neutral or alkaline.
(*b*) To its solution add silver nitrate solution and warm.
(*c*) To its solution add a little mercury (II) chloride solution and warm.
(*d*) Heat a portion of the solid.
(*e*) To its solution add dilute potassium manganate (VII) solution.
(*f*) To a small portion of its solution add, carefully, bromine water until a faint permanent yellow colour appears, then add silver nitrate solution.

OP 3

Using a solution of the potassium salt, **HB**, carry out the following experiments:

(*a*) Add silver nitrate solution, followed by dilute nitric acid.
(*b*) Add ammonia solution to the result of (*a*).
(*c*) Add mercury (II) nitrate solution, followed by more **HB**.
(*d*) Add mercury (I) nitrate solution.
(*e*) Add a very dilute solution of **HB** to mercury (I) nitrate solution.
(*f*) Add a few drops of copper (II) sulphate solution, and then a larger volume.
(*g*) Warm the result of (*f*) with sodium bisulphite solution.
(*h*) Add a few drops of cobalt nitrate solution, and shake the mixture with 3-methylbutan-1-ol. Pour the mixture into a larger volume of water.
(*i*) Add 1 drop of iron (III) chloride solution and a few drops of ether.
(*j*) Pour a drop or two of **HB** on to some solid ammonium chloride, then add a drop of iron (III) chloride solution.

Describe what happens, and comment on the nature of **HB**, and of the reactions you observe.

OP 4

Carry out the following experiments with substance **HC**:

(*a*) Heat it in a hard-glass test tube.
(*b*) Heat it with a little calcium hydroxide.
(*c*) To its solution add silver nitrate solution, followed by dilute nitric acid.
(*d*) To the solution of the residue from (*a*) add silver nitrate solution, followed by dilute nitric acid.
(*e*) To its solution add a little iron (III) chloride solution, followed by concentrated hydrochloric acid.
(*f*) To its solution in nitric acid add ammonium molybdate solution and warm.

Describe what you observe, and deduce what you can about **HC**.

OP 5

Carry out the following experiments with the salt, **E**, which you are not required to identify, and record your results and observations. Give what explanations you can.

(*a*) Heat it.
(*b*) Heat it with a little sodium hydroxide solution.
(*c*) Treat it with a little concentrated hydrochloric acid.
(*d*) Add to its solution in water a little concentrated hydrochloric acid and a piece of granulated zinc, or some zinc powder.
(*e*) To its solution in water add a little potassium iodide solution and a little dilute hydrochloric acid. Boil gently for a time.
(*f*) Add to its solution in water a little hydrogen peroxide and acidify with dilute sulphuric acid.

OP 6

Record what happens when you carry out the following experiments with substance **HY**:

(*a*) Heat it.
(*b*) Heat it with anhydrous sodium acetate.
(*c*) To its solution in dilute sulphuric acid add a few drops of potassium manganate (VII) solution, and warm.
(*d*) To its solution in water add silver nitrate solution.
(*e*) To its solution in water add ammonia solution gradually until no further change is seen. To the resulting solution add a few drops of silver nitrate solution, and warm.

Say what you think **HY** is, and give what explanations you can of the reactions you have observed.

OP 7

Carry out the following experiments with substance **Z**:

(*a*) Dissolve a little in cold water.
(*b*) Warm the solution from (*a*).
(*c*) Add sodium hydroxide solution drop by drop to the solution from (*b*) until there is an excess of alkali.
(*d*) Add a little hydrogen peroxide to the solution from (*c*), and warm.
(*e*) Acidify the solution from (*d*) with dilute sulphuric acid. Add a layer of ether, and another drop of hydrogen peroxide.
(*f*) To its solution in water add barium chloride solution and dilute hydrochloric acid.
(*g*) Carry out a flame test on **Z**.

Identify **Z**, and explain the reactions as far as you can.

OP 8

Carry out the following experiments with substance **N**:

(*a*) Examine its solubility in water.
(*b*) To a small portion add dilute nitric acid and warm until no further reaction takes place.
(*c*) To separate portions of the product from (*b*) add (i) sodium hydroxide solution and warm, and (ii) ammonia solution.
(*d*) Dissolve a small quantity of **N** in sodium thiosulphate solution. To separate portions of this solution add (i) sodium hydroxide solution and warm, and (ii) dilute sulphuric acid and warm.

Record your observations. State what you think substance **N** to be, and suggest what explanations you can for the reactions which have taken place.

OP 9

Carry out the following experiments with substance **P**:

(*a*) Warm a little solid **P** with potassium iodide solution.
(*b*) Warm a little solid **P** with manganese (II) sulphate solution containing 1 or 2 drops of silver nitrate solution.
(*c*) Warm a little solid **P** with silver nitrate solution.
(*d*) Warm a little solid **P** with silver nitrate solution and a little pyridine.
(*e*) Warm a little solid **P** with potassium manganate (VII) solution and dilute sulphuric acid.
(*f*) Dissolve a little **P** in cold water, and add barium chloride solution and dilute hydrochloric acid.
(*g*) Dissolve a little **P** in water, boil for a few minutes, and then add barium chloride solution and dilute hydrochloric acid.

(*h*) Warm a little solid **P** with sodium hydroxide solution, then cool. Add a layer of ether, followed by potassium chromate solution and dilute sulphuric acid.

Identify **P**, and explain the reactions you observe.

Appendices

APPENDIX I. PREPARATION OF COMMON LABORATORY REAGENTS

In this book, the standard concentration chosen for dilute acids and bases is twice molar (2M). Most other reagents are one-tenth molar (0·1M). When a solid reagent is being dissolved, there is usually no difficulty about the quantity required, though care should be taken to find out the degree of hydration of the solid being used and to allow for it. A number of compounds, however, including the common acids, are purchased as liquids or solutions in water, and have to be diluted. Tables 36 and 37 give the approximate concentrations of the purchased materials, and the volumes required to make up 2 litres of reagent.

Table 36. Concentrations of commercial materials

Concentrated hydrochloric acid	1·19 g cm^{-3}	12M
Concentrated nitric acid	1·42 g cm^{-3}	16M
Concentrated sulphuric acid	1·84 g cm^{-3}	18M
Glacial acetic acid	1·05 g cm^{-3}	17M
'0·880' ammonia	0·88 g cm^{-3}	15M
'0·90' ammonia	0·90 g cm^{-3}	10M

Table 37. Preparation of laboratory solutions

2M hydrochloric acid	400 cm^3 of concentrated acid in 2 litres of solution
2M nitric acid	250 cm^3 of concentrated acid in 2 litres of solution
2M sulphuric acid	230 cm^3 of concentrated acid in 2 litres of solution
2M acetic acid	225 cm^3 of glacial acid in 2 litres of solution
2M ammonia	215 cm^3 of '0·880' ammonia in 2 litres of solution

Remember, when diluting sulphuric acid, always to add the acid in a slow stream to water with good mixing, never vice versa.

Two other compounds usually purchased in solution are:

Hydrogen peroxide: '20 volume' is approximately 1·8M when fresh.

Sodium hypochlorite: 15% w/v is approximately 2M when fresh.

The following is a list of reagents which are not used in 0·1M solution, or whose preparation needs special comment.

Alizarin	0·1 g in 100 cm^3 of ethanol.
2M ammonium carbonate	add 160 g of solid 'ammonium carbonate' to 140 cm^3 of '0·880' ammonia, and dilute to 1 litre.
2M ammonium chloride	dissolve 107 g per litre of aqueous solution.
0·1M butanedione dioxime	11·6 g in 1 litre of ethanol.
Fehling's solution 'A'	dissolve 70 g of copper (II) sulphate pentahydrate per litre of aqueous solution, and add a few drops of concentrated sulphuric acid.
Fehling's solution 'B'	dissolve 350 g of sodium potassium tartrate and 100 g of sodium hydroxide per litre of aqueous solution.
Fehling's solution	mix equal volumes of 'A' and 'B' just before use.
0·1M iodine solution	dissolve 25·4 g of iodine and 60 g of potassium iodide per litre of aqueous solution.
Litmus	boil 10 g of solid with 200 cm^3 water and filter.
Magneson I	0·1 g with 1 g of sodium hydroxide in 100 cm^3 of water.

Methyl orange	1 g in 500 cm^3 of ethanol, then add 500 cm^3 of water.
Phenolphthalein	2 g in 600 cm^3 of ethanol, then add 400 cm^3 of water.
Potassium antimonate (V)	dissolve 20 g in 500 cm^3 of boiling water, cool, add 30 cm^3 of 2M potassium hydroxide, and dilute to 1 litre.
2M potassium hydroxide	dissolve 112 g per litre of aqueous solution.
0·5M potassium iodide	dissolve 83 g per litre of aqueous solution.
0·05M silver nitrate	dissolve 8·5 g per litre of aqueous solution.
1·5M sodium carbonate	dissolve 159 g of anhydrous sodium carbonate per litre of aqueous solution.
2M sodium hydroxide	dissolve 80 g per litre of aqueous solution.
Starch solution	make 2 g into a paste with cold water, add 0·01 g of mercury (II) chloride and boil with 1 litre of water, then cool.
Uranyl zinc acetate	dissolve 10 g of uranyl acetate dihydrate in 15 cm^3 of 2M acetic acid, dilute to 50 cm^3, and mix with a solution of 30 g of zinc acetate dihydrate in 10 cm^3 of 2M acetic acid diluted to 50 cm^3. Add a little sodium chloride, leave to stand, then filter.

If a solution of a tin (IV) compound is required, it is suggested that 0·1M ammonium chlorostannate (IV) be used. This overcomes the problems encountered in attempting to prepare a stable solution of tin (IV) chloride.

Purification of water

The method employed for the purification of water in a particular laboratory should be based on a cost analysis: in low-hardness areas, water softening by ion-exchange (e.g. a Permutit 'Deminrolit' unit) is cheaper, but in hard-water areas distillation using an electric immersion heater (e.g. a Manesty water still) may be cheaper.

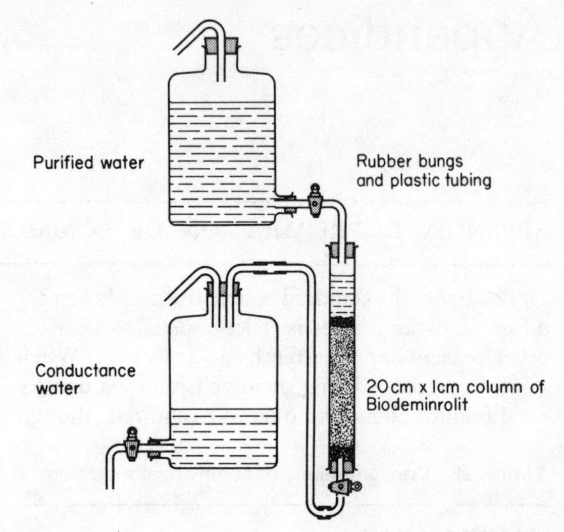

Fig. 46. Apparatus for the production of conductance water.

Conductance water of satisfactory quality is obtained by passing purified water through a column of 'Indicator Biodeminrolit' (Permutit Co. Ltd.) which changes from caviar to khaki when exhausted. A convenient set-up is shown in Fig. 46. Prepare conductance water the day before required by passing purified water through the column at 5 cm^3 per minute. If the apparatus stands in the laboratory, protect the contents with soda-lime tubes. Elsewhere protection against carbon dioxide is not worth while, as it will be rapidly absorbed in normal experimental conditions.

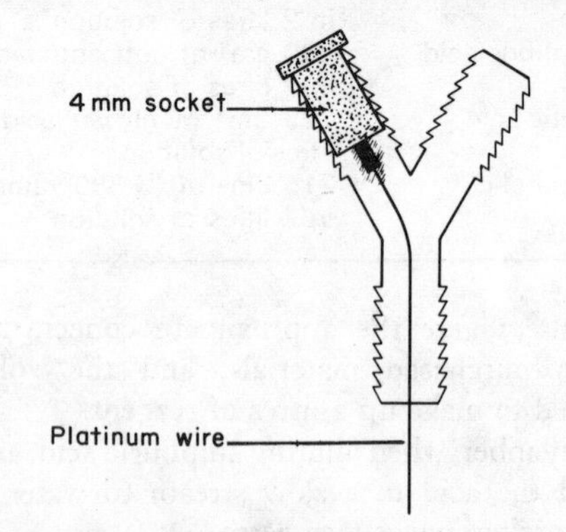

Fig. 47. A hydrogen electrode.

Preparation of a hydrogen electrode

A length of platinum wire is soldered to a 4 mm socket and the soldered joint coated with rubber solution or similar material to prevent chemical attack. The 4 mm socket is inserted in one arm of a 'Y' pattern polypropylene connector of outside diameter 10–12 mm (Xlon Products Ltd., see Fig. 47). The lower part of the platinum wire must now be coated with platinum black electrolytically.

Clean two electrodes with acid, wash well with water and support in an electrolyte consisting of 0·5 g platinum (IV) chloride plus 0·01 g lead acetate dissolved in 50 cm³ of 2M hydrochloric acid. Electrolyse for 15 minutes using 4 volts D.C. and reversing the current every half minute. To remove occluded gases repeat the electrolysis procedure but using dilute sulphuric acid as electrolyte.

The electrodes should be stored in pure water when not in use.

APPENDIX II. THE MATHEMATICS OF REACTION KINETICS

The mathematics of reaction kinetics are a frequent source of confusion because of the variety of ways in which the rate law of a reaction can be expressed. A rate law expression will have two variables, one will be *time* and the other a *changing concentration*, and will take the general form:

'*the rate of change of concentration during a reaction depends on the concentration of reactants*'.

However the *variable concentration* can either be that of a reactant in which case for a simple reaction $A \longrightarrow$ products the rate law expression becomes:

$$\frac{-d[A]}{dt} = k[A]^n$$

or the variable concentration can be that of a product (x), in which case the rate law expression becomes:

$$+\frac{dx}{dt} = k(a - x)^n$$

These expressions can be integrated for different values of n and the results are given in Table 40. The order of a reaction can also be deduced from the half-life times ($t_{\frac{1}{2}}$) which are the times taken for the concentration (A_0) of a reactant to be halved. A graph of log $t_{\frac{1}{2}}$ against log A_0 should be a straight line of gradient ($1-n$). For a fuller account see FURSE, A. J. and POORE, A. D., *School Science Review*, 156, **53** (182), 1971.

Table 38

Order of reaction n	*Differential form of rate law*	*Integrated form of rate law*	*Simple graphical plot*
0	$\frac{-d[A]}{dt} = k$	$kt = [A]_0 - [A]_t$	$[A]_t$ against t
1	$\frac{-d[A]}{dt} = k[A]$	$kt = 2{\cdot}303 \log_{10} \frac{[A]_0}{[A]_t}$	$\log_{10} [A]_t$ against t
2	$\frac{-d[A]}{dt} = k[A]^2$	$kt = \frac{1}{[A]_t} - \frac{1}{[A]_0}$	$\frac{1}{[A]_t}$ against t
0	$\frac{dx}{dt} = k$	$kt = x$	x against t
1	$\frac{dx}{dt} = k(a - x)$	$kt = 2{\cdot}303 \log_{10} \frac{a}{a - x}$	$\log_{10} (a - x)$ against t
2	$\frac{dx}{dt} = k(a - x)^2$	$kt = \frac{x}{a(a - x)}$	$\frac{x}{a - x}$ against t

APPENDIX III. APPARATUS AND SUPPLIERS

Most chemicals, ion exchange resins and materials for chromatography in this book are available from British Drug Houses Ltd.

The interchangeable ground-glass joint (standard taper) glassware recommended is the Quickfit 27 BU set. The following additional items are needed for some experiments:

Item	Code
Drying tube, angled B14	MF23/1
Air condenser B14/B14	C2/11
Screw-cap adaptors B14	ST51/13
Adaptor, multiple with two necks parallel B14/B14	MA1/11
Receiver adaptor, bends with vent B14/B14	RA2/11
Flask, round-bottom short-neck B14	FR25/1S

250 ml flask set

Item	Code
Flask, round-bottom short-neck B24	FR250/3S
Column, Dufton fractionation B24/B24	FC2/13
Adaptor, reduction B14/B24	DA13
Adaptor, socket to cone with 'T' connection B19/B24	MF18/3
Funnel, cylindrical type B19/B19	D1/22

Fractionation set

Item	Code
Column, Dufton fractionation B19/B19	FC1/12
Adaptor, reduction B14/B19	DA12
Adaptor, expansion B19/B14	XA21

General apparatus

Apparatus	Supplier
Analytical balance	L. Oertling Ltd., Cray Valley Works, ORPINGTON, Kent, BR5 2HA
Rough balance and hand bellows	Ferris and Co. Ltd., Kenn Road, Hillside Road, BRISTOL BS5 7PE
Plastic stereo models for organic chemistry	Rinco Instrument Co. Inc., 503, South Prairie Street, GREENVILLE, Illinois, U.S.A.
Space-filling models, ionic lattice models	Catalin Ltd., WALTHAM CROSS, Essex.

Trimetric apparatus

Apparatus	Supplier
General (E-MIL)	H. J. Elliott Ltd., Treforest Industrial Estate, PONTYPRIDD, Glam.
Pipette filler	Griffin and George Ltd., Frederick Street, BIRMINGHAM 1.
pH meter and electrodes	Chandos Intercontinental, High Street, New Mills, STOCKPORT, Cheshire.
Calomel electrode RJ23	Electronic Instruments Ltd., RICHMOND, Surrey.

Physical chemical apparatus

Apparatus	Supplier
Washburn and Read apparatus, silicone rubber seal CL-926	Gallenkamp & Co. Ltd., Technico House, Christopher Street, LONDON, E.C.2.
Valve voltmeter 1M-18U	Heath (Gloucester) Ltd., GLOUCESTER GL2 6EE

Gas syringes and oven	W. G. Flaig and Sons Ltd., Exelo Works, Margate Road, BROADSTAIRS, Kent
Prepared layers for T.L.C.	Camlab, CAMBRIDGE CB4 1HI
Eastman chromatogram kit, model 104	Kodak Ltd., Research Chemical Sales Division, KIRKBY, Liverpool
Water-bath and stirrers	Grant Instruments (Cambridge) Ltd., Barrington, CAMBRIDGE CB2 5QZ
CT50 Conductivity bridge	Grayshaw Instruments, 126, Sandgate High Street, FOLKESTONE, Kent.
Conductivity dip cell (E7591/B)	Pye Unicam Ltd., York Street, CAMBRIDGE CB1 2PX
Polypropylene bottles	Xlon Products Ltd., 323a, Kennington Road, LONDON, S.E.11
Multi-pour plastic beakers	Arnold R. Horwell Ltd., 2, Grangeway, Kilburn High Road, LONDON, N.W.6.

Radiochemical apparatus

Scaler (Type 102ST)	Panax Equipment Ltd., Holmethorpe Industrial Estate, REDHILL, Surrey.
G.M. tubes for liquids (M6Ha) G.M. tube for solids (ST3)	20th-Century Electronics Ltd., King Henry's Drive, New Addington, CROYDON CR9 0BG
Ionization chamber dosimeter	R. A. Stephen and Co. Ltd., Miles Road, MITCHAM CR4 3YP
Vinyl adhesive tape with the radioactive symbol	Jencons Ltd., HEMEL HEMPSTEAD, Hertfordshire.
Minerals	R. F. D. Parkinson Ltd., Doulting, SHEPTON MALLET, Somerset.
Nuclear emulsion plates	Ilford Ltd., ILFORD, Essex.
X-Ray film	Kodak Ltd. (Industrial Sales Division), Kodak House, Kingsway, LONDON WC2B 6TG

Literature

The following free booklets are available:

B.D.H. publications on 'Ion exchange resins', etc. (British Drug Houses Ltd., Poole, Dorset)

Camlab guide to thin-layer chromatography (Camlab (Glass) Ltd., Cambridge)

Unilever educational booklets (Unilever Education Section, Unilever House, Blackfriars, London, E.C.4)

Whatman technical bulletins on chromedia (H. Reeve Angel & Co. Ltd., 9 Bridewell Place, London, E.C.4)

APPENDIX IV. SOME NOTES ON NOMENCLATURE

In recent years there have been several moves to encourage all chemists to use systematic names for chemical compounds. I.U.P.A.C. (the International Union of Pure and Applied Chemistry) have made recommendations, but these have not been universally accepted. One of the major problems is that the existing literature, with its many unsystematic names, will continue to be referred to for many years to come. Likewise the chemical industry is, in general, conservative in its attitude. It is likely to be quite a long time before we see no bottles labelled 'acetic acid', for example.

In the United Kingdom there have been (in 1972) recommendations published by the Association for Science Education that all names should be systematic. The chief advantage is that the name will enable a person who is familiar with the system to write the formula at once. Some of the names are, unfortunately, much more cumbersome than the present ones (benzoic acid becomes benzenecarboxylic acid, citric acid becomes 2-hydroxypropane-1,2,3-tricarboxylic acid) but they are meaningful.

Because these systematic names are unlikely to become generally used in the next year or two, we have not completely changed over in this book, though we have abandoned acetone for propanone, urea for carbamide and permanganate for manganate (VII), as well as a number of less familiar names. We have retained oxalic acid, acetic acid, formic acid and hypochlorite, amongst commonly used names.

Some trivial names and their systematic equivalents are listed below:

Acetaldehyde
Ethanal
Acetamide
Ethanamide
Acetanilide
N-Phenylethanamide
Acetic acid
Ethanoic acid
Acetone
Propanone
Acetylacetone
Pentan-2,4-dione
Acetylene
Ethyne
Adipic acid
Hexane-1,6-dioic acid
n-Amyl alcohol
Pentan-1-ol
Aniline
Phenylamine
Aniline hydrochloride
Phenylammonium chloride
Anisole
Methoxybenzene
Benzil
1,2-Diphenylethanedione
Benzyl alcohol
Phenylmethanol
t-Butyl alcohol
2-Methylpropan-2-ol
Catechol
Benzene-1,2-diol
Chloroform
Trichloromethane
Dibenzalacetone
1,5-Diphenylpentan-1,4-dien-3-one
Diethyl ether
Ethoxyethane
Dimethylglyoxime
Butanedione dioxime
Ethyl acetoacetate
Ethyl 3-oxobutanoate
Ethylenediamine
Ethane-1,2-diamine
Ethylene glycol
Ethane-1,2-diol
Formaldehyde
Methanal
Formic acid
Methanoic acid
Glycerol
Propane-1,2,3-triol
Hydrobenzoin
1,2-Diphenylethanediol
Iodoform
Tri-iodomethane

Isoamyl alcohol
3-Methylbutan-1-ol
Oxalic acid
Ethanedioic acid
Quinol
Benzene-1,4-diol
Quinone
Cyclohexa-2,5-dien-1,4-dione
Rcsorcinol
Benzene-1,3-diol
Salicylic acid
2-Hydroxybenzenecarboxylic acid
Sebacic acid
Decane-1,10-dioic acid
Sulphanilic acid
4-Aminobenzenesulphonic acid
Tartaric acid
2,3-Dihydroxybutane-1,4-dioic acid
Toluene
Methylbenzene
Urea
Carbamide
Xylene
Dimethylbenzene

APPENDIX V. KEY TO THE OBSERVATIONAL PROBLEMS

OP 1: **A** is calcium acetate

OP 2: **HA** is sodium phosphite

OP 3: **HB** is potassium thiocyanate

OP 4: **HC** is ammonium sodium hydrogen orthophosphate

OP 5: **E** is ammonium metavanadate

OP 6: **HY** is nickel formate

OP 7: **Z** is chromium (III) potassium sulphate

OP 8: **N** is copper (I) iodide

OP 9: **P** is potassium peroxodisulphate

Index

(figures in bold type are **preparations**)

Table of Atomic Weights, 1975 (Based on the Assigned Relative Atomic Mass of $^{12}C = 12$)

The following values apply to elements as they exist naturally on earth and to certain artificial elements. When used with due regard to footnotes, they are considered reliable to ± 1 in the last digit, or ± 3 if that digit is in bold type.

Name	Symbol	At. No.	A_r	Name	Symbol	At. No.	A_r
Actinium	Ac	89	227·0278[z]	Mercury	Hg	80	200·5**9**
Aluminium	Al	13	26·98154	Molybdenum	Mo	42	95·94
Americium	Am	95	(243)	Neodymium	Nd	60	144·2**4**[x]
Antimony	Sb	51	121·7**5**	Neon	Ne	10	20·17**9**[y]
Argon	Ar	18	39·94**8**[w,x]	Neptunium	Np	93	237·0482[z]
Arsenic	As	33	74·9216	Nickel	Ni	28	58·70
Astatine	At	85	(210)	Niobium	Nb	41	92·9064
Barium	Ba	56	137·33[x]	Nitrogen	N	7	14·0067
Berkelium	Bk	97	(247)	Nobelium	No	102	(259)
Beryllium	Be	4	9·01218	Osmium	Os	76	190·2[x]
Bismuth	Bi	83	208·9804	Oxygen	O	8	15·999**4**
Boron	B	5	10·81[w,y]	Palladium	Pd	46	106·4[x]
Bromine	Br	35	79·904	Phosphorus	P	15	30·97376
Cadmium	Cd	48	112·41[x]	Platinum	Pt	78	195·0**9**
Caesium	Cs	55	132·9054	Plutonium	Pu	94	(244)
Calcium	Ca	20	40·08[x]	Polonium	Po	84	(209)
Californium	Cf	98	(251)	Potassium	K	19	39·098**3**
Carbon	C	6	12·011[w]	Praseodymium	Pr	59	140·9077
Cerium	Ce	58	140·12[x]	Promethium	Pm	61	(145)
Chlorine	Cl	17	35·453	Protactinium	Pa	91	231·0359[z]
Chromium	Cr	24	51·996	Radium	Ra	88	226·0254[x,z]
Cobalt	Co	27	58·9332	Radon	Rn	86	(222)
Copper	Cu	29	63·54**6**[w]	Rhenium	Re	75	186·207
Curium	Cm	96	(247)	Rhodium	Rh	45	102·9055
Dysprosium	Dy	66	162·5**0**	Rubidium	Rb	37	85·467**8**[x]
Einsteinium	Es	99	(254)	Ruthenium	Ru	44	101·0**7**[x]
Erbium	Er	68	167·2**6**	Rutherfordium	Rf	104	—
Europium	Eu	63	151·96[x]	Samarium	Sm	62	150·4[x]
Fermium	Fm	100	(257)	Scandium	Sc	21	44·9559
Fluorine	F	9	18·998403	Selenium	Se	34	78·9**6**
Francium	Fr	87	(223)	Silicon	Si	14	28·085**5**
Gadolinium	Gd	64	157·2**5**[x]	Silver	Ag	47	107·868[x]
Gallium	Ga	31	69·72	Sodium	Na	11	22·98977
Germanium	Ge	32	72·5**9**	Strontium	Sr	38	87·62[x]
Gold	Au	79	196·9665	Sulphur	S	16	32·06[w]
Hafnium	Hf	72	178·4**9**	Tantalum	Ta	73	180·947**9**
Hahnium	Ha	105	—	Technetium	Tc	43	(97)
Helium	He	2	4·00260[x]	Tellurium	Te	52	127·6**0**[x]
Holmium	Ho	67	164·9304	Terbium	Tb	65	158·9254
Hydrogen	H	1	1·0079[w]	Thallium	Tl	81	204·3**7**
Indium	In	49	114·82[x]	Thorium	Th	90	232·0381[x,z]
Iodine	I	53	126·9045	Thulium	Tm	69	168·9342
Iridium	Ir	77	192·2**2**	Tin	Sn	50	118·6**9**
Iron	Fe	26	55·84**7**	Titanium	Ti	22	47·9**0**
Krypton	Kr	36	83·80[x,y]	Tungsten	W	74	183·8**5**
Lanthanum	La	57	138·905**5**[x]	Uranium	U	92	238·029[x,y]
Lawrencium	Lr	103	(260)	Vanadium	V	23	50·941**4**
Lead	Pb	82	207·2[w,x]	Xenon	Xe	54	131·30[x,y]
Lithium	Li	3	6·94**1**[w,x,y]	Ytterbium	Yb	70	173·0**4**
Lutetium	Lu	71	174·97	Yttrium	Y	39	88·9059
Magnesium	Mg	12	24·305[x]	Zinc	Zn	30	65·38
Manganese	Mn	25	54·9380	Zirconium	Zr	40	91·22[x]
Mendelevium	Md	101	(258)				

[w] Element for which known variations in isotopic composition in normal terrestrial material prevent a more precise atomic weight being given.

[x] Element for which geological specimens are known in which the element has an anomalous isotopic composition.

[y] Element for which substantial variations in atomic weight can occur in commercially available material because of change of isotopic composition.

[z] Element for which the atomic weight value is that of the radioisotope of longest half-life.

[Table reprinted from *Pure and Applied Chemistry*, 1976, **47**, 80–81, by permission of the International Union of Pure and Applied Chemistry.]

Table of The Long Form of the Periodic Table

When used in conjunction with the Table of Atomic Weights and footnotes opposite the values given for atomic weights are considered reliable to ±1 in the last digit, or ±3 if that digit is in bold type.

	Ia	IIa	IIIa		IVa	Va	VIa	VIIa	VIII			Ib	IIb	IIIb	IVb	Vb	VIb	VIIb	0
1	1 H 1·0079																		2 He 4·00260
2	3 Li 6·941	4 Be 9·01218												5 B 10·81	6 C 12·011	7 N 14·0067	8 O 15·9994	9 F 18·998403	10 Ne 20·179
3	11 Na 22·98977	12 Mg 24·305												13 Al 26·98154	14 Si 28·0855	15 P 30·97376	16 S 32·06	17 Cl 35·453	18 Ar 39·948
4	19 K 39·0983	20 Ca 40·08	21 Sc 44·9559		22 Ti 47·90	23 V 50·9414	24 Cr 51·996	25 Mn 54·9380	26 Fe 55·847	27 Co 58·9332	28 Ni 58·70	29 Cu 63·546	30 Zn 65·38	31 Ga 69·72	32 Ge 72·59	33 As 74·9216	34 Se 78·96	35 Br 79·904	36 Kr 83·80
5	37 Rb 85·4678	38 Sr 87·62	39 Y 88·9059		40 Zr 91·22	41 Nb 92·9064	42 Mo 95·94	43 Tc —	44 Ru 101·07	45 Rh 102·9055	46 Pd 106·4	47 Ag 107·868	48 Cd 112·41	49 In 114·82	50 Sn 118·69	51 Sb 121·75	52 Te 127·60	53 I 126·9045	54 Xe 131·30
6	55 Cs 132·9054	56 Ba 137·33	57 La 138·9055	LANTHANIDES	72 Hf 178·49	73 Ta 180·9479	74 W 183·85	75 Re 186·207	76 Os 190·2	77 Ir 192·22	78 Pt 195·09	79 Au 196·9665	80 Hg 200·59	81 Tl 204·37	82 Pb 207·2	83 Bi 208·9804	84 Po —	85 At —	86 Rn —
7	87 Fr —	88 Ra 226·0254	89 Ac 227·0278	ACTINIDES	104 Rf —	105 Ha —	106	107	108	109	110	111	112	113	114	115	116	117	118

Lanthanide Series	57 La 138·9055	58 Ce 140·12	59 Pr 140·9077	60 Nd 144·24	61 Pm —	62 Sm 150·4	63 Eu 151·96	64 Gd 157·25	65 Tb 158·9254	66 Dy 162·50	67 Ho 164·9304	68 Er 167·26	69 Tm 168·9342	70 Yb 173·04	71 Lu 174·97
Actinide Series	89 Ac —	90 Th 232·0381	91 Pa 231·0359	92 U 238·029	93 Np 237·0482	94 Pu —	95 Am —	96 Cm —	97 Bk —	98 Cf —	99 Es —	100 Fm —	101 Md —	102 No —	103 Lr —